Methods for Serum-Free Culture of Epithelial and Fibroblastic Cells

Cell Culture Methods for Molecular and Cell Biology

David W. Barnes, David A. Sirbasku, and Gordon H. Sato, *Editors*

Volume 1: Methods for Preparation of Media, Supplements, and Substrata for Serum-Free Animal Cell Culture

Volume 2: Methods for Serum-Free Culture of Cells of the Endocrine System

Volume 3: Methods for Serum-Free Culture of Epithelial and Fibroblastic Cells

Volume 4: Methods for Serum-Free Culture of Neuronal and Lymphoid Cells

Methods for Serum-Free Culture of Epithelial and Fibroblastic Cells

Editors

David W. Barnes

Department of Biological Sciences
University of Pittsburgh
Pittsburgh, Pennsylvania

David A. Sirbasku

Department of Biochemistry
and Molecular Biology
University of Texas Medical School
Houston, Texas

Gordon H. Sato

W. Alton Jones Cell Science Center
Lake Placid, New York

Alan R. Liss, Inc., New York

Address all Inquiries to the Publisher
Alan R. Liss, Inc., 150 Fifth Avenue, New York, NY 10011

Library of Congress Cataloging in Publication Data

Main entry under title:

Methods for serum-free culture of epithelial and fibroblastic cells.
(Cell culture methods for molecular and cell biology; v. 3)
Includes index.
1. Serum-free culture media. 2. Cell culture.
3. Epithelial cells. 4. Fibroblasts. I. Barnes,
David W. (David William), 1949- . II. Sirbasku,
David A. (David Andrew), 1941- . III. Sato, Gordon.
IV. Series
QH585.5.S45M48 1984 619 84-7910
ISBN 0-8451-3802-2

Contents

Contributors ix

Contents of Volumes 1, 2, and 4 xiii

Preface xix

SERUM-FREE CULTURE OF EPITHELIAL CELLS

1 **Growth of Primary and Established Kidney Cell Cultures in Serum-Free Media**
Mary Taub 3

2 **Hormonally Defined, Serum-Free Medium for a Proximal Tubular Kidney Epithelial Cell Line, LLC-PK_1**
Milton H. Saier, Jr. 25

3 **Serum-Free Organ Culture of Embryonic Mouse Metanephros**
Ellis D. Avner, William E. Sweeney, Jr., and Demetrius Ellis 33

4 **Primary Culture of Hepatocytes**
H.L. Leffert, K.S. Koch, and H. Skelly 43

5 **Selective Growth of Human Small Cell Lung Cancer Cell Lines and Clinical Specimens in Serum-Free Medium**
Desmond N. Carney, Martin Brower, Virginia Bertness, and Herbert K. Oie 57

6 **Primary Tissue Cultures of Human Colon Carcinomas in Serum-Free Medium: An In Vitro System for Tumor Analysis and Therapy Experiments**
Jürgen van der Bosch 73

7 **Growth and Differentiation of Human Bronchogenic Epidermoid Carcinoma Cells in Serum-Free Media**
Kaoru Miyazaki, Hideo Masui, and Gordon H. Sato 83

8 **Serum-Free Cell Culture of A431 Human Epidermoid Carcinoma**
David W. Barnes 95

9 **Growth and Differentiation of Embryonal Carcinoma Cells in Defined and Serum-Free Media**
Angie Rizzino 107

10 **α_2-Macroglobulin, a Contaminant of Commercially Prepared Pedersen Fetuin: Isolation, Characterization, and Biological Activity**
David S. Salomon, Kathryn B. Smith, Ilona Losonczy, Mozeena Bano, William R. Kidwell, Giulio Alessandri, and Pietro M. Gullino 125

SERUM-FREE CULTURE OF FIBROBLASTIC CELLS

11 **On Deciding Which Factors Regulate Cell Growth**
Arthur B. Pardee, Paul V. Cherington, and Estela E. Medrano 157

12 **Purification of Pituitary and Brain Fibroblast Growth Factors and Their Use in Cell Culture**
Denis Gospodarowicz 167

13 **Preparation of Bovine Pituitary Fibroblast Growth Factor**
Sandra K. Lemmon and Ralph A. Bradshaw. 199

14 **Preparation of Pituitary Acidic FGF**
Angelo A. Gambarini, Mari C.S. Armelin, and Hugo A. Armelin 209

15 **Growth of SV40 Balb/c-3T3 Cells in Serum-Free Culture Medium**
G.A. Rockwell 221

16 **Use of Hormone-Toxin Conjugates and Serum-Free Media for the Isolation and Study of Cell Variants in Hormone Responses**
Nobuyoshi Shimizu 233

17 **Growth of Human Fibroblast Cultures in Serum-Free Media**
Richard G. Ham 249

18 **Serum-Free Cell Culture for Growth of NIH 3T3 and 10T1/2 Mouse Embryo Fibroblast Cell Lines, SV40 Virus Propagation, and Selection of SV40-Transformed Cells**
Lin-Chang Chiang, Janet Silnutzer, James M. Pipas, and David W. Barnes 265

Index 277

Contributors

Giulio Alessandri, Laboratory of Pathophysiology, National Cancer Institute, National Institutes of Health, Bethesda, MD 20205 [**125**]

Hugo A. Armelin, Instituto de Quimica, Universidade de São Paulo, São Paulo, Brazil [**209**]

Mari C.S. Armelin, Instituto de Quimica, Universidade de São Paulo, São Paulo, Brazil [**209**]

Ellis D. Avner, Division of Pediatric Nephrology, Children's Hospital of Pittsburgh, Pittsburgh, PA 15213 [**33**]

Mozeena Bano, Laboratory of Pathophysiology, National Cancer Institute, National Institutes of Health, Bethesda, MD 20205 [**125**]

David W. Barnes, Department of Biological Sciences, University of Pittsburgh, Pittsburgh, PA 15260 [**xix,95,265**]

Virginia Bertness, NCI-Navy Medical Oncology Branch, National Cancer Institute, National Institutes of Health and the National Naval Medical Center, Bethesda, MD 20814 [**57**]

Ralph A. Bradshaw, Department of Biological Chemistry, California College of Medicine, University of California at Irvine, Irvine, CA 92717 [**199**]

Martin Brower, Division of Medical Oncology, Rochester General Hospital, Rochester, NY 14621 [**57**]

Desmond N. Carney, NCI-Navy Medical Oncology Branch, National Cancer Institute, National Institutes of Health and the National Naval Medical Center, Bethesda, MD 20814 [**57**]

Paul V. Cherington, Laboratory of Tumor Biology, Sidney Farber Cancer Institute and Department of Pharmacology, Harvard Medical School, Boston, MA 02115 [**157**]

Lin-Chang Chiang, Department of Biological Sciences, University of Pittsburgh, Pittsburgh, PA 15260 [**265**]

Demetrius Ellis, Division of Pediatric Nephrology, Children's Hospital of Pittsburgh, Pittsburgh, PA 15213 [**33**]

Angelo A. Gambarini, Instituto de Quimica, Universidade de São Paulo, São Paulo, Brazil [**209**]

The number in brackets is the opening page number of the contributor's article.

Denis Gospodarowicz, Cancer Research Institute and Departments of Medicine and Ophthalmology, University of California Medical Center, San Francisco, CA 94143 **[167]**

Pietro M. Gullino, Laboratory of Pathophysiology, National Cancer Institute, National Institutes of Health, Bethesda, MD 20205 **[125]**

Richard G. Ham, Department of Molecular, Cellular, and Developmental Biology, University of Colorado, Boulder, CO 80309 **[249]**

William R. Kidwell, Laboratory of Pathophysiology, National Cancer Institute, National Institutes of Health, Bethesda, MD 20205 **[125]**

K.S. Koch, Department of Medicine, Division of Pharmacology, School of Medicine, University of California at San Diego, La Jolla, CA 92093 **[43]**

H.L. Leffert, Department of Medicine, Division of Pharmacology, School of Medicine, University of California at San Diego, La Jolla, CA 92093 **[43]**

Sandra K. Lemmon, Department of Biological Sciences, Carnegie-Mellon University, Pittsburgh, PA 15213 **[199]**

Ilona Losonczy, Laboratory of Tumor Immunology and Biology, National Cancer Institute, National Institutes of Health, Bethesda MD 20205 **[125]**

Hideo Masui, Department of Biology, University of California at San Diego, La Jolla, CA 92093 **[83]**

Estela E. Medrano, Laboratory of Tumor Biology, Sidney Farber Cancer Institute and Department of Pharmacology, Harvard Medical School, Boston, MA 02115 **[157]**

Kaoru Miyazaki, Division of Enzymology, Institute for Protein Research, Osaka University, Osaka 565, Japan **[83]**

Herbert K. Oie, NCI-Navy Medical Oncology Branch, National Cancer Institute, National Institutes of Health and the National Naval Medical Center, Bethesda, MD 20814 **[57]**

Arthur B. Pardee, Laboratory of Tumor Biology, Sidney Farber Cancer Institute and Department of Pharmacology, Harvard Medical School, Boston, MA 02115 **[157]**

James M. Pipas, Department of Biological Sciences, University of Pittsburgh, Pittsburgh, PA 15260 **[265]**

Angie Rizzino, Eppley Institute for Research in Cancer and Allied Diseases, University of Nebraska Medical Center, Omaha, NE 68105 **[107]**

Graham A. Rockwell, Unit of Cellular and Molecular Biology, C.S.I.R.O., North Ryde, N.S.W., Australia **[221]**

Milton H. Saier, Jr., Department of Biology, The John Muir College, University of California at San Diego, La Jolla, CA 92093 **[25]**

David S. Salomon, Laboratory of Tumor Immunology and Biology, National Cancer Institute, National Institutes of Health, Bethesda, MD 20205 **[125]**

Gordon H. Sato, W. Alton Jones Cell Science Center, Lake Placid, NY 12946 **[xix,83]**

Nobuyoshi Shimizu, Department of Molecular and Cellular Biology, University of Arizona, Tucson, AZ 85721 and Department of Molecular Biology, Keio University School of Medicine, Tokyo 160, Japan **[233]**

Janet Silnutzer, Department of Biological Sciences, University of Pittsburgh, Pittsburgh, PA 15260 [**265**]

David A. Sirbasku, Department of Biochemistry and Molecular Biology, University of Texas Medical School, Houston, TX 77225 [**xix**]

H. Skelly, Department of Medicine, Division of Pharmacology, School of Medicine, University of California at San Diego, La Jolla, CA 92093 [**43**]

Kathryn B. Smith, Laboratory of Tumor Immunology and Biology, National Cancer Institute, National Institutes of Health, Bethesda MD 20205 [**125**]

William E. Sweeney, Jr., Division of Pediatric Nephrology, Children's Hospital of Pittsburgh, Pittsburgh, PA 15213 [**33**]

Mary Taub, Department of Biochemistry, State University of New York at Buffalo, Buffalo, NY 14214 [**3**]

Jürgen van der Bosch, Forschungsinstitut Borstel, Institut für Experimentelle Biologie und Medizin, 2061 Borstel, Federal Republic of Germany [**73**]

Contents of Volumes 1, 2, and 4

Volume 1: Methods for Preparation of Media, Supplements, and Substrata for Serum-Free Animal Cell Culture

METHODS FOR PREPARATION OF BASAL NUTRIENT MEDIA

1 **Formulation of Basal Nutrient Media**
Richard G. Ham

2 **Preparation and Use of Serum-Free Culture Media**
Charity Waymouth

3 **Preparations and Uses of Lipoproteins to Culture Normal Diploid and Tumor Cells Under Serum-Free Conditions**
Denis Gospodarowicz

METHODS FOR PREPARATION OF MITOGENIC PEPTIDES

4 **Preparation of Human Platelet-Derived Growth Factor**
Elaine W. Raines and Russell Ross

5 **Purification of Multiplication-Stimulating Activity**
Lawrence A. Greenstein, S. Peter Nissley, Alan C. Moses, Patricia A. Short, Yvonne W.-H. Yang, Lilly Lee, and Matthew M. Rechler

6 **Preparation of Guinea Pig Prostate Epidermal Growth Factor**
Jeffrey S. Rubin and Ralph A. Bradshaw

7 **Purification of Human Epidermal Growth Factor From Urine**
C. Richard Savage, Jr. and Robert A. Harper

8 **Isolation of Growth Factors From Human Milk**
Yuen W. Shing and Michael Klagsbrun

9 **Purification of Type β Transforming Growth Factors From Non-neoplastic Tissues**
Anita B. Roberts, Charles A. Frolik, Mario A. Anzano, Richard K. Assoian, and Michael B. Sporn

10 **Preparation of Endothelial Cell Growth Factor**
Thomas Maciag and Robert Weinstein

METHODS FOR PREPARATION OF SUBSTRATA

11 **Use of Basic Polymers as Synthetic Substrata for Cell Culture**
Wallace L. McKeehan

12 **Preparation of Cellular Fibronectin**
Kenneth M. Yamada and Steven K. Akiyama

13 **Isolation of Laminin**
Steven R. Ledbetter, Hynda K. Kleinman, John R. Hassell, and George R. Martin

14 **Isolation of Chondronectin**
Hugh H. Varner, A. Tyl Hewitt, and George R. Martin

15 **Human Serum Spreading Factor (SF): Assay, Preparation, and Use in Serum-Free Cell Culture**
Janet Silnutzer and David W. Barnes

16 **Purification of Epibolin From Human Plasma**
K.S. Stenn

17 **Preparation of Extracellular Matrices Produced by Cultured Bovine Corneal Endothelial Cells and PF-HR-9 Endodermal Cells: Their Use in Cell Culture**
Denis Gospodarowicz

18 **Analysis of Basement Membrane Synthesis and Turnover in Mouse Embryonal and Human A431 Epidermoid Carcinoma Cells in Serum-Free Medium**
David S. Salomon, Lance A. Liotta, Mounanandham Panneerselvam, Victor P. Terranova, Atul Sahai, and Paula Fehnel

19 **Cell Attachment and Spreading on Extracellular Matrix-Coated Beads**
Shing Mai and Albert E. Chung

Volume 2: Methods for Serum-Free Culture of Cells of the Endocrine System

1 **Growth of GH_3, a Rat Pituitary Cell Line, in Serum-Free, Hormone-Supplemented Medium**
Izumi Hayashi

2 **Growth of Adrenocortical Cell Cultures in Serum-Free Medium**
Michael H. Simonian and Mark L. White

3 **Primary Culture of Testicular Somatic Cells**
J.P. Mather and D.M. Phillips

4 **Isolation and Growth of Adult Human Prostatic Epithelium in Serum-Free, Defined Medium**
Mukta M. Webber, Donna M. Chaproniere-Rickenberg, and Robert E. Donohue

5 **Growth of Functional Primary and Established Rat Ovary Cell Cultures in Serum-Free Medium**
Joseph Orly

6 **Growth of Rat Mammary Tumor Cells in Serum-Free, Hormone-Supplemented Medium**
Tamiko Kano-Sueoka

7 **Growth of Normal Mammary Epithelium on Collagen in Serum-Free Medium**
William R. Kidwell, Mozeena Bano, and David S. Salomon

8 **Isolation and Serum-Free Cultivation of Mammary Epithelial Cells Within a Collagen Gel Matrix**
Walter Imagawa, Yasuhiro Tomooka, Jason Yang, Raphael Guzman, James Richards, and Satyabrata Nandi

9 **Serum-Free Culture of the Isolated Whole Mammary Organ of the Mouse: A Model for the Study of Differentiation and Carcinogenesis**
Mihir R. Banerjee and Michael Antoniou

10 **Growth of Human Mammary Epithelial Cells in Monolayer Culture**
Martha Stampfer

11 **Definition of Hormones and Growth Factors Required for Optimal Proliferation and Expression of Phenotypic Responses in Human Breast Cancer Cells**
Marc E. Lippman

12 **Serum-Free Cell Culture of MCF7 Human Mammary Carcinoma**
David W. Barnes

13 **General Methods for Isolation of Acetic Acid- and Heat-Stable Polypeptide Growth Factors for Mammary and Pituitary Tumor Cells**
Tatsuhiko Ikeda, David Danielpour, Peter R. Galle, and David A. Sirbasku

Volume 4: Methods for Serum-Free Culture of Neuronal and Lymphoid Cells

SERUM-FREE CULTURE OF NEURONAL CELLS

1 **Culture Methods for Growth of Neuronal Cell Lines in Defined Media**
Jane E. Bottenstein

2 **Preparation of a Chemically Defined Medium for Purified Astrocytes**
Richard S. Morrison and Jean de Vellis

3 **Growth and Differentiation of Pheochromocytoma Cells in Chemically Defined Medium**
R. Goodman

4 **Differentiated Mouse Fetal Hypothalamic Cells in Serum-Free Medium**
A. Faivre-Bauman, J. Puymirat, C. Loudes, and A. Tixier-Vidal

5 **Regulation of Pigmentation and Proliferation in Cultured Melanocytes**
John M. Pawelek

6 **Neuron-Glia Interaction in Mammalian Brain: Preparation and Quantitative Bioassay of a Neurotrophic Factor (NTF) From Primary Astrocytes**
Wilfried Seifert and Hans Werner Müller

7 **Preparation and Assay of Nerve Growth Factor**
Thomas L. Darling and Eric M. Shooter

SERUM-FREE CULTURE OF LYMPHOID CELLS

8 **Production and Purification of Interleukin-2 for the Initiation and Maintenance of T-Cell Lines**
Diane Mochizuki and James D. Watson

9 **Methods for Production and Purification of Human T-Cell Growth Factor**
M.G. Sarngadharan, R.C. Ting, and R.C. Gallo

10 **Preparation of Thymosins**
Teresa L.K. Low and Allan L. Goldstein

11 **Culture of Lymphocytes and Hemopoietic Cells in Serum-Free Medium**
N.N. Iscove

12 **Growth of Lymphoid Cells in Serum-Free Medium**
Frederick J. Darfler and Paul A. Insel

13 **Serum-Free Cultivation of Plasmacytomas and Hybridomas**
Hiroki Murakami

14 **Culture of Human Lymphocytes in Serum-Free Medium**
John Mendelsohn, Alendry Caviles, Jr., and Janice Castagnola

15 **Studies of Growth and Differentiation of Human Myelomonocytic Leukemia Cell Lines in Serum-Free Medium**
Theodore R. Breitman, Beverly R. Keene, and Hiromichi Hemmi

16 **Serum-Free Growth of SP2/0-AG-14 Hybridomas**
Kathelyn Sue Steimer

Preface

In this volume are described methods for preparation of serum-free, hormonally defined media for transformed and untransformed fibroblasts and a number of types of differentiated epithelial cells. Investigations of tissue-specific functional expressions by epithelial cells are one of the most active areas of research in molecular and cell biology, and the development of serum-free, defined media for these cell types has had considerable impact on the field. Striking examples of the new types of studies accomplished by culturing epithelial cells in defined media are those in this volume describing the functional state of cultured kidney tubule epithelium. Different combinations of hormones and other factors can be shown to promote expression of either proximal or distal tubule functions selectively. Serum-free culture methods also are described for organ culture of developing kidney. Studies of hormone-responsive transport mechanisms as well as studies of the factors altering the states of kidney cell differentiation are open to new approaches using the methods outlined.

In addition to kidney, methods for preparing primary cultures of normal, functional hepatocytes are described; these will provide interested workers with a starting point for further investigation of the complex regulation of growth and gene expression in these cells in a defined environment. Also examined in this volume is the serum-free culture of epithelia from the lung; in this case, defined media have been used successfully to grow cell types that are readily lost when cultured in serum-containing medium. Well-defined cultures of lung epithelium from human and other species will facilitate the study of carcinogenesis in this tissue. Chapters describing growth of both human lung and colon carcinoma point out the experimentally-proven potential of the use of serum-free medium in the growth of human tumor cells from biopsy samples for drug screening.

Additional detailed methods are given in the first section for the serum-free culture of two cell types that may be of interest to a broad audience. The A431 human epidermoid carcinoma line, widely used recently in studies of epidermal growth factor effects and growth factor-receptor interactions, is finding use in other types of studies, including those dealing with extracellular matrix production (See Volume 1). Methods for growth of several lines of differentiating embryonal carcinoma, an obvious choice for a model system in the study of control of cell differentiation, are also reviewed. For those investigators requiring formulations for differentiated epithelial cell types not described in these volumes, application of the methods presented with lung, colon, and kidney tissue will provide a useful starting point. Information provided in Volume 1 of this series will provide methods of preparation of some of the basic components of the defined media used with the epithelial cells.

In the second part of this volume, methods for serum-free culture of fibroblasts are detailed, with an emphasis on those systems and techniques applicable to studies of growth control and malignancy in vitro. Included is an introductory chapter (reprinted from Prog. Clin. Biol. Res., Vol. 66) that provides an overview of the use of serum-free cell culture in the study of proliferation in vitro, as well as detailing serum-free culture conditions for lines of transformed and untransformed fibroblasts. With methods described in this section, a variety of applications are possible. Several approaches are described for the use of serum-free cell culture as a selection process for isolating cell variants. Also presented in this volume are methods for isolation and purification of brain- and pituitary-derived fibroblast growth factors. Methods for preparation of both the acidic and basic forms of these polypeptide factors are described in detail, and provide the interested investigator not only with access to those preparations, but also with outlines of how one goes about attempting the successful isolation of new factors that may be required for cell growth.

Applications of the methods described here are direct for several of the major cell types of the body. With some thought and experimentation, these formulations, and the methods of approach that they represent, should be applicable to other cell types as well.

David W. Barnes
David A. Sirbasku
Gordon H. Sato

Serum-Free Culture of Epithelial Cells

Methods for Serum-Free Culture of Epithelial and Fibroblastic Cells,
pages 3–24

1
Growth of Primary and Established Kidney Cell Cultures in Serum-Free Media

Mary Taub

Hormonally defined serum-free media have been developed for established kidney epithelial cell lines and primary kidney cell cultures. These hormonally defined media permit a number of studies to be done with cultured kidney cells that were previously impeded because of serum in the culture medium. In vitro studies have been primarily concerned with a limited number of established kidney epithelial cell lines, including Madin Darby canine kidney (MDCK) and LLC-PK_1. These cell lines closely resemble several types of kidney tubule cells in vivo, which reabsorb solutes into the blood (Tables I and II). The availability of hormonally defined media now permits an examination of the mechanisms by which the growth and transport functions of these cells are regulated by hormones. The hormonally defined serum-free media also permit additional types of kidney tubule cells to be studied in vitro, by facilitating the growth of the appropriate cells in primary cultures. In this manner additional renal transport systems can be examined in vitro. The cells affected in a number of medically related problems of the kidney can now be maintained in primary culture. Thus the alterations that occur in particular medical disease states can be examined in vitro. Primary kidney cultures in hormonally defined media should also prove to have practical importance. Kidney tubule cells in vivo produce such products as erythropoietin, which potentially could be made and readily purified from primary kidney cultures in defined medium.

In this chapter the methods for the growth of several established kidney epithelial cell lines and primary cultures in defined medium are discussed in detail. These culture systems include the MDCK cell line, primary baby mouse

Department of Biochemistry, State University of New York at Buffalo, Buffalo, New York 14214

TABLE I. Properties of MDCK Cells

1. Morphology: [Leighton et al., 1970]
   Domes
   Form polarized monolayers
   Adjacent cells form tight junctions
2. Transport properties:
   Amiloride-sensitive Na^+ channel, $K_1 = 5 \times 10^{-5}$ M [Rindler et al., 1979b]
   Furosemide-sensitive Na^+, K^+ cotransport [McRoberts et al., 1982]
   Transepithelial salt and water transport [Misfeldt et al, 1976]
      1.0 mV potential, Amiloride-sensitive [Cereijido et al., 1978]
      84-ohm/cm^2 resistance
3. Hormone responses:
   Arginine-vasopressin-sensitive cAMP production [Rindler et al., 1979a]
   Norepinephrine-sensitive cAMP production [Taub et al., 1979]
   α and β receptors [Meier and Insel, 1982]
   DOC receptors [Ludens et al., 1978]
4. Growth properties:
   Form nodules in baby but not adult nude mice [Stiles et al., 1976]
   Grow in defined medium [Taub et al., 1979]

kidney epithelial cells, and primary rabbit kidney proximal tubule cells. The procedures described for these three culture systems can be used as the starting point for growing other kidney cells serum-free.

METHODS TO GROW ESTABLISHED KIDNEY CELL LINES IN HORMONE-SUPPLEMENTED SERUM-FREE MEDIUM

Growth of the MDCK Cell Line in Defined Medium

The MDCK cell line can be grown in medium K-1, a hormone-supplemented serum-free medium, at the same rate as in serum-supplemented medium [Taub et al., 1979]. The components of this hormonally defined serum-free medium are described in Table III. A serum-free basal medium (DME/F12—a 50:50 mixture of Dulbecco's modified Eagle's medium and Ham's F12) is supplemented with insulin (5 μg/ml), transferrin (5 μg/ml), prostaglandin E_1 (PGE_1, 25 ng/ml), triiodothyronine (T_3, 5×10^{-12} M), and hydrocortisone (5×10^{-8} M). This hormonally defined medium (medium K-1) permits MDCK cells to grow over the long term. Medium K-1 also permits MDCK colonies to form at low plating densities, and it thus may be used for genetic studies.

MDCK monolayers are prepared for growth in medium K-1 as follows. Subconfluent or newly confluent MDCK monolayers are washed twice with phosphate-buffered saline (PBS), and then once with 0.1% trypsin, 0.3%

TABLE II. Properties of LLC-PK$_1$

1. Transport properties:
   Domes [Hull et al., 1976]
   Na^+-dependent sugar transport [Rabito and Ausiello, 1980; Mullin et al., 1980]
   Polarized amino acid transport [Rabito and Karish, 1982]
   Transepithelial glucose and salt transport [Misfeldt and Sanders, 1981]
   2.8-mV potential (phlorezin-sensitive)
   211-ohm/cm^2 resistance
2. Hormone responses: [Goldring et al., 1978]
   Arginine-vasopressin-sensitive cAMP production
   Calcitonin-sensitive cAMP production
3. Growth properties:
   Do not form tumors in immunosuppressed animals [Hull et al., 1976]
   Grow in defined medium [Chuman et al., 1982]

ethylene diamine tetraacetate (EDTA) in PBS. Excess trypsin is removed by aspiration, so that the monolayer is covered by a film of trypsin. The MDCK cells are then incubated at 37°C until the cells on the dish round up. The trypsin period should be minimized, so as to reduce damage to the cells. Trypsin action is then stopped by adding to the cell layer an equivalent amount of 0.5% soybean trypsin inhibitor in PBS. The cells are removed from the dish and subjected to low-speed centrifugation. The cell pellet is washed twice, with serum-free medium, and finally is resuspended in serum-free medium. The MDCK cells are then plated directly into plastic dishes containing medium K-1. The cultures are maintained at 37°C in a humidified 95% air/5% CO_2 incubator.

MDCK cells in medium K-1 will attach directly to the plastic surface and will grow without the use of any additional substratum. High plating efficiencies will be achieved, provided a) that the MDCK cells plated in the dishes originate from subconfluent cultures, and b) that the trypsinization period is minimized. When MDCK cells are plated into culture dishes containing medium K-1 at 10^4 cells per 60-mm dish, the cells will grow at 1.5 doublings per day. The growth of MDCK cells under these conditions is dependent upon the presence of each of the five supplements in medium K-1. The individual deletion of any one of the five supplements from the medium (in particular PGE_1 and transferrin) has an inhibitory effect on cell growth. At higher plating densities (greater than 10^4 cells per 60-mm dish) supplements such as PGE_1 have a much less dramatic effect on cell growth.

After MDCK cell cultures have become confluent, dome formation is observed (Fig. 1). The formation of domes in defined medium is dependent upon the presence of PGE_1, T_3, and hydrocortisone. MDCK monolayers that

TABLE III. Components in Hormonally Defined Serum-Free Media for Cultured Kidney Epithelial Cells

	Medium cell optimized for:			
	Medium K-1	Medium K-2	Medium RK-1	Medium RK-2
Medium supplements	(MDCK)	($LLC\text{-}PK_1$)	(primary rabbit kidney proximal tubule cells)	(primary rabbit kidney cells, unpurified)
Insulin	5 μg/ml	25 μg/ml	5 μg/ml	5 μg/ml
Transferrin	5 μg/ml	10 μg/ml	5 μg/ml	5 μg/ml
Hydrocortisone	5×10^{-8} M	2×10^{-7} M	5×10^{-8} M	5×10^{-8} M
Triiodothyronine	5×10^{-12} M	10^{-9} M	—	5×10^{-12} M
Prostaglandin E_1	25 ng/ml	—	—	—
Arginine-vasopressin	—	10 μU/ml	—	—
Epidermal growth factor	—	—	—	10 ng/ml
Cholesterol	—	10^{-8} M	—	—
Reference	Taub et al. [1979]	Chuman et al. [1982]	Chung et al. [1982]	Chung et al. [1982]

Basal medium (used for all four defined media): A 50:50 mixture of Dulbecco's modified Eagle's medium (DME) and Ham's F12 supplemented with 1) 15 mM HEPES buffer, 2) 1.2 mg/ml sodium bicarbonate, 3) 192 IU/ml penicillin and 200 μg/ml streptomycin, and 4) 5×10^{-8} M selenium (Difco).

form domes can be maintained for many days at confluency, given routine medium changes. Alternatively, the monolayers can be subcultured into medium K-1 and passaged for months in the absence of serum. New stock cultures should be initiated on a monthly basis from frozen ampules of MDCK cells. In this way the possibility is avoided that genetic changes occur in MDCK cell populations during long-term growth in medium K-1.

Medium K-1 can also be used as a cloning medium for MDCK cells. MDCK cells plated at low densities (less than 2×10^2 cells per cm^2) form colonies after a 1-week growth period in medium K-1. As observed with MDCK cell growth at higher densities, MDCK colony formation is dependent upon the presence of each of the five supplements in medium K-1. Thus the effect of individual hormone deletion (or addition) on MDCK plating efficiency can be used as a sensitive assay of the growth-stimulating effects of the medium

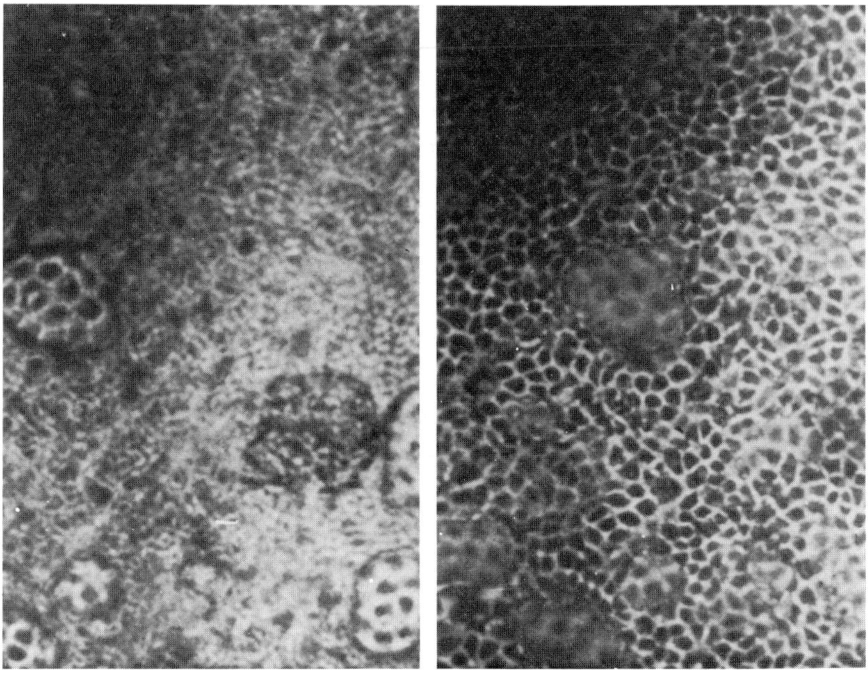

Fig. 1. Dome formation by MDCK cells. MDCK cells were grown to confluency in Dulbecco's modified Eagle's medium supplemented with 10% fetal calf serum. When domes were observed in the culture dish, the cells in one microscope field were photographed in two different focal planes. From Taub et al. [1979], with permission.

supplements. The MDCK colonies that form in medium K-1 can be isolated from the culture dishes with cloning rings. Trypsin action in the rings may be inhibited by the addition of soybean trypsin inhibitor directly to the cloning rings.

MDCK monolayers may be used for a number of growth and functional studies in defined medium. The mechanisms by which hormones and other agents induce dome formation may be examined. Confluent monolayers of MDCK cells may be used for studies concerning membrane transport and hormone receptors. In order to examine the effects of hormones on these functions, the hormone supplements in the defined medium may be manipulated, so that the effects of particular hormones may be studied. Synergistic interaction among hormones can also be analyzed.

Growth of Other Established Kidney Cell Lines in Defined Medium

The hormonally defined serum-free medium for MDCK cells can be used as a means to grow other established kidney epithelial cell lines and primary cell cultures in the absence of serum. Table IV illustrates the growth properties of several other established kidney epithelial cell lines in medium K-1, following a serum preincubation. The LLC-MK_2 monkey kidney and the Madin Darby bovine kidney (MDBK) cell lines both grow in medium K-1 at an equivalent or greater rate than in medium supplemented with 10% fetal calf serum. In contrast the BSC-1 cell line, and an epithelial subline of normal rat kidney (NRK) cells, grow at a very low rate in K-1 medium compared with serum-supplemented medium. Consequently, the hormonally defined serum-free medium for MDCK cells is *not* optimal for the growth of a number of kidney epithelial cell lines in the absence of serum.

However, the hormonally defined medium for MDCK cells can be used as a starting point in the development of serum-free culture media for such kidney epithelial cell lines. For example, medium K-1 for MDCK was not an optimal growth medium for the LLC-PK_1 cell line. Unlike MDCK this pig kidney cell line has several of the transport properties of the proximal tubule (Table II). Nevertheless, medium K-1 was a useful starting point in the development of the defined medium for LLC-PK_1 cells. LLC-PK_1 cells were shown to grow in response to four of the five supplements in the MDCK medium (i.e., insulin, transferrin, T_3, and hydrocortisone). These four supplements were finally added to the defined medium developed for LLC-PK_1 (medium K-2, Table III). In addition, arginine-vasopressin and cholesterol were also

TABLE IV. Group of Kidney Cell Lines in Medium K-1

Cell type		Growth rate (doublings/day)	
Cell line	Species of origin	SF-DME/F12 + FCS	Medium K-1
LLC-MK_2 (7)	Monkey	1.6	1.6
MDBK (8)	Bovine	0.8	1.0
BSC-1	Monkey	0.7	0.4
NRK (15)	Rat	1.3	0.1
RAG (16)	Rat	1.6	1.2
MDCK (3)	Canine	1.5	1.5

Cells were inoculated at 2.5×10^4/60-mm dish into dishes containing SF-DME/F12 supplemented with 10% fetal calf serum. The next day the medium was removed by aspiration, and the cells were washed three times with SF-DME/F12 and then incubated with medium K-1 or SF-DME/F12 supplemented with 10% fetal calf serum. The growth rate was estimated from daily cell counts in duplicate dishes over a 5-day period. From Taub and Sato [1979], with permission.

added to the LLC-PK_1 medium. Arginine-vasopressin increases cAMP synthesis in LLC-PK_1 and thus may act by a pathway that is similar to that initiated by PGE_1 in MDCK cells. The hormonally defined medium for LLC-PK_1 permits the long-term growth of LLC-PK_1 cells at the same rate as in serum-supplemented medium. LLC-PK_1 cells also form domes at confluency in this defined medium.

GROWTH OF PRIMARY CULTURES OF KIDNEY EPITHELIAL CELLS IN HORMONALLY DEFINED, SERUM-FREE MEDIUM

General Approach

The use of hormone-supplemented serum-free medium has alleviated a number of problems previously encountered with kidney primary cell cultures. Included among such problems are the low yield of kidney epithelial cells obtained in the primary cultures, and extensive overgrowth of the primaries by fibroblasts. As a consequence the reproducibility of results may be poor. Such problems may be attributed to the procedures used in preparing kidney cells for primary culture, as well as the cell culture conditions. Tissue culture media containing serum are usually not optimal for the long growth and survival of primary kidney epithelial cell cultures [Taub and Sato, 1980]. However, such serum-supplemented media do support the growth of kidney fibroblasts.

The use of primary kidney cell cultures has been limited not only by the presence of fibroblasts, but also by the presence of numerous types of epithelial cells in the cultures. Primary cultures of kidney epithelial cells have usually been initiated from cells obtained from kidney slices. Because of the complex structure of the kidney, the cells in such slices originate from a number of nephron segments. Consequently, the cell cultures that have been obtained from such slices are only of limited use in studies concerned with the functional properties of particular cell types in the kidney.

The problems may be alleviated by the use of hormonally defined serum-free media for primary kidney cell cultures. Such hormonally defined serum-free media can be developed by means of several approaches. Hormonally defined serum-free media may be developed for particular established kidney epithelial cell lines. These media can then be applied towards primary kidney cell cultures [Taub et al., 1979]. Alternatively, hormonally defined serum-free media can be developed directly for the primary kidney cells of interest. The growth requirements of particular epithelial cell types in the kidney can be determined (particular renal cell types can be purified for these purposes). These studies may then facilitate the development of media that permit the selective growth of particular types of kidney cells in primary cultures, without fibroblast overgrowth.

Primary Cultures of Kidney Epithelial Cells in Medium K-1

Baby mouse kidney primaries in medium K-1. Medium K-1, the hormone-supplemented serum-free medium developed for the MDCK cell line, can be used for the selective growth and/or maintenance of primary cultures of kidney epithelial cells without fibroblast overgrowth. The method for preparing baby mouse kidney primary cells for growth in medium K-1 is described below; it can be applied to kidney primaries derived from other species [Taub and Sato, 1980].

The primaries are initiated from the kidneys of the 10-day-old mice (Balb/c). To obtain the kidneys, the mice are killed by cervical dislocation (forceps are applied to the neck, and the tail is pulled). To insure sterility, the mice are then doused in 70% ethanol. The kidneys are removed with sterile forceps and are placed in a 60-mm-diameter culture dish. The kidneys are washed several times with serum-free culture medium that contains antibiotics (192 IU/ml penicillin and 200 μg/ml streptomycin), so as to reduce the probability of contamination. The kidneys are then transferred into another 60-mm dish and are minced into 5-mm-diameter pieces (a curved-blade scissors, #M5400, V. Mueller Co., is useful for such purposes).

The kidney tissue is now ready to be dissociated, so as to obtain isolated tubules in suspension. Towards these ends, the chunks of kidney tissue are transferred into a sterile 10-ml culture tube containing 1 mg/ml collagenase (class IV, Worthington/Millipore Corp.) and 1 mg/ml soybean trypsin inhibitor in serum-free medium. The tissue is incubated for 15–20 min at 37°C, with occasional shaking. After this incubation period, the tube is shaken vigorously so as to dislodge any dissociated nephron segments from the chunks of kidney tissue. The chunks of tissue are allowed to settle to the bottom of the tube.

The serum-free medium, which contains the nephron segments (tubules and glomeruli) in suspension, is transferred into another culture tube. The isolated tubules and glomeruli in this suspension are transferred into another culture tube and are washed twice by centrifugation. These nephron segments are finally resuspended into serum-free medium.

The remaining kidney tissue in the other culture tube is resuspended into serum-free medium containing 1 mg/ml collagenase and 1 mg/ml soybean trypsin inhibitor as before. This material is incubated 15–20 min at 37°C. Dissociated tubules and glomeruli are again removed and washed as described above. This process is continued until the appropriate number of tubules and glomeruli are obtained in suspension.

Primary kidney epithelial cell cultures can be initiated with either a suspension of kidney cells or isolated nephron segments. A suspension of kidney cells can be obtained by subjecting the nephron segments briefly to trypsin treatment. Trypsin action is stopped with either soybean trypsin inhibitor (as described above for kidney cell lines) or serum. Then the cells are washed twice by centrifugation and resuspended in defined medium for cell culture. Prolonged incubation periods with trypsin not only are deleterious to the subsequent attachment and growth of the primary kidney cells, but also may cause cell lysis. In general, higher yields of cells in culture are obtained when isolated renal tubules, rather than cell suspensions, are used.

Baby mouse kidney cells are plated into culture dishes at a high density in order to obtain confluent monolayers (more than 5×10^4 cells per 35-mm-diameter dish are plated). The kidney cells attach during the first day in culture. Subsequently, cells that are primarily epithelial in morphology grow at 0.6 doublings per day, either in medium K-1 or in medium supplemented with 10% fetal calf serum. Confluent monolayers of baby mouse kidney epithelial cells are obtained after a 1-week growth period in medium K-1. The primary kidney epithelial cell cultures also become confluent in serum-supplemented medium. However, confluent monolayers of kidney epithelial cells maintained in serum-supplemented medium eventually become overgrown with fibroblasts, unlike monolayers in K-1.

The characteristics of confluent monolayers of baby mouse kidney cells in medium K-1 are summarized in Table V. The cells form domes. Amiloride-sensitive sodium uptake, a distal tubule marker, is observed, whereas sodium-dependent α-methylglucoside uptake, a proximal tubule marker, is not observed [Taub and Sato, 1980]. Brush border enzymes including leucine aminopeptidase and alkaline phosphatase are also present. Consequently, the baby mouse kidney primaries exhibit characteristics similar to those observed in the MDCK cell line.

Baby mouse kidney cells plated at lower densities (less than 2.5×10^4 cells per 35-mm dish) grow to form individual colonies. Colony formation can be observed after a 1-week culture period. The colonies that form in medium K-1 are uniformly epithelial in morphology, whereas fibroblast colonies form in serum-supplemented medium. The formation of epithelial colonies in medium K-1 is dependent upon the presence of each of the five supplements in this medium [Taub and Sato, 1980]. The individual deletion of PGE_1 or transferrin from the medium has a more dramatic inhibitory effect on epithelial colony formation than the individual deletion of any one of the other three supplements. In this respect, the growth requirements of the baby mouse kidney cells resemble those of MDCK.

TABLE V. Functional Properties of Cultured Kidney Epithelial Cells in Defined Medium: Comparison of Primaries With MDCK Cells

Cell type	Primary cultures of rabbit kidney proximal tubule cells	Primary baby mouse kidney	MDCK
Growth medium	Medium RK-1	Medium K-1	Medium K-1
Morphology	Epithelial	Epithelial	Epithelial
Dome formation	+	++	+++
Amiloride sensitive of $^{22}Na^+$ uptake	—	Yes	Yes
(Amiloride conc. required for 50% inhibition of 1.1 mM $^{22}Na^+$ uptake)	—	5×10^{-7} M	5×10^{-4} M
Na^+-dependent α-methyl-glucoside (αMG) uptake (initial rate of uptake)	Yes 0.18 nmoles αMG accum./min/mg protein	None detected	None detected
γ-Glutamyl transpeptidase activity (nmoles p-nitro-phenol released/mg protein)	2.7	—	2.0
Leucine amino peptidase activity (nmoles p-nitro-anilide released/min/mg protein)	12.0	1.77	0.5
Alkaline phosphatase activity (nmoles p-nitro-phenol released/min/mg protein)	0.75	—	0.2
Hormones that stimulate cAMP production	Parathyroid hormone	—	Arginine-vasopressin; isoproterenol

Primary cultures derived from purified proximal tubules were grown in medium RK-1 for 2 weeks [Chung et al., 1982]. MDCK and primary baby mouse kidney cell cultures were grown to confluency in medium K-1 [Taub and Sato, 1980]. The frequency of dome formation was determined by scanning 30 microscope fields. The cultures were classified as having a high frequency (+++), a moderate frequency (++), or a low frequency (+) of domes if more than 75%, 25%, or 5%, respectively, of the microscope fields had domes. The uptake of 1 mM ^{14}C-α-methylglucoside (αMG) was determined over a 10-min period at 23°C as described by Chung et al. [1982]. The inhibitory effect of amiloride on ouabain-insensitive $^{22}Na^+$ uptake was similarly determined with 1 mM NaCl at 23°C, as described by Taub and Saier [1979]. Leucine aminopeptidase activity, alkaline phosphatase activity, and γ-glutamyl trans-peptidase activity were measured by assaying the hydrolysis of L-leucine-p-nitroanilide,p-nitrophenyl phosphate, and L-γ-glutamyl-p-nitroanilide, respectively [Chung et al., 1982].

Baby mouse kidney colonies composed of fibroblasts can be grown in serum-free medium. When baby mouse kidney cells are plated at a low density into serum-free medium supplemented with 5 μg/ml insulin and 5 μg/ml transferrin, colonies consisting of fibroblasts are obtained as well as epithelial colonies [Taub and Sato, 1980]. PGE_1 and hydrocortisone have an inhibitory effect on the formation of the baby mouse kidney fibroblast colonies. Thus the absence of baby mouse kidney fibroblast colonies in medium K-1 may be explained by the presence of growth-inhibiting supplements in the medium, as well as by the absence of supplements required for optimal fibroblast growth.

Adult mouse kidney primaries in medium K-1. Adult mouse kidney primary cultures also can be grown in serum-free medium. However, another procedure is used for preparing the kidney tissue for primary culture. The collagenase digestion procedure described above for baby mouse kidney primaries results in only very low yields of adult mouse kidney cells (the collagenase treatment has deleterious effects on the adult mouse kidney cells). A faster procedure, which results in higher yields of cells in culture, is as follows.

The kidney from an adult mouse is placed in a 60-mm-diameter culture dish and is washed with serum-free culture medium. The kidney is decapsulated and transferred into another culture dish. Kidney cortical slices are removed with a sterile 7-ml Dounce tissue homogenizer containing culture medium. The kidney tissue is then subjected to six strokes of a loose pestle (type B), which releases nephron segments (tubules and glomeruli) into solution. The nephron segments are transferred to a 10-ml sterile centrifuge tube and are washed several times by centrifugation. The tubules and glomeruli are then treated for 2 min at 37°C with 0.125 mg/ml collagenase (class IV, Millipore/Worthington) and 0.025 mg/ml soybean trypsin inhibitor in serum-free medium. The collagenase treatment is stopped by centrifugation of the material. The tubules and glomeruli are again washed by centrifugation and are plated into culture dishes containing medium K-1. As observed with baby mouse kidney primary cells, these adult mouse kidney cells will grow to form confluent monolayers of epithelial cells interspersed with domes.

Growth and maintenance of primary kidney cultures derived from a number of species in medium K-1. Primary cultures of kidney epithelial cells derived from a number of species can be similarly grown and/or maintained in medium K-1. Human embryonic, adult bovine, rat, canine, and even wolf kidney epithelial cells have been prepared by the collagenase digestion procedure described for baby mouse kidney cells. These cells attach and grow in culture dishes containing medium K-1, with varying efficiencies.

The primary kidney epithelial cell cultures derived from these species can be maintained for a number of weeks in medium K-1, with routine medium changes. The growth of kidney fibroblasts derived from a number of species is inhibited in medium K-1. However, PGE_1 and hydrocortisone are not necessarily growth-inhibiting to all fibroblasts. Indeed the growth of diploid human fibroblasts is reportedly stimulated by hydrocortisone.

Growth of Different Subpopulations of Renal Cells in Defined Medium

Summary of available methodologies. The kidney cells derived from particular nephron segments can be purified and grown in hormonally defined serum-free medium. Several of the available methods for purifying particular types of renal cells are described below. A number of cell culture studies using such purified renal cells indicate that different types of epithelial cells in a particular kidney differ with regard to their growth requirements for hormones. Such differences are not surprising considering that the cells in the various nephron segments differ dramatically with regard to their metabolic capacities [Schmidt and Guder, 1976], as well as their hormonal responses [Morel, 1981]. Primary cultures of kidney epithelial cells derived from unpurified preparations of rabbit renal tubules not only grew in response to insulin, transferrin, and hydrocortisone, but also exhibited dramatic growth responsiveness to epidermal growth factor (EGF) and T_3. However, when the proximal tubule cells from the rabbit kidney were purified and cultured, the dramatic growth response to EGF and T_3 was no longer observed [Chung et al., 1982]. This study suggested that different subpopulations of rabbit kidney tubule cells differ with regard to their growth responsiveness to medium supplements. The studies of Wilson and Horster [1982] with microdissected renal tubules similarly suggested that differences exist in the growth requirements of the cells in the thick ascending limb of Henle and the cortical collecting duct, for example.

Thus the manipulation of the growth supplements in serum-free media (e.g., hormones, growth factors) may facilitate the selective growth of particular types of cells in the kidney. In addition, the modification of the basal medium is also an important factor in the development of selective media for the different cells types in the kidney. Methodologies for the development of optimal basal media for particular types of differentiated cells have been described by Ham [1982] and by Ham and McKeehan [1979]. The choice of the appropriate substratum for cell attachment may also be important in facilitating the selective growth of the cells in the different nephron segments.

In order to develop appropriate hormonally defined media, the rabbit kidney cells derived from particular nephron segments must be purified for subsequent cell culture studies. A number of procedures for purifying the different cell types in the kidney have been described. Particular types of kidney cells have been purified from single-cell suspensions on the basis of their differences in density from other kidney cell types. For example, the cells from the thick ascending limb of Henle's loop have been purified from a suspension of rabbit renal medullary cells with a Ficoll gradient [Eveloff et al., 1980]. Proximal tubule cells from the rat kidney have similarly been purified from a suspension of trypsinized cells by means of Ficoll density gradient centrifugation, as well as by free-flow electrophoresis [Kreisberg et al., 1977a,b]. However, the trypsin treatment that is used to obtain the single-cell suspensions is deleterious to the subsequent growth and attachment of the tubule epithelial cells. Consequently, when these procedures are followed, only a low yield of cultured cells is obtained.

Higher yields of cultured cells are obtained when renal tubules rather than isolated cells are used directly for cell culture. Particular types of renal tubules and/or glomeruli have been purified from suspensions of nephron segments with sieves [Chung et al., 1982; Burlington and Cronkite, 1973; Quadracci and Stricker, 1970]. These purified nephron segments can be put into culture as described below.

In order to insure that the cell types of interest are being accurately characterized in such in vitro studies, primary cultures derived from appropriate microdissected nephron segments can also be initiated [Horster, 1980]. Thus if the supplements that stimulate the growth of primary cultures of purified proximal tubule cells are identified, the effects of these supplements on microdissected tubules put into culture can be similarly examined.

Primary cultures of rabbit kidney proximal tubule cells. Proximal tubules from the rabbit kidney are prepared for cell culture as described by Chung et al. [1982] and as summarized in Figure 2. An adult male New Zealand White rabbit (6–7 lb) is killed by cervical dislocation. A kidney is removed with the renal artery intact (the kidney on the left side has the larger artery of the two kidneys). This kidney is placed in a sterile dish and washed several times with serum-free medium. A sterile 18-gauge needle (with the end blunted with a file) is inserted into the renal artery. The needle is held in place with a sterile suture (extra strong carpet thread will do). With a 50-ml syringe the kidney is perfused, first with PBS and then with an iron oxide solution (the iron oxide solution is prepared as described in Table VI). During the perfusion, iron oxide particles become associated with the glomeruli but do not enter the tubules. As a result of the perfusion, the kidney appears grey-black in color.

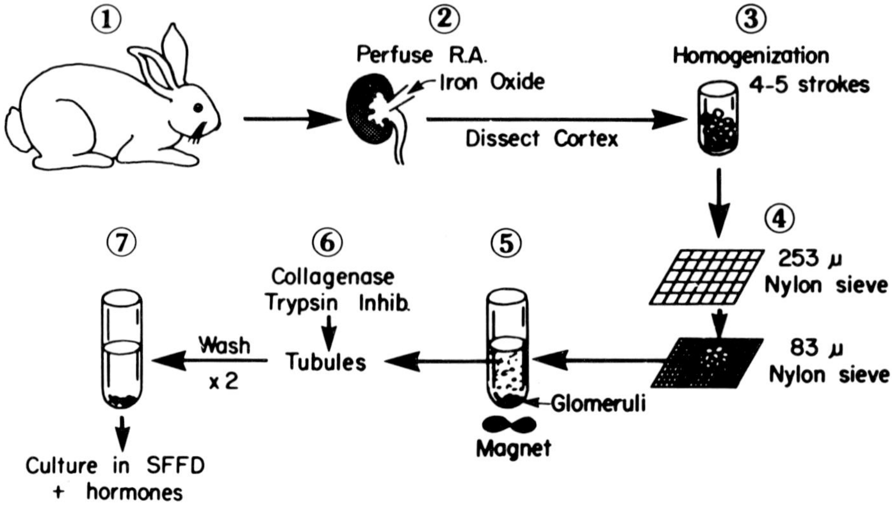

Fig. 2. Method for preparing primary cultures of rabbit kidney proximal tubule cells. Details are described in Chung et al. [1982].

TABLE VI. Preparation of Iron Oxide Solution

1. Sodium hydroxide (2.6 g) and potassium nitrate (20 g) were dissolved in 100 ml of oxygen-saturated water.
2. Ferrous sulphate (9 g) is dissolved in 100 ml of oxygen-saturated water. The solution made in step 1 is added to the ferrous sulphate solution.
3. The mixture is boiled for 20 min.
4. The resulting precipitate of magnetic iron oxide is washed 5–10 times as follows. The precipitate is resuspended in water and is then attracted to the bottom of the wash flask with a strong magnet. The wash water is then decanted.
5. The iron oxide is resuspended in several liters of 0.9% NaCl. This concentrated iron oxide solution is then autoclaved for future use. Immediately prior to use, the iron oxide solution can be diluted in PBS. See Cook and Pickering [1958] for details.

The perfused kidney is carefully washed in a 100-mm-diameter dish and decapsulated. The kidney is then transferred into another 100-mm dish and is washed several times with serum-free medium. The cortex is removed, with sterile, curved-blade dissecting scissors. The cortical slices obtained in this manner are transferred into a sterile 7-ml Dounce tissue homogenizer containing serum-free medium. This cortical material is then subjected to six strokes of a type B (loose) pestle. The resulting suspension of nephron segments is transferred into a sterile 50-ml centrifuge tube.

In order to purify the proximal tubules from this suspension of nephron segments, the suspension is passaged through a series of two 9-cm-diameter sieves (the sieves are constructed from nylon mesh, 253 μm or 83 μm in diameter, obtained from Tetko, Elmsford, NY). The first sieve in the series (253 μm) removes large aggregates of nephrons, whereas the second sieve (83 μm) separates the distal tubules from proximal tubules and glomeruli. The distal tubules are narrow in diameter and pass through the sieve. After the tubule suspension has been passed over the sieves, 300-500 ml of PBS is washed through the sieves to facilitate the separation of the tubules and glomeruli.

The purified proximal tubules and glomeruli are removed from the top of the second sieve, with a sterile spatula. This material is transferred into a sterile 50-ml centrifuge tube containing 25 ml serum-free medium and a sterile magnetic stirring bar. The tubules are incubated for 5 min at 37°C, which permits the glomeruli covered with iron oxide to become associated with the stir bar. To insure the removal of the glomeruli, the purified proximal tubules are transferred into another 50-ml culture tube containing serum-free medium and a stir bar.

The suspension of purified proximal tubules is transferred into another 50-ml centrifuge tube containing 0.050 mg/ml collagenase (class IV, Worthington/Millipore) and 0.050 mg/ml soybean trypsin inhibitor (Gibco). The tubules are incubated for 2 min at 37°C. The reaction is stopped by centrifugation for 5 min in a desktop centrifuge. The tubules are again washed by centrifugation and resuspended in serum-free medium. In order to quantitate the amount of renal material in suspension, the protein content is determined by the method of Bradford [1976], or the DNA content is assayed by the method of Labarca and Paigen [1980]. The tubule suspension is then ready for cell culture.

The purified proximal tubules are suspended at 0.5 mg protein per ml into serum-free medium supplemented with 5 μg/ml insulin, 5 μg/ml transferrin, and 5×10^{-8} M hydrocortisone (medium RK-1, Table III). In order to insure uniformity of the suspension, the tubules are stirred with a magnetic stirring bar. This suspension of purified proximal tubules is inoculated into culture dishes at 2 ml per 35-mm dish, with continued stirring. The culture dishes containing the purified proximal tubules are then placed in a 5% CO_2/95% air humidified incubator.

The proximal tubules attach to the dishes within the first day of the culture period. The medium should be changed after the first day, so as to remove unattached tubules, and medium changes should occur routinely every 4 or 5 days thereafter. After 3-4 days in culture, extensive outgrowth of epithelial cells from the proximal tubules is observed. After 2 weeks in culture, confluent monolayers of kidney epithelial cells are observed in medium RK-1 without fibroblast overgrowth (Fig. 3). In contrast, fibroblast overgrowth is observed with serum-supplemented medium. One hundred confluent monolayers of cells in 35-mm dishes containing RK-1 can be obtained from a single proximal tubule preparation.

Two-week-old primary proximal tubule cell cultures exhibit a number of functions typical of the proximal tubule (Table IV). The monolayers exhibit sodium-dependent α-methylglucoside uptake, which is sensitive to inhibition by phlorizin and ouabain. A transport system with these properties is distinctive of the proximal tubule of the kidney. The primary rabbit proximal tubule cultures also exhibit sodium-dependent phosphate transport (Taub, unpublished observation), brush border enzymes, and parathyroid hormone-sensitive cyclic AMP production [Chung et al., 1982]. Cyclic AMP production is not stimulated by either arginine-vasopressin or calcitonin. Consequently, these primary cultures have the hormone responses typical of proximal tubule cells and can be used to study hormonal regulation of membrane transport functions.

Primary cultures of proximal tubule cells derived from other species. Primary cultures of kidney proximal tubule cells derived from other species can be similarly prepared for growth in hormonally defined serum-free medium. However, several modifications in methodology may be important. The kidneys from small animals such as the rat can be perfused in situ via the aorta. Both kidneys are perfused simultaneously by this means, so that more material can be obtained. Sieves of 253-μm and 83-μm mesh are appropriate for separating the nephron segments from the rat kidney, for example. However, the mesh size of the two sieves used in the separation of the nephron segments can vary with the kidney of interest. The hormone supplements added to the culture medium may also differ when the proximal tubule cells from different species are cultured. For example, medium K-1 for MDCK cells may be important to use for the selective growth of the epithelial cells in mouse and rat kidney proximal tubule cell populations, rather than fibroblasts. PGE_1, a component in medium K-1, is required to inhibit the growth of mouse kidney fibroblasts, for example [Taub and Sato, 1980].

Primary cultures of rabbit kidney epithelial cells derived from microdissected nephron segments. Primary cultures of kidney epithelial cells can also be initiated from microdissected nephron segments [Horster, 1980; Cohen and Fine, 1982]. A 1-mm slice of renal tissue is removed from a rabbit kidney and is transferred into a dish containing chilled medium. Appropriate nephron segments are individually isolated by microdissection

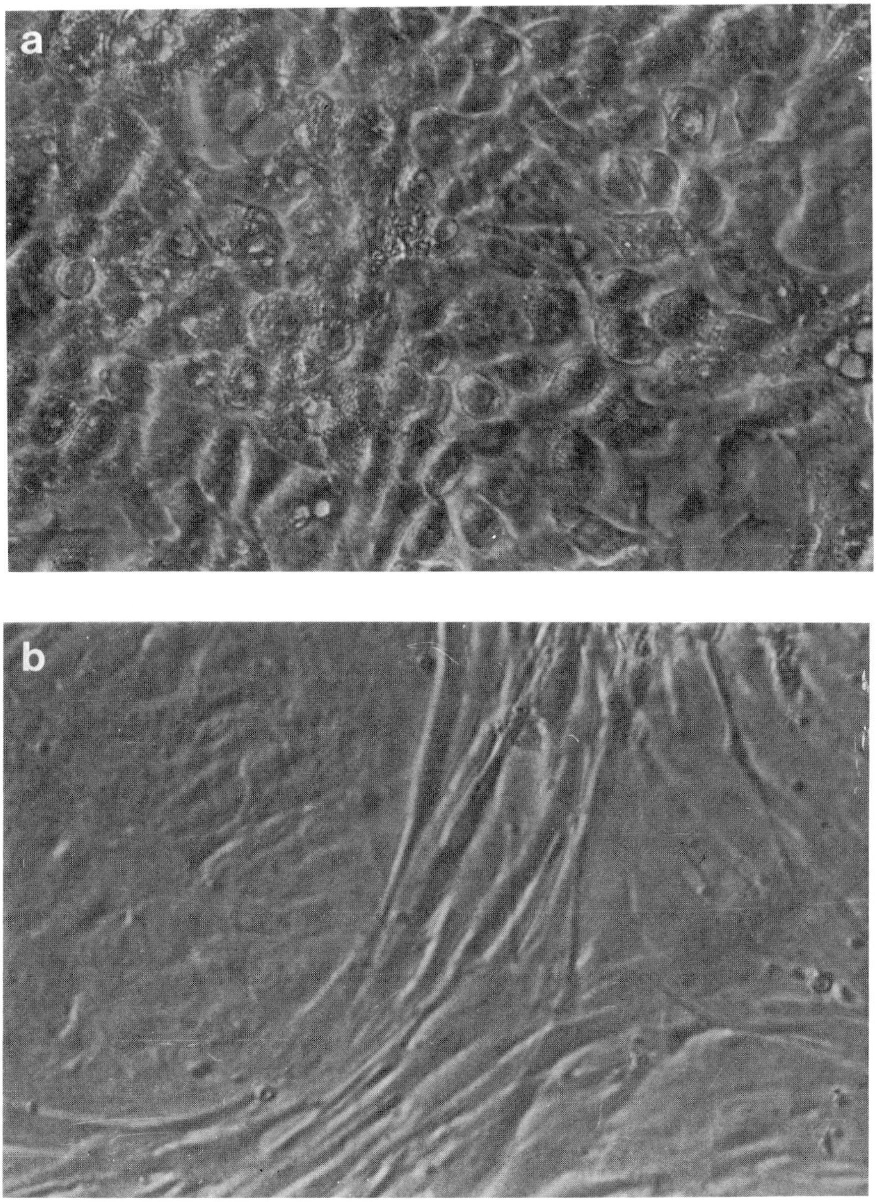

Fig. 3. Primary cultures of rabbit kidney proximal tubule cells in medium RK-1 (a) and in medium supplemented with 10% fetal calf serum (b). Three weeks later representative microscope fields of the cultures were photographed. From Chung et al. [1982], with permission.

(without use of any chemical disaggregation) from the fresh renal slice. The dissected nephron segments are transferred in a drop of medium into Terasaki microtiter wells (Falcon) containing appropriate hormonally defined serum-free medium [Cohen and Fine, 1982]. The monolayers can also be grown on millipore filters (THWP 2500) that permit microscopic visualization or on plasma-coated dishes [Horster, 1979]. Wilson and Horster [1982] pierce the nephron segments with forceps at intervals to stimulate cell migration and proliferation.

Hormone-supplemented serum-free media have not yet been designed for the optimal growth of primaries derived from such microdissected rabbit nephron segments. However, rabbit cortical collecting tubules and proximal convoluted tubules have been grown in medium K-1 for MDCK [Cohen and Fine, 1982] and in medium K-2 for LLC-PK_1 cells [Chuman et al., 1982], respectively. Furthermore Horster [1980] and Wilson and Horster [1982] have examined the effects of hormone supplements on the growth of epithelial cells from several nephron segments in a quantitative manner. Thus, optimal hormonally defined media may be developed for microdissected kidney tubule cells in the future.

SUMMARY

Hormonally defined serum-free media have been developed for established kidney epithelial cell lines and primary kidney cell cultures. A defined medium was first developed for the established kidney epithelial cell line Madin Darby canine kidney (MDCK) [Taub et al., 1979]. This hormonally defined medium, medium K-1, has been used to grow a number of different types of kidney epithelial cells in culture. Although medium K-1 is not optimal for all cultured kidney cells, this medium has been a useful starting point in developing defined media for other established kidney cell lines such as LLC-PK_1 [Chuman et al., 1982]. These cell lines can then be used to study hormonal regulation of renal functions

Hormonally defined media such as K-1 have also been used for the growth of primary cultures of kidney epithelial cells [Taub et al., 1979; Taub and Livingston, 1981]. Previous studies with kidney primary cultures in serum-supplemented medium have been limited by the many different types of epithelial cells in the cultures, as well as by fibroblast overgrowth. The hormonally defined culture media have been used to selectively grow primary kidney epithelial cell cultures without fibroblast overgrowth [Taub and Sato, 1980]. Several approaches have been taken in this regard. The hormonally defined serum-free media developed for established kidney epithelial cell lines have been used for kidney primaries [Taub et al., 1979]. In addition hormonally defined serum-free media have been developed directly for the kidney primary cells of interest [Chung et al., 1982]. Such hormonally defined serum-free media have facilitated the development of primary kidney cell culture systems that retain the functional properties of kidney tubule cells. Such primary cell culture systems should prove to be invaluable in validating the physiological significance of studies with established kidney cell lines. Such primary culture systems should also facilitate the study of renal cell types not previously examined in vitro.

A number of the kidney cell lines available for study have retained properties that are characteristic of the transporting epithelial cells in the kidney tubule. The Madin Darby canine kidney cell line is typical of such kidney epithelial cell lines. The MDCK cells possess a morphological polarity in culture that is unique to transporting epithelial cells. The mucosal surface of the cells in the monolayer uniformly faces the culture medium, and their serosal surface faces the culture dish; adjacent cells in the culture are interconnected by tight junctions [Leighton et al., 1969]. As a result, a continuous sheet of MDCK cells is formed at confluency, which is similar to a frog skin or a toad bladder. Multicellular domes or hemicysts may form in such confluent monolayers of MDCK cells (Fig. 1). The formation of these domes (groups of cells slightly elevated from the tissue culture dish surface) may be explained by the transport of salt and water across the cell layer (Fig. 4). The transported salt and water trapped between the cell layer and the dish, give rise to hydrostatic pressure and dome formation.

The transepithelial salt and water transport by MDCK cells has been measured by Misfeldt et al. [1976] and Cereijido et al. [1978]. Confluent mono-

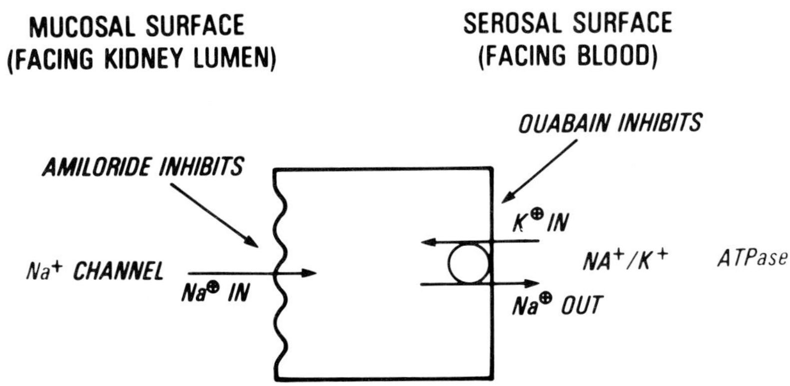

Fig. 4. MDCK: A model of transporting kidney epithelial cells.

layers of MDCK cells on a collagen-coated permeable support were put into an Ussing flux chamber. A spontaneous transepithelial potential difference (PD) that was sensitive to inhibition by amiloride was observed. In addition a transepithelial resistance and water flux were measured. MDCK cells had a low resistance (84 ohm/cm^2), typical of the leaky epithelial cells in the proximal tubule. However, MDCK cells had biochemical properties typical of the distal tubule and the loop of Henle. Included among these properties is amiloride-sensitive Na^+ transport [Rindler et al., 1979b], furosemide-sensitive chloride transport [McRoberts et al., 1982], and arginine-vasopressin-sensitive cyclic AMP production [Rindler et al., 1979a] (Table I).

Dome formation and transepithelial solute transport has similarly been observed in the LLC-PK_1 cell line. A transepithelial PD of 2.8 mV and a resistance of 211 ohm/cm^2 was observed when D-glucose was present [Misfeldt and Sanders, 1981]. The addition of phlorizin to the mucosal bath or phoretin to the serosal bath caused a decrease in the PD. These studies suggested that transepithelial sugar transport occurs in LLC-PK_1 as is observed in the proximal tubule of the kidney. Labeled flux studies similarly indicated that LLC-PK_1 cells possess a sodium-dependent sugar transport system that is typical of proximal tubule cells [Rabito and Ausiello, 1980; Mullin et al., 1980]. The growth and functional properties of LLC-PK_1 are summarized in Table II.

Primary kidney cell culture systems, being obtained directly from the animal, similarly retain functional properties of kidney tubule cells. The kidney primaries can be initiated from the animal of interest. Thus, if previous biochemical studies have been done with rat kidney material, for example, rat kidney primary cultures can be initiated for studying the same phenomena in vitro. Similarly, if a medical disease state of the kidney has previously been examined in vivo, the rat being used as the animal model, for example, then rat kidney primary cultures can be initiated for in vitro studies of the disease.

The use of hormonally defined serum-free culture media is particularly advantageous for these in vitro studies. The primary cells of interest can be grown by choosing the appropriate hormone supplements. The hormonal and nutritional environment of the cells can be readily manipulated, in a reproducible manner. The direct effects of these factors on cell function can be examined, whereas previously hormonal effects were often masked by the serum.

ACKNOWLEDGMENTS

The author acknowledges Gloria Viola for preparation of the manuscript. The work was funded by NIH grant 1 ROI CA28112-04 and NIH Career Development Award 1 KO4 CA00888-01 to M. Taub.

REFERENCES

Bradford M (1976): A rapid and sensitive method for the quantitation of microgram quantities of protein utilizing the principle of protein-dye binding. Anal Biochem 72:248-254.

Burlington H, Cronkite EP (1973): Characteristics of cell cultures derived from renal glomeruli. Proc Soc Exptl Biol Med 142:143-149.

Cereijido M, Robbins ES, Dolan WJ, Rotunno CA, Sabatini DD (1978): Polarized monolayers formed by epithelial cells on a permeable and translucent support. J Cell Biol 77:853-880.

Chuman L, Fine LG, Cohen AI, Saier MH (1982): Continuous growth of proximal tubular kidney epithelial cells in hormone-supplemented serum-free medium. J Cell Biol 94:506-510.

Chung SD, Alavi N, Livingston D, Hiller S, Taub M (1982): Characterization of primary rabbit kidney cultures that express proximal tubule functions in a hormonally defined medium. J Cell Biol 95:118-126.

Cohen AI, Fine L (1982): Personal communication.

Cook WF, Pickering GW (1958): A rapid method for separating glomeruli from rabbit kidney. Nature 182:1103-1104.

Eveloff J, Haase W, Kinne R (1980): Separation of renal medullary cells: Isolation of cells from the thick ascending limb of Henle's loop. J Cell Biol 87:672-681.

Goldring SR, Dayer JM, Ausiello DA, Krane SM (1978): A cell strain cultured from porcine kidney increases cyclic AMP content upon exposure to calcitonin or vasopressin. Biochem Biophys Res Com 83:434-440.

Ham RG (1982): Importance of the basal nutrient medium in the design of hormonally defined medium. In Sato GH, Pardee AB, Sirbasku DA (eds): Cold Spring Harbor Conferences on Cell Proliferation Vol 9, Book A. Cold Spring Harbor, New York: Cold Spring Harbor Laboratory, pp 39-60.

Ham RG, McKeehan WL (1979): Media and growth requirements. Methods Enzymol 58:44-93.

Horster M (1979): Primary culture of mammalian nephron epithelia. Requirements for cell outgrowth and proliferation from defined explanted nephron segments. Plügers Arch 382:209-215.

Horster M (1980): Hormonal stimulation and differential growth response of renal epithelial cells cultivated in vitro from individual nephron segments. Int J Biochem 12:29-35.

Hull RN, Cherry WR, Weaver GW (1976): The origin and characteristics of a pig kidney cell, strain LLC-PK_1. In Vitro 12:670-677.

Kreisberg J, Pitts AM, Pretlow TG (1977a): Separation of proximal tubule cells from suspensions of rat kidney cells in density gradients of Ficoll in tissue culture medium. Amer J Path 86:591-602.

Kreisberg JL, Sachs G, Pretlow TG, McGuire RA (1976): Separation of proximal tubule cells from suspensions of rat kidney cells by free-flow electrophoresis. J Cell Physiol 93:169-172.

Labarca C, Paigen K (1980): A simple, rapid, and sensitive DNA assay procedure. Anal Biochem 102:344-352.

Leighton J, Brada Z, Estes LW, Justh G (1969): Secretory activity and oncogenicity of a cell line (MDCK) derived from canine kidney. Science 158:472-473.

Leighton J, Estes LW, Mansukhani S, Brada Z (1970): A cell line derived from normal dog kidney (MDCK) exhibiting qualities of papillary adenocarcinoma and of renal tubular epithelium. Cancer 26:1022-1028.

Ludens JH, Vaughn DA, Mawe RC, Fanestil DD (1978): Specific binding of deoxycorticosterone by canine kidney cells in culture. J Ster Biochem 9:17-21.

McRoberts JA, Erlinger S, Rindler MJ, Saier MH (1982): Furosemide-sensitive salt transport in the Madin-Darby canine kidney cell line. J Biol Chem 254:2260-2266.
Meier KE, Insel PA (1982): Clonal variation in the expression of catecholamine receptors in MDCK cells. J Cell Biol 95:416a.
Misfeldt DS, Sanders MJ (1981): Transepithelial transport in cell culture: D-glucose transport by a pig kidney cell line (LLC-PK_1). J Membr Biol 59:13-18.
Misfeldt DS, Hamamoto ST, Pitelka DR (1976): Transepithelial transport in cell culture. Proc Natl Acad Sci USA 73:1212-1216.
Morel F (1981): Sites of hormone action in the mammalian nephron. Amer J Physiol 240:F159-F164.
Mullin JM, Weibel J, Diamond L, Kleinzeller A (1980): Sugar transport in the LLC-PK_1 renal epithelial cell line: Similarity to mammalian kidney and influence of cell density. J Cell Physiol 104:375-389.
Quadracci LJ, Stricker GE (1970): Growth and maintenance of glomerular cells in vitro. Proc Soc Exptl Biol Med 135:947-950.
Rabito CA, Ausiello DA (1980): Na^+ dependent sugar transport in a cultured epithelial cell line from pig kidney. J Mem Biol 54:31-38.
Rabito CA, Karish MV (1982): Polarized amino acid transport by an epithelial cell line of renal origin (LLC-PK_1). The basolateral systems. J Biol Chem 257:6802-6808.
Rindler MJ, Chuman LM, Shaffer L, Saier MH (1979a): Retention of differentiated properties in an established dog kidney epithelial cell line (MDCK). J Cell Biol 81:635-648.
Rindler MJ, Taub M, Saier MH (1979b): Uptake of $^{22}Na^+$ by cultured dog kidney cells (MDCK). J Biol Chem 254:11431-11439.
Schmidt U, Guder WG (1976): Sites of enzyme activity along the nephron. Kidney Int 9:233-242.
Stiles CD, Desmond W, Chuman LM, Sato G, Saier MH Jr (1976): Growth control of heterologous tissue culture cells in the congenitally athymic nude mouse. Cancer Res 36:3300-3305.
Taub M, Livingston D (1981): The development of serum-free hormone-supplemented media for primary kidney cultures and their use in examining renal functions. Ann NY Acad Sci 372:406-421.
Taub M, Saier MH Jr (1979): Regulation of $^{22}Na^+$ transport by calcium in an established kidney epithelial cell line. J Biol Chem 254:11440-11444.
Taub M, Sato G (1979): Growth of kidney epithelial cells in hormone supplemented, serum-free medium. J Supramol Struct 11:207-216.
Taub M, Sato G (1980): Growth of functional primary cultures of kidney epithelial cells in defined medium. J Cell Physiol 105:369-378.
Taub M, Chuman L, Saier MH, Sato G (1979): Growth of Madin Darby canine kidney epithelial cell (MDCK) line in hormone-supplemented serum-free medium. Proc Natl Acad Sci USA 76:3338-3342.
Wilson PD, Horster MF (1982): Differential response to hormones of defined distal nephron epithelia in culture. Amer J Physiol (in press).

Methods for Serum-Free Culture of Epithelial and Fibroblastic Cells,
pages 25-31

2
Hormonally Defined, Serum-Free Medium for a Proximal Tubular Kidney Epithelial Cell Line, LLC-PK_1

Milton H. Saier, Jr.

Little is known concerning the in vivo conditions and factors controlling growth and expression of the differentiated state in kidney epithelial cells. Nor are the developmental conditions that give rise to distinct types of epithelial cells in the different segments of the kidney tubule understood. In order to provide information about these important biological problems, we have focused our studies on established cell lines derived from kidney tissue [Saier, 1981; McRoberts et al., 1981]. Of these kidney epithelial lines, the best characterized is the Madin-Darby canine kidney (MDCK) line [Gaush et al., 1966; Leighton et al., 1969, 1970; Handler et al., 1980]. Our investigations have defined the in vivo [Stiles et al., 1976a] and in vitro [Stiles et al., 1976b; Taub et al., 1979] growth properties of these cells. Differentiated characteristics of these cells have also been studied [Rindler et al., 1979a,b; Taub and Saier, 1979], and the polar nature of an MDCK cell epithelium has been established by functional studies [Misfeldt et al., 1976; Cereijido et al., 1978] as well as structural studies [Rabito et al., 1978; U et al., 1979, 1980].

Whereas the MDCK cell line appears to derive from the collecting duct or distal tubule, recent evidence suggests that another established cell line, designated LLC-PK_1 [Hull, 1976], may have been isolated from the proximal tubule of the porcine kidney. These cells form continuous epithelial sheets in vitro and retain Na^+-dependent glucose uptake activity [Rabito et al., 1980].

Department of Biology, The John Muir College, University of California at San Diego, La Jolla, California 92093

Although an increasing number of laboratories are initiating studies with the LLC-PK_1 cells, little is yet known concerning the growth properties of these cells or about primary epithelial cells derived from the proximal tubule. We therefore initiated comparative growth studies with this line, the first of which are reported here. It is shown that while the LLC-PK_1 cells will not grow continuously in the serum-free medium that supports rapid MDCK cell growth, a few qualitative and quantitative modifications of the medium allow continuous growth of these proximal tubular cells. The qualitative changes include substitution of vasopressin for prostaglandin E_1 and the inclusion of cholesterol in the growth medium. Some of the growth agents influence cell morphology and expression of specific differentiated characteristics of the cells. On the basis of the results reported, we tentatively conclude that the factors and conditions supporting growth of proximal tubular epithelial cells are basically similar to, but differ slightly from, those supporting growth of epithelia from the distal segment or collecting duct.

MATERIALS AND METHODS

Origins of cells, sources of hormones, growth conditions, media, and procedures for the isolation of primary cultures of rabbit proximal tubular epithelial cells have all been described [Chuman et al., 1982].

RESULTS

Hormonally Defined Media for LLC-PK_1 and MDCK Cells

Table I summarizes the composition of the hormone-supplemented serum-free medium for the LLC-PK_1 cells and compares this medium with that for the MDCK cells [Taub et al., 1979; Chuman et al., 1982]. Both cell lines require transferrin, insulin, and selenium for growth, as do numerous other cell lines [Rizzino et al., 1979; Hayashi and Sato, 1976]. Additionally, the two lines are growth-stimulated by hydrocortisone and triiodothyronine, both of which presumably act at the nuclear level. An adenylate cyclase-stimulating hormone (vasopressin in the case of the LLC-PK_1 line; prostaglandin E_1 in the case of the MDCK cells) is also required. Only the LLC-PK_1 cells showed growth stimulation with cholesterol. Epidermal growth factor was also noticeably stimulatory in the presence of transferrin and insulin but had little effect in the complete defined medium. The quantitative differences in concentration of those agents that are required by both cell lines may not be of particular significance, since the two studies were not conducted under strictly parallel conditions.

The growth dependencies of the LLC-PK_1 cells on individual hormones were studied as a function of concentration [Chuman et al., 1982]. None of the required factors inhibited growth at concentrations that exceeded those that gave optimal growth. Growth stimulation was frequently observed at concentrations that were several orders of magnitude lower than those reported in Table I.

The LLC-PK_1 cells have been grown continuously for over 6 months with weekly passages at a 1-to-10 dilution in this serum-free medium (Table I). They grow slightly slower in the defined medium, with a doubling time of 26.4 h, compared to a doubling time in serum (7.5% horse serum, 2.5% fetal calf serum) of 21.6 h. The plating efficiency of the serum-free cells is 70%, whereas the serum grown cells routinely plate with 97% efficiency. Cells grown in serum-containing and serum-free hormone-supplemented media form hemicysts with approximately equal frequency. The hemicysts were nearly identical with respect to number, size, and morphology (data not shown).

The growth requirement for vasopressin could not be replaced by prostaglandins or other adenylate cyclase-stimulating hormones. Arg-vasopressin stimulated cell growth to the same extent as did lys-vasopressin, and human calcitonin, reported to be active in LLC-PK_1 cells [Goldring et al., 1978], had little or no effect. Other hormones tested did not enhance cyclic AMP production to the extent observed with vasopressin (data not shown). However, the vasopressin growth dependency could be partially replaced by isobutylmethylxanthine (a potent inhibitor of cyclic AMP phosphodiesterase) or dibutyryl cyclic AMP. These two agents caused cellular morphological changes analogous to

TABLE I. Hormone-Defined Media for Two Kidney Epithelial Cell Lines

	LLC-PK_1	MDCK
Transferrin	25 μg/ml	5 μg/ml
Insulin	10 μg/ml	5 μg/ml
Selenium	5×10^{-8} M	5×10^{-8} M
Hydrocortisone	2×10^{-7} M	5×10^{-8} M
Triiodothyronine	10^{-9} M	10^{-9} M
Vasopressin	10 μU/ml	
Prostaglandin E_1		25 ng/ml
Cholesterol	10^{-8} M	

Optimal concentrations of the growth factors are indicated. The hormone-free basal medium for both cell lines was DME/F-12.

those caused by vasopressin. The cells appeared more flattened and assumed a less ragged, more healthy appearance.

The LLC-PK_1 cells were examined by electron microscopy after growth in either serum-containing medium or in the complete serum-free hormone-defined medium. The cells grown under these two conditions appeared similar. Normal junctional complexes were present in both cases. The only striking difference noted was the appearance of the brush border. Cells grown in the hormone-supplemented medium possessed more dense but shorter microvilli than were observed for the serum-grown cells.

MDCK cells were found to grow in the medium defined for the LLC-PK_1 cells, but no hemicyst formation was observed. This observation correlates with the fact that vasopressin stimulates cyclic AMP production in MDCK cells much less efficiently than do the E prostaglandins. The LLC-PK_1 cells did not grow continuously in the serum-free medium that supports MDCK cell growth.

Use of Serum-Free Medium to Isolate Primary Cultures of Kidney Epithelial Cells

The hormone-supplemented serum-free medium for the LLC-PK_1 cells was used to isolate primary cultures of epithelial cells, free of fibroblasts, from microdissected segments of proximal kidney tubules. Proximal tubules were identified with certainty, and care was taken to insure that no transfer of floating cells occurred. Cells were seen to proliferate from the broken ends of the proximal tubular segment and to spread out concentrically over the Millipore filter within 48 h. In 10 days the area covered by the monolayer was more than 100 times greater than that covered by the initial explant, indicating that proliferation rather than migration of cells had occurred. Light microscope examination indicated the cells to be somewhat flattened and elongated cuboidal epithelial cells. Although they grew on the membrane filter as a monolayer, they were often present in collections of two or more cells in thickness, and occasionally they were arranged as very small papillary growths four to six cells deep. Scattered cells had large clear cytoplasmic vacuoles. No fibroblasts were seen by light or electron microscopy. These results establish that the epithelial cells that proliferated were of proximal tubular origin.

To test whether the serum-free medium defined for the LLC-PK_1 cells was specific in supporting the growth of proximal tubular cells, explants of dissected cortical collecting tubules were studied. Cells from these tubule segments also proliferated in this medium. Thus, while the medium does support the growth of proximal tubular cells, it cannot be regarded as being selective for this cell type.

Ultrastructural study revealed the free surfaces of the cells to possess widely and irregularly separated microvilli. The cells were attached to one another by tight junctions at the apical surface; they interdigitated with one another on the lateral surfaces by complex infoldings. The nuclei were located in central or basal portions of the cells. Organelles, consisting mainly of elongated mitochondria and cisternae of endoplasmic reticulum, were randomly distributed throughout the cytoplasm; numerous free ribosomes constituted the bulk of the cytoplasm. A basement membrane underlying the basal portion of the cells was not too evident.

It is interesting that primary cultures of rabbit kidney epithelial cells have been isolated, free of fibroblasts, in a serum-free medium containing only insulin, transferrin, and hydrocortisone [Chung et al., 1982]. The cells were derived from isolated proximal tubules and exhibited many of the differentiated functions expected of proximal tubular epithelial cells. Thus, the primary cultures formed hemispherical clones, actively accumulated methyl α-glucoside by a Na^+-dependent mechanism, and exhibited adenylate cyclase activity that was responsive to parathyroid hormone. Certain enzyme activities, such as leucine aminopeptidase and alkaline phosphatase, characteristic of proximal tubular epithelia, were lower in cells grown in cultures than in the original proximal tubules.

DISCUSSION

We have shown that the LLC-PK_1 cell line, a kidney epithelial line of presumed proximal tubular origin, can be maintained in continuous growth in the absence of serum [Chuman et al., 1982]. Growth depends on several agents as listed in Table I. The growth requirements of the LLC-PK_1 cells are similar to those of the MDCK cells studied previously, except that vasopressin substitutes for prostaglandin as an adenylate cyclase-stimulating hormone, and cholesterol is required. The specificity of this medium is indicated by the facts that a) the medium supports rapid growth of the LLC-PK_1 cells in the absence of serum; b) the medium supports hemicyst formation with a frequency similar to that observed in serum-containing medium; and c) the medium has been successfully used to isolate epithelial cells, free from fibroblasts, from dissected proximal tubules of rabbit kidney. Because the growth rate of LLC-PK_1 cells in serum exceeded that in defined medium, it can be anticipated that an additional factor will be found that stimulates cell division. This suggestion is substantiated by the fact that the primary cultures of proximal tubular epithelial cells, although polarized with apical membrane microvilli and tight junctions, did not fully differentiate in vitro into cells with a dense, microvillous apical brush border, characteristic of proximal tubular epithelia. The medium should nevertheless prove useful in defining a) the cell surface receptor properties of the $LLC-PK_1$ cells, b) the growth requirements of these cells in molecular terms, and c) the hormonal dependencies for expression of differentiated biochemical traits. It should also be of use in the isolation of epithelial cells from proximal tubules of a variety of mammalian species.

REFERENCES

Bar-Khayan Y, Cohen AH, Houston B, Trizna W, Fine LG (1981): In vivo "culture" of a transporting renal epithelial monolayer. Kidney Int 19:465-470.

Cereijido M, Robbins ES, Dolan WJ, Rotunno CA, Sabatini DD (1978): Polarized monolayers formed by epithelial cells on a permeable and translucent support. J Cell Biol 77:853-880.

Chuman L, Fine LG, Cohen LG, Saier MH Jr (1982): Continuous growth of proximal tubular kidney epithelial cells in hormone-supplemented serum-free medium. J Cell Biol 94:506-510.

Chung SD, Alavi N, Livingston D, Hiller S, Taub M (1982): Characterization of primary rabbit kidney cultures that express proximal tubule functions in a hormonally defined medium. J Cell Biol 95:118-126.

Gaush CR, Hard WL, Smith TF (1966): Characterization of an established line of canine kidney cells (MDCK). Proc Soc Exp Biol Med 122:931-935.

Goldring SR, Dayer JM, Ausiello DA, Krane SM (1978): A cell strain cultured from porcine kidney increases cyclic AMP content upon exposure to calcitonin or vasopressin. Biochem Biophys Res Commun 83:434-440.

Handler JS, Perkins FM, Johnson JP (1980): Studies of renal cell function using cell culture techniques. Am J Physiol 238:F1-F9.

Hayashi I, Sato GH (1976): Replacement of serum by hormones permits growth of cells in a defined medium. Nature (Lond) 259:132-134.

Hull RN, Cherry WR, Weaver GW (1976): The origin and characteristics of a pig kidney cell strain, $LLC-PK_1$. In Vitro (Rockville) 12:670-677.

Leighton J, Brada Z, Estes W, Justh G (1969): Secretory activity and oncogenicity of a cell line (MDCK) derived from canine kidney. Science 163:472-473.

Leighton J, Estes LW, Mansukhani S, Brada Z (1970): A cell line derived from normal dog kidney (MDCK) exhibiting qualities of papillary adenocarcinoma and of renal tubular epithelium. Cancer 26:1022-1028.

McRoberts JA, Taub M, Saier MH Jr (1981): The Madin-Darby canine kidney (MDCK) cell line. In Sato G (ed): "Functionally Differentiated Cell Lines." New York: Alan R. Liss, pp 117-139.

Misfeldt DS, Hamamoto ST, Pitelka DR (1976): Transepithelial transport in cell culture. Proc Natl Acad Sci USA 73:1212-1216.

Rabito CA, Ausiello DA (1980): Na^+-dependent sugar transport in a cultured epithelial cell line from pig kidney. J Membr Biol 54:31-38.

Rabito C, Tchao R, Valentich JD, Leighton J (1978): Distribution and characteristics of occluding junctions in a monolayer of a cell line (MDCK) derived from canine kidney. J Membr Biol 43:351-365.

Rindler MJ, Chuman LM, Shaffer L, Saier MH Jr (1979a): Retention of differentiated properties in an established dog kidney epithelial cell line (MDCK). J Cell Biol 81:635-648.

Rindler MJ, Taub M, Saier MH Jr (1979b): Uptake of $^{22}Na^+$ by cultured dog kidney cells (MDCK). J Biol Chem 254:11431-11439.

Rizzino A, Rizzino H, Sato G (1979): Defined media and the determination of nutritional and hormonal requirements of mammalian cells in culture. Nutrition Rev 37:369-378.

Saier MH Jr (1981): Growth and differentiated properties of a kidney epithelial cell line (MDCK). Am J Physiol 240:C106-C109.

Stiles CD, Desmond W, Chuman LM, Sato G, Saier MH Jr (1976a): Growth control of heterologous tissue culture cells in the congenitally athymic nude mouse. Cancer Res 36:1353-1360.

Stiles CD, Desmond W, Chuman LM, Sato G, Saier MH Jr (1976b): Relationship of cell growth behavior in vitro to tumorigenicity in athymic nude mice. Cancer Res 36:3300-3305.

Taub M, Saier MH Jr (1979): Regulation of $^{22}Na^+$ transport by calcium in an established kidney epithelial cell line. J Biol Chem 254:11440-11444.

Taub M, Chuman LM, Saier MH Jr, Sato G (1979): The growth of Madin-Darby canine kidney epithelial cell (MDCK) line in hormone-supplemented, serum free medium. Proc Natl Acad Sci USA 76:3338-3342.

U HS, Saier MH Jr, Ellisman MH (1979): Tight junction formation is closely linked to the polar redistribution of intramembranous particles in aggregating MDCK epithelia. Exp Cell Res 122:384-392.

U HS, Saier MH Jr, Ellisman MH (1980): Tight junction formation in the establishment of intramembranous particle polarity in aggregating MDCK cells. Exp Cell Res 128:223-237.

Methods for Serum-Free Culture of Epithelial and Fibroblastic Cells, pages 33–41

3
Serum-Free Organ Culture of Embryonic Mouse Metanephros

Ellis D. Avner, William E. Sweeney, Jr., and Demetrius Ellis

Successful organ culture of embryonic tissue permits controlled in vitro study of the complex processes of organogenesis. Over the past 30 years, various organ culture systems have been designed to examine aspects of mammalian metanephric development [Grobstein, 1953; Saxén et al., 1968; Emura and Tanaka, 1972; Ekblom et al., 1981a]. Such models have clear limitations in the study of renal morphogenesis or the biochemical and endocrinological basis of normal renal development in that they mechanically separate metanephric blastema from ureteric bud and require animal serum or tissue extracts of undefined chemical composition for in vitro growth.

To overcome these limitations, we have developed a whole-organ mouse metanephric culture system in which advanced organotypic renal proximal tubular and epithelial glomerular differentiation occur in serum-free, hormone-supplemented medium [Avner et al., 1982]. This chapter 1) will describe our methods for primary tissue isolation and culture, medium preparation and storage, and tissue harvesting and processing; and 2) will discuss the potential applications of serum-free metanephric organ culture in the study of mammalian renal development.

METHODS

Primary Tissue Isolation and Culture

Pregnant Swiss-Webster albino mice are sacrificed by cervical dislocation at 13.0 ± 0.4 days gestation. Embryos are aseptically removed from uteri and

Department of Pediatrics (Nephrology), University of Pittsburgh School of Medicine, Pittsburgh, Pennsylvania 15213 (E.D.A., D.E.) and Division of Pediatric Nephrology, Children's Hospital of Pittsburgh, Pittsburgh, Pennsylvania 15213 (E.D.A., W.E.S., D.E.)

placed in a sterile solution of equal volumes of Dulbecco's modified Eagle's medium (DMEM) and Ham's F12 medium (HF12) at 4°C. Gestational age is based on timed breedings, and precise gestational age of fetal tissues is confirmed by formal developmental staging of embryos by established methods [Gruneberg, 1943; Theiler, 1972]. Under 18 × magnification in a laminar flow hood, embryos are opened with an abdominal incision with microsurgical instruments and the paired metanephroi are identified posterior to fetal gonads and inferior to the adrenal glands. Whole metanephroi are extracted, trimmed, and transferred onto a 0.8 μm Millipore filter sitting atop a Trowell-type organ culture assembly [Trowell, 1959]. The assembly consists of a double-welled organ culture dish (Falcon No. 3037, Oxnard, CA) with an outer well filled with an absorbant ring soaked in double-distilled water and a center well covered by a fitted metal grid that supports the filter and tissue. Culture medium is added to the center well and the level is adjusted to soak the bottom of the filter without submerging the tissue. The complete assemblies are incubated at 36 ± 0.2°C and 95% humidity in the mixed air-5% CO_2 environment of a water-jacketed incubator. Culture medium is completely changed every 48 h.

The culture medium is a modification of Medium K1 initially developed to support the growth of Madin-Darby canine kidney (MDCK) cells and primary cultures of kidney epithelial cells [Taub et al., 1979; Taub and Sato, 1980]. It consists of equal volumes of DMEM and HF12 supplemented with 10 mM HEPES buffer, sodium bicarbonate (1.1 mg/ml), 10 nM $Na_2 \cdot SeO_35H_2O$, insulin (5 μg/ml), prostaglandin E_1 (25 ng/ml), triiodothyronine (3.2 pg/ml), and transferrin (5 μg/ml). For experiments requiring longer than 72 h of incubation, mycostatin (50 U/ml), penicillin (50 U/ml), and gentamicin (50 μg/ml) are added to the culture medium.

Medium Preparation and Storage

DMEM and HF12 (FlowLab, McLean, VA) are prepared as a 50:50 solution in 100 ml aliquots, protected from light, and stored at −20°C. The solution may be stored in this form for 3 months without loss of growth-supporting activity in the organ culture system.

Transferrin and all hormones were initially purchased as premeasured, lyophilized, membrane-filtered products prepared in sterile vials, particularly for cell culture use (Collaborative Research, Waltham, MA). For the organ culture system, all such products could be stored without loss of activity for 6 months at 4°C (lyophilized) or 2 months at −20° (reconstituted solutions). We found such prepared reagents reliable and convenient to use, though relatively expensive.

After preliminary studies defined the final composition of our growth medium, we utilized standard lyophilized stock chemicals for all basal medium supplements in subsequent experiments (Sigma Chemical, St. Louis). Transferrin and all hormones could be stored in a dessicator at $-20°C$ for up to 6 months. Following reconstitution, solutions were sterilized by passage through a series of 0.22-μm filters and stored in aliquots at $-20°C$. Reconstituted transferrin and hormone solutions were stable for 2.5 months when stored in this fashion. Solutions used for reconstitution of the various products were as follows: a) transferrin and hydrocortisone—triple-distilled H_2O; b) insulin—0.1 N HCl; c) triiodothyronine—0.2 N NaOH; and d) prostaglandin E_1—95% ETOH. Complete medium (DMEM:F12 with all supplements) was prepared freshly with each culture change. Transferrin and hormone mixtures were not active in fully prepared medium for longer than 48 h even when stored at $-4°C$.

Tissue Harvesting and Processing

Although tissue can be separated from filter substrate by chemical dissociation or vigorous washing, we found simple mechanical removal most effective for intact organ preservation. A microspatula with finely ground edges can be carefully passed between flattened explant and filter, and tissue can be lifted in toto from the culture assembly. The harvested, intact organ can then be processed for morphological and immunological study with minor modifications of standard methods [Avner et al., 1982, 1983b,d].

For biochemical assays, cellular viability measurements, or isolation and subculture of cells from intact explants, further chemical or mechanical tissue dispersion is required. Mechanical dispersion is best accomplished with a Wheaton micro tissue grinder (Wheaton Scientific, Millville, NJ). Homogenized tissue is then suitable for micro protein measurements with a modified Lowry method [Hartree, 1972], or assay of enzymes such as γ-glutamyltranspeptidase by sensitive fluorometric methods [Smith et al., 1979; Avner et al., 1983e].

Mild chemical tissue dispersion is required to obtain intact cells for viability measurements and secondary subculture. Best results have been obtained in our system with a combination of pronase and collagenase. Following washes in phosphate-buffered saline at 4°C, individual explants are gently minced with microdissection scissors and then placed in a buffered solution of 0.25% pronase (Calbiochem-Behring, LaJolla, CA) and 0.075% type IV collagenase (Sigma Chemical Co., St. Louis). The tissue solution is stirred for 45 min at 4°C, and the resultant cell suspension is centrifuged at 100g for 10 min. Cells are washed in basal medium at 4°C, recentrifuged, and resuspended in the

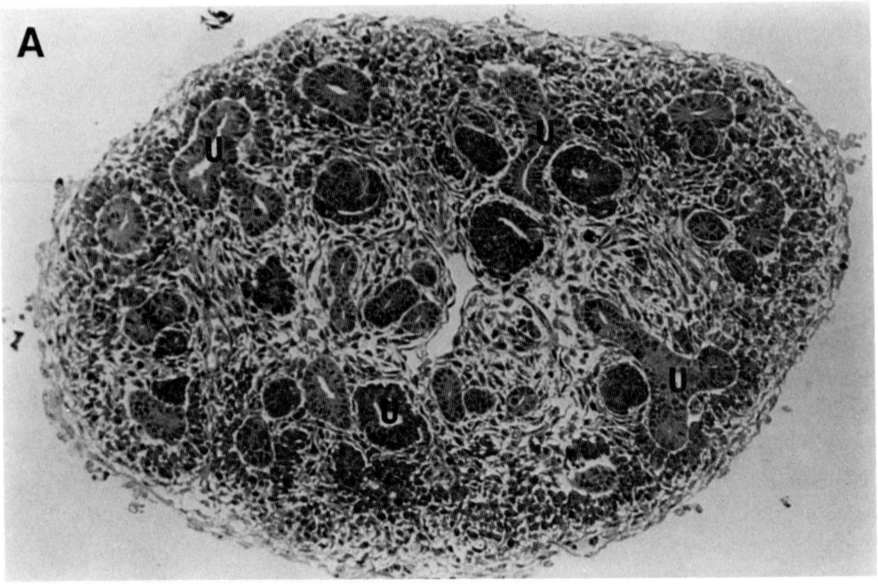
A
U
U
U
U

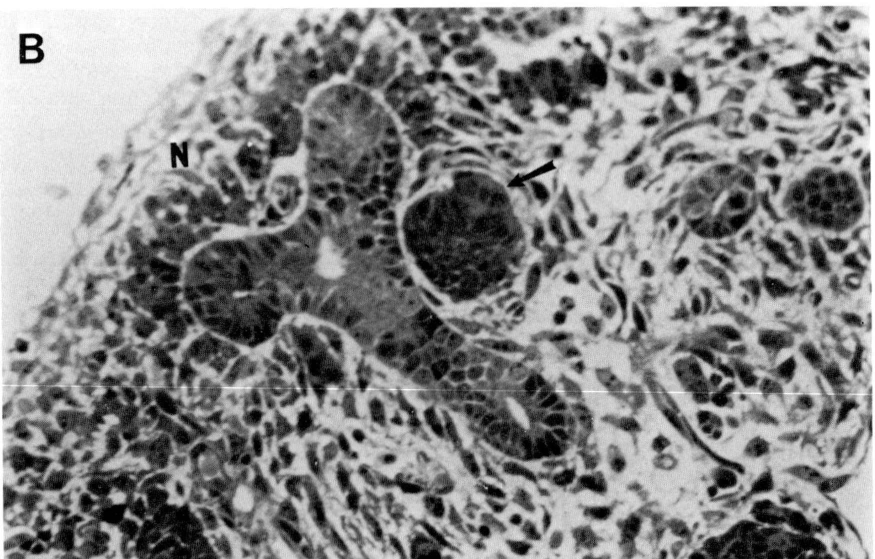
B
N

desired volume of medium for subculture or other studies. Cellular viability can be rapidly assessed by standard trypan blue dye exclusion or neutral red dye uptake methods [Patterson, 1979].

RESULTS

Morphogenesis

At the time of primary tissue isolation and explanation into organ culture (13 ± 0.4 days gestation), the embryonic metanephros is at an early stage of tubular development (Fig. 1A,B). Distal ramifications of the ureteric bud are present, and condensations of nephrogenic cells surround actively inducing ampullary areas. Early stages of normal tubular development including oval cell masses and primitive S-shapes can be identified amidst areas of undifferentiated mesenchyme. No endothelial elements or vascular structures are present. Under the culture conditions employed, metanephric development continues beyond the early S-shaped tubule (Fig. 2A,B). Nephrogenesis proceeds through normal stages of tubular development with the formation by 72–96 h in culture of discrete proximal tubules with characteristic ultrastructural features including elongated microvilli, complete epithelial junctional complexes, organized cytoplasmic organelles, and an advanced apical endocytic system. In addition, by 96–120 h of culture, completely devascularized glomeruli form. These structures consist of a core of elongated epithelial cells surrounded by a well-formed capsule of squamous epithelial cells. Organ culture glomeruli have characteristic ultrastructural features that include mature foot processes surrounding islands of basement membrane and developing slit diaphragms [Avner et al., 1982]. There is no in vitro differentiation beyond the 120-h stage.

Growth and Viability

During culture, the explants elongate and flatten along the filter substrate. While differentiation of tissue through defined morphological stages permits qualitative assessment of growth and viability, quantitative assessment is best obtained by measurement of protein biosynthesis and membrane dye uptake and exclusion. At the time of explantation of primary tissue into culture, protein content per explant is 9.5 ± 0.7 μg (mean $\pm$ SE). Following the first

Fig. 1. Embryonic metanephros at the time of explantation (13 ± 0.4 days gestation). A. Low magnification shows distal ramifications of the branching ureteric bud (U) in transverse and cross-section amidst undifferentiated mesenchyme and early tubular forms (toluidine blue $\times$ 120). B. High magnification shows a cap of condensed nephric blastema (N) around an actively inducing ampullary area and a primitive nephrogenic cell mass (arrow) (toluidine blue $\times 420$).

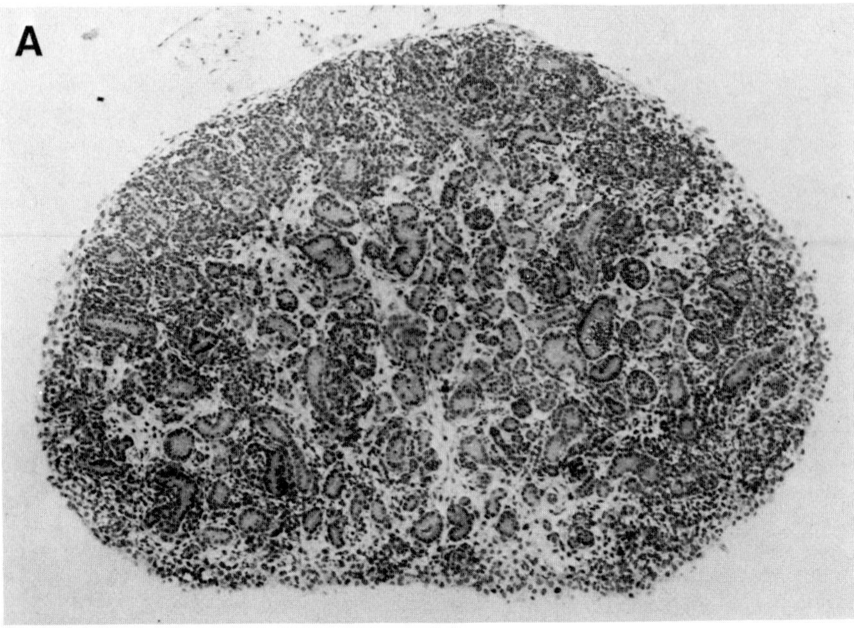
A

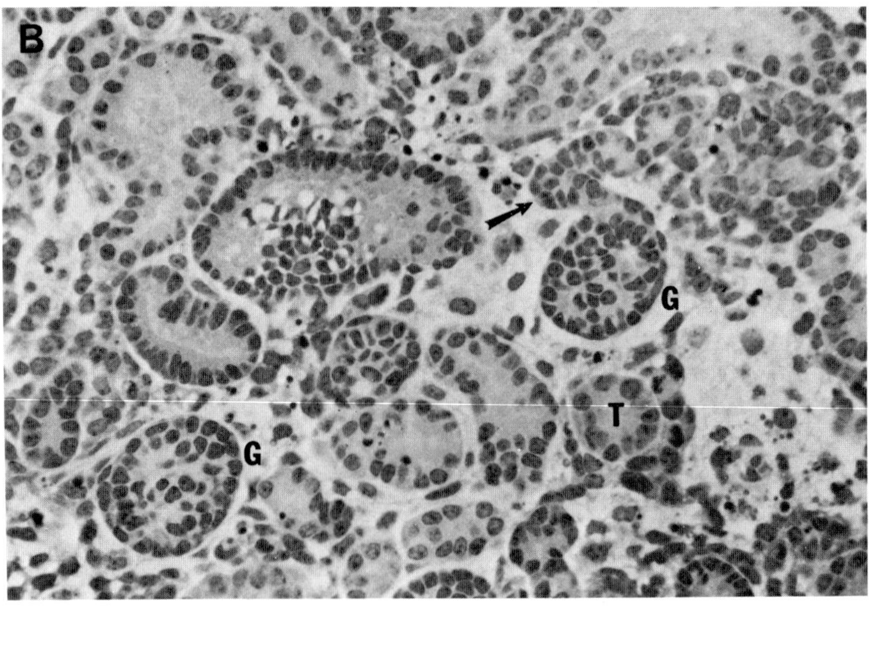
B
G
T
G

24 h of culture, protein content per explant increases linearly to a maximal value of 43.4 $\pm$ 3 μg by 144 h in vitro. At the time of explantation of primary tissue into culture, cellular viability is 95 $\pm$ 2%. Cellular viability shows only minor decreases during the culture period and reaches a value of 85 $\pm$ 3% by 144 h in vitro. Beyond 144 h in culture, tissue rapidly deteriorates morphologically, and protein biosynthesis and cell viability sharply decrease.

DISCUSSION AND APPLICATIONS

Under the conditions described, intact embryonic mouse metanephros undergoes organotypic development in serum-free, hormone-supplemented medium. The whole organ metanephric system was initially developed in medium supplemented with 20% donor bovine serum [Avner et al., 1983b]. Variations in tissue behavior in vitro led to analysis of protein and hormonal composition of serum batches that promoted optimal differentiation. It was reasoned that if the growth factors present in serum were defined and then added to a nutritionally rich serum-free medium, then an intact organ in culture might produce its own stromal scaffolding to guide differentiation without requiring serum components. Medium K-1, composed of equal volumes of DMEM and HF12 supplemented with 10 mM HEPES buffer, sodium bicarbonate 1.1 mg/ml, 10 nM $Na_2 \cdot SeO_3\ 5H_2O$, insulin (5 μg/ml), prostaglandin E_1 (25 ng/ml), triiodothyronine (3.2 pg/ml), hydrocortisone (18 ng/ml), and transferrin (5 μg/ml) [Taub et al., 1979] approximated the nutritional and hormonal profile of the serum that promoted optimal growth in the metanephric organ culture system. Thus Medium K-1 was used in our initial studies of serum-free organ culture, and the medium was subsequently modified to promote maximal growth, differentiation, and viability in the metanephric system. The final medium promoting optimal metanephric development in whole-organ culture differs from K-1 only in that it completely eliminates hydrocortisone. When hydrocortisone was added to the medium in concentrations of 18 ng/ml to 5 μg/ml, no alterations in growth, differentiation, or viability were noted. However, the addition of greater than 10 μg/ml of hydrocortisone to the medium induced striking tubular alterations in the developing nephrons with the production of cystic tubular malformations. This finding has been exploited to create an organ culture model of cystic renal maldevelopment [Avner et al., 1983a].

Fig. 2. Metanephric explant following 6 days of culture. A. Low magnification shows a panorama of advanced tubular and glomerular structures (hematoxylin $\times$80). B. High magnification highlights the differentiated glomeruli (G) and tubules (T), and shows a convoluted proximal tubule attached to its glomerulus (arrow) (hematoxylin $\times$520).

Although some differentiation in the system occurs with deletion of any of the four growth supplements (insulin, prostaglandin E_1, triiodothyronine, transferrin) from the basal medium, maximal growth, differentiation, and viability occur only when all are present simultaneously. The deletion of prostaglandin E_1 or transferrin has the most severe inhibitory effect on morphogenesis. This suggests that prostaglandin E_1 may have a role in renal differentiation as well as renal growth, and confirms previous observations that transferrin is necessary for differentiation of renal tissue in organ culture [Ekblom et al., 1981b].

The currently described organ culture model overcomes the two major limitations of previous metanephric culture systems: a) the requirement for separation and recombination of metanephric blastema with inducing tissue; and b) the growth requirement for animal serum or tissue extracts. Under the culture conditions described, embryonic metanephros undergoes advanced in vitro cytodifferentiation of proximal tubular and glomerular epithelial elements that parallels the ultrastructural differentiation of the mammalian nephron in vivo [Larsson, 1975; Reeves et al., 1978]. By maintaining the normal structural relationships of epithelial and mesenchymal elements in experimental organogenesis without vascularization or perfusion, the model permits controlled study of normal and abnormal renal morphological, immunological, and biochemical development [Avner et al., 1983c,d,e].

ACKNOWLEDGMENTS

This work has been supported in part by Basil O'Connor Starter Research grant no. 5-349 from the March of Dimes Birth Defects Foundation to E.D.A.

REFERENCES

Avner ED, Ellis D, Temple T, Jaffee R (1982): Metanephric development in serum-free organ culture. In Vitro 18:675–682.

Avner ED, Sweeney WE, Piesco NP, Ellis D (1983a): Cystic metanephric development in serum-free organ culture. Kidney Int 23:179 (abstract).

Avner ED, Villee DB, Schneeberger EE, Grupe WE (1983b): An organ culture model for the study of metanephric development. J Urol 129:660–664.

Avner ED, Sweeney WE, Ellis D (1983c): Cyst formation in metanephric organ culture induced by cis-dichlorodiammineplatinum (II). Experientia 39:74–76.

Avner ED, Jaffe R, Temple T, Ellis D, Chung AE (1983d): The development of renal basement membrane glycoproteins in metanephric organ culture. Lab Invest 48:263–268.

Avner ED, Ellis D, Sweeney WE, Rankin B, Curthoys NP (1983e): The ontogeny of γ-glutamyltranspeptidase in metanephric organ culture. Pediatr Res 17:345A (abstract).

Ekblom P, Meittinen A, Virtanen I, Wahlström T, Dawnay A, Saxén L (1981a): In vitro segregation of the metanephric nephron. Dev Biol 84:88–95.

Ekblom P, Thesleff I, Miettinen A, Saxén L (1981b): Organogenesis in a defined medium supplemented with transferrin. Cell Differ 10:281-288.

Emura M, Tanaka T (1972): Development of endothelia and erythroid cells in the mouse metanephrogenic mesenchyme in culture with the fetal liver. Dev Growth Differ 14: 237-246.

Grobstein C (1953): Inductive epithelio-mesenchymal interaction in cultured organ rudiments of the mouse. Science 118:52-55.

Gruneberg H (1943): The development of some external features in mouse embryos. J Hered 34:88-92.

Hartree EF (1972): Determination of protein: A modification of the Lowry method that gives a linear photometric response. Anal Biochem 48:422-427.

Larsson L (1975): The ultrastructure of the developing proximal tubule in the rat kidney. J Ultrastruct Res 51:119-139.

Patterson MK (1979): Measurement of growth and viability of cells in culture. In Jakoby WB, Pastan IH (eds): "Methods in Enzymology," Vol 58: "Cell Culture." New York: Academic, pp 141-151.

Reeves W, Caulfield JP, Farquhar MG (1978): Differentiation of epithelial foot processes and filtration slits: Sequential appearance of occluding junctions, epithelial polyanion, and slit membranes in developing glomeruli. Lab Invest 39:90-100.

Saxén L, Koskimies O, Miettinen H, Rapola J, Wartiovaara J (1968): Differentiation of kidney mesenchyme in an experimental model system. Adv Morphogen 7:251-291.

Smith GD, Ding JL, Peters TJ (1979): A sensitive fluorimetric assay for γ-glutamyltransferase. Anal Biochem 100:136-139.

Taub M, Sato G (1980): Growth of functional primary cultures of kidney epithelial cells in defined medium. J Cell Physiol 105:369-378.

Taub M, Chuman L, Saier M, Sato G (1979): Growth of Madin-Darby canine kidney epithelial cell (MDCK) line in hormone-supplemented, serum-free medium. Proc Natl Acad Sci USA 76:3338-3342.

Theiler K (1972): "The House Mouse." New York: Springer-Verlag, pp 87-118.

Trowell OA (1959): The culture of mature organs in a synthetic medium. Exp Cell Res 16:-118-147.

Methods for Serum-Free Culture of Epithelial and Fibroblastic Cells,
pages 43–55

4
Primary Culture of Hepatocytes

H.L. Leffert, K.S. Koch, and H. Skelly

Methods for establishing primary monolayer fetal and adult rat hepatocyte cultures are presented. Techniques are emphasized rather than rationale or methodological validation. Typical differentiated properties of these cell cultures are summarized in Table I. Media preparations (Tables II and III), buffer formulations (Table IV), and plating procedures (Tables V and VI) are outlined with explanatory details provided in the text. Lastly, protocols are given to set up both fetal and adult hepatocytes for growth reinitiation assays, under chemically defined conditions. The protocols were originally published elsewhere [Leffert and Paul, 1972, 1973; Leffert, 1974a,b; Koch and Leffert, 1974, 1979; Leffert and Weinstein, 1976; Leffert et al., 1977, 1978].

MEDIA AND DIGESTION BUFFER CONSTRUCTION

Twice-concentrated (×2) medium is prepared from stock solutions (Table II) as described in Table III. Amino acid, vitamin, and antibiotic stock solutions (Table II) are prepared in the appropriate portions and stored in darkness at −20°C until use. The antimycotic stock solution is stored in darkness at room temperature. L-Ornithine usually is present but is not necessary for fetal liver cultures. Arginine is added after plating only under special conditions (see, e.g., Leffert and Paul [1972, 1973] and Koch and Leffert [1974]).

A 100-ml amount of ×1 medium at pH 7.4 is freshly prepared as follows: 45 ml sterile ×2 medium; L-ornithine, if desired (see Table II); serum (0–30 ml); and sterile water are combined to make a final volume of 100 ml. To

Department of Medicine, Division of Pharmacology, University of California at San Diego, La Jolla, California 92093

TABLE I. Differentiated Properties of Primary Monolayer Rat Liver Cell Cultures

	Fetal	Adult
Enzymes		
Alcohol dehydrogenase [EC 1.1.1.1]	−	+[b]
Gluconeogenic	+	+[b]
Glutathione S-transferase B [EC 2.5.1.18]	U[a]	+[b]
Glycogenolytic	+	+
Microsomal $P_{448,450}$	+	+
Pyruvate kinase [EC 2.7.1.40]		
Fetal (type III)	+	+[b]
Adult (type I)	−	+[b]
Tyrosine aminotransferase [EC 2.6.1.5]	+	+
Urea "cycle" (ornithine transcarbamylase [EC 2.1.3.3])	+	+
Secretory proteins		
Albumin	+	+
α_1-Fetoprotein	+[b]	+[b]
Fibrinogen	U	+
Haptoglobin	+	+
Hemopexin	+	+
Transferrin	U	+
Very low density lipoprotein	−[c]	+[b,c]
Ultrastructure[c]		
Bile canaliculi	+	+
Desmosomes	+	+
Glycogen granules	+	+
Peroxisomes	+	+
Tight junctions	+	+

[a]Unknown.
[b]Growth state-dependent [Leffert and Sell, 1974; Leffert et al., 1978, 1982; Lad et al., 1982].
[c]Unpublished observations (D. Weinstein, K. Dempo, K. Miyai, S. Sell, and H.L. Leffert).

further minimize the possibility of contamination, 0.25 ml of antibiotic stock is added.

For the adult system, "complete medium" contains L-ornithine, 15% (v/v) serum, and 10 μg/ml each of insulin, hydrocortisone, and inosine. "Basal medium" is complete medium minus hormone and purine supplements. Stock solutions (5 ml) of hormone or purine (1 mg/ml) are freshly prepared in polypropylene test tubes in 0.15 M NaCl.

One drop of 6 N HCl dissolves insulin in the 5-ml solution. Supplements, 1 ml/100 ml medium, are added and the medium is sterilized through a 0.20-μm Nalgene filter.

TABLE II. Stock Solutions for Arginine-Free Medium

Amino acids (4 liters)	
L-Glycine	6.00 g
L-Histidine HCL	8.40 g
L-Isoleucine	20.96 g
L-Leucine	20.96 g
L-Lysine	29.24 g
L-Methionine	6.00 g
L-Ornithine	16.80 g[a]
L-Phenylalanine	13.20 g
L-Serine	8.40 g
L-Threonine	19.04 g
L-Tryptophan	3.20 g
L-Valine	18.72 g
Vitamins (2 liters)	
Choline chloride	0.80 g
Folic acid	0.80 g
Nicotinamide	0.80 g
Pantothenic acid	0.80 g
Pyridoxal HCl	0.80 g
Thiamine	0.80 g
Riboflavin	0.80 g
Antibiotics (330 ml)	
Penicillin G	30×10^6U
Streptomycin sulfate	6.0 g
Antimycotic (1 liter)	
n-Butyl-p-hydroxybenzoate	0.1 g

[a]Can be omitted or added from a standard sterile stock solution (332 mg/100 ml water; add 2.5 ml/100 ml × 1 medium).

Digestion buffer (Table IV) is stored at −20°C until use. Fresh collagenase, 3 mg/ml, is added to buffer at 37°C; after dissolving, the solution is sterilized (as above) and preincubated at 37°C for 1-4 h prior to use. For adult liver tissue digestion only, prior to sterilization, 100 ml of buffer containing enzyme is supplemented with 1 ml each of the insulin, hydrocortisone, and inosine stock solutions.

CELL ISOLATION AND PLATING

Detailed procedures appear in Table V (fetal cells: Leffert and Paul [1972]) and Table VI (adult cells: Leffert et al. [1977]). All steps are performed by

TABLE III. Construction of 8 Liters of Twice-Concentrated (×2) Medium

1. Thaw: 320 ml amino acid stock
   160 ml vitamin stock
   40 ml antibiotic stock
2. Dissolve 1.152 g L-tyrosine in 500 ml boiling water. Add 3.2 g $MgSO_4 \cdot 7H_2O$. Cool to room temperature.
3. Dissolve 0.748 g L-cystine in 200 ml water (at 60°C) containing 7 ml 1 N NaOH. Cool to room temperature.
4. Dissolve 4.2 g $CaCl_2 \cdot 2H_2O$, 6.4 g KCl, and 102.4 g NaCl in 1,500 ml water.
5. Add amino acid stock to #4 using stirring bar. Add:
   24 ml 1% (w/v) phenol red
   72 g glucose (anhydrous)
   2.3 g NaH_2PO_4
   0.112 g inositol
   16 ml 0.01% (w/v) $FeCl_3$
   40 ml antibiotic stock
   16 ml antimycotic
   160 ml vitamin stock
   Let stir for 5–10 min.
6. Add 9.34 g L-glutamine and 1.76 g sodium pyruvate to #5. Stir.
7. Add cooled solutions (#2, #3) to #6. Stir.
8. Bring to 7,200 ml with water; bubble CO_2 for about 1 h while adding 800 ml water containing 59.7 g $NaHCO_3$.
9. Filter through 0.22-μm Millipore GS filter in a 142-mm tripod stand.
10. Dispense into sterile bottles and store in the dark at 4°C for up to 6 weeks.

sterile technique. The stirring speed, i.e., rheostat knob setting, for all tissue digestion is 30–35. For fetal cultures, four animals are used and usually yield 20–40 fetuses, giving rise to about 1–3 × 10^8 cells. For adult cultures, two livers are used, giving rise to about 2 × 10^8 cells. For both protocols the time from surgery to plating should be about 3 h or less. At plating, greater than 95% of isolated cells are hepatocytes with 70–95% viability as judged by exclusion of 0.05% (w/v) trypan blue at 1 min.

Seeding densities, i.e., numbers of cells per ml medium, and volume of media per dish vary with experimental design. For the fetal system, usual parameters for 35-, 50-, and 90-mm-diameter dishes are 1.25 × 10^5 cells/ml, 2 ml; 1.5 × 10^5 cells/ml, 5 ml; and 2 × 10^5 cells/ml, 10 ml, respectively. For the adult system, 5 × 10^5 cells/ml, 2 ml, are plated per 35-mm dish. "Healthy" cultures show spread cellular aggregates upon the surface of the dish within 24 h and flattening (nonrefractile nuclear and cytoplasmic morphology under phase optics) within 48 h (see Fig. 1). Medium changes are performed as needed but are not necessary. In this laboratory, functional fetal cultures plated without L-ornithine survive for 6–8 weeks without a medium change. Functional adult cultures plated with L-ornithine survive for 2–3

TABLE IV. Buffer Solutions

Digestion buffer (1 liter)	
KCL	0.224 g
NaCl	7.587 g
$NaH_2PO_4 \cdot H_2O$	0.138 g
D-glucose	1.802 g
Phenol red	0.0012 g
Insulin[a]	0.010 g
Hydrocortisone-succinate[a]	0.010 g
Inosine[a]	0.010 g
(Adjust pH to 7.4)	
Tris buffer (3 liters)	
$Ca^{2+}Mg^{2+}$-free	750 ml
Deionized water	2,250 ml
15% (w/v) $CaCl_2 \cdot 2H_2O$	2 ml
15% (w/v) $MgCl_2$	2 ml
(Adjust pH to 7.4)	
$Ca^{2+}Mg^{2+}$-free concentrate (2 liters)	
NaCl	64.0 g
KCl	3.04 g
Na_2HPO_4 (anhyd.)	0.08 g
Tris base	24.0 g
1 N HCl	160 ml
(Adjust pH to 7.4)	
$Ca^{2+}Mg^{2+}$-free Tris buffer (2 liters)	
$Ca^{2+}Mg^{2+}$-free concentrate	500 ml
Deionized water	1,500 ml
(Adjust pH to 7.4)	
Trypsin concentrate (1 liter)	
Difco 1:250 trypsin	25 g
Deionized water	1.0 liter
Steps	
1. Dissolve 8 h at 4°C with stirring; let settle.	
2. Filter supernatant at 21°C using Whatman No. 1 paper.	
3. Sterilize with 0.22 μm GS Millipore and AP2512450 prefilter.	
4. Store 60-ml portions at −20°C.	
Trypsin stock solution for cell counts (0.6 liter)	
Trypsin concentrate	60 ml
1% (w/v) phenol red	2 ml
(Filter through 0.45-μm Nalgene filter)	
Sterile $Ca^{2+}Mg^{2+}$-free Tris buffer	540 ml
Store 10-ml portions at −20°C.	

[a]Freshly added for adult liver cell isolation only.

weeks without a medium change and beyond 4 weeks with weekly media changes (beginning at 10–12 days after plating) into fresh medium supplemented with 1% (v/v) serum or with 50 ng/ml each of insulin, glucagon, and epidermal growth factor. Under these plating conditions, long-term viability and differentiated function are not further improved by precoating the dishes with any one (or a mixture) of the following proteins: fibronectin, type IV collagen, ceruloplasmin, laminin, and fetuin.

CELL COUNTS

At plating, titrations are performed with 0.5 ml of cell suspension added to 9.5 ml of Isoton (Scientific Products, #B3157-11) and counted electronically

TABLE V. Fetal Rat Liver Cell Culture Plating Procedure

1. Lightly etherize pregnant rat. Immobilize the animal on its back and soak entire abdomen with 95% (v/v) ethanol.
2. Make a midline incision with a sterile scissors, cutting first skin and then muscle layers; do not perforate intestines, which lie under muscle layer.
3. Remove fetuses one at a time from embryonic sacs. Cut umbilical cord and decapitate immediately. Place about 10-15 fetuses in a sterile 9-cm dish.
4. Dissect out intact livers sterilely and put about 15 into a 5-cm dish containing 5 ml ornithine-minus medium at 37°C supplemented with 10% (v/v) serum. Inspect each liver to ensure absence of nonhepatic tissue and blood clots.
5. Cut each liver into its four separate "lobes"; remove medium by aspiration and wash once with 5 ml of fresh medium.
6. Pool washed, cut tissue and incubate in fresh collagenase-supplemented digestion buffer (1-2 ml/liver), for 8-10 min at 37°C in a 125-ml Erlenmeyer flask containing a magnetic stirring bar. Set rheostat = 30-35.
7. Tilt flask to allow undigested tissue to settle and remove cell suspension, with a sterile Pasteur pipette, placing it into a "storage" flask containing 10 ml medium (maintain on ice).
8. Repeat tissue digestion steps 6 and 7 a total of 4-5 times, pooling each subsequent harvest. Add an equal volume of fresh ice-cold medium to the final cell suspension and swirl gently. Dispense 20-40 ml (by pouring) into sterile 50-ml plastic tubes.
9. Centrifuge 10 sec at full speed and apply brake immediately thereafter until rotor stops.
10. Remove supernatant liquid by suction and discard; wash pellet once with 10 ml of serum-free medium (blow fluid rapidly against sidewall to disrupt pellet and then take up the cell suspension; blow it out gently against sidewall). Repeat step 9.
11. Discard second supernatant liquid. Resuspend pellets as in step 10, pool (store on ice), and titer a 0.5-ml aliquot.
12. Pipette cell suspensions into appropriate media and plate using a flow syringe.

(threshold at 10 μm, attenuation and aperture settings of 1 and 256, respectively, for Counter Model Fn; or, for Model Zf, threshold at 10 μm, and 1/aperture current and 1/amplification settings of 16 and 1, respectively). Hemocytometer counts are within $\pm 10\%$ of corrected electronic counts when these settings are used.

Table IV lists buffer solutions employed for measurements of attached cell numbers. Dishes are washed twice with Tris buffer, pH 7.4 (a volume equal to initial medium volume). Then an equal volume of $Ca^{2+}Mg^{2+}$-free Tris buffer with trypsin (10 ml trypsin stock solution per 180 ml $Ca^{2+}Mg^{2+}$-free Tris chloride at pH 7.4) is added and the dishes are incubated for 30-40 min at 37°C. A 2-ml (#122402010) or 5-ml (#122405010) glass hand pipette (Bellco) equipped with a rubber bulb (American Scientific Products, #R5000-1) is used for detaching the cells. Vigorous pipetting, about 20 times, over the dish surface usually reduces cell aggregation. The cell suspension or an aliquot of it is brought up to 10 ml with Isoton and counted.

TABLE VI. Adult Rat Liver Cell Culture Plating Procedure

1. Lightly etherize rat. Immobilize the animal on its back and soak entire abdomen with 95% ethanol.
2. Make a midline incision extending to the sternum, with lateral incisions to form flaps, freely exposing the entire peritoneum. Cut the ligaments attaching the liver to the diaphragm and stomach. Exsanguinate the animal by cardiac puncture, removing 5-8 ml blood, and push the intestinal contents aside, everting them to expose portal vessels.
3. Clamp portal vein (to the left of splanchnic and gastric venous inputs) together with the celiac trunk (small pulsating arterial system dorsal to everted portal system). Clamp inferior vena cava above the renal venous inputs but below the hepatic vein. After the clamps are in place cut the vena cava below the clamp.
4. Insert a 20-ml plastic syringe equipped with a 23-gauge needle (bevel up) and perfuse 40 ml hormone, purine, and collagenase-supplemented "digestion buffer" for 2-3 min through the liver. Complete tissue blanching; a light tan color and a swollen liver indicate adequate perfusion.
5. Excise perfused liver, placing it in a 9-cm plastic petri dish containing 15 ml of basal medium supplemented with 10% (v/v) serum. Remove all extraneous tissue and blood clots if present. Quickly transfer the liver to dishes with 15 ml media supplemented with 5% serum (1st dish) and 1% serum (2nd dish) in order to wash thoroughly. Aspirate the last dishful of medium, leaving the tissue damp. Dissect the tissue (two livers) into 5 × 8 mm pieces. *Do not mince.*
6. Incubate cut tissue at 37°C with 35-40 ml of hormone, purine, and collagenase-supplemented digestion buffer for 8-10 min in a 125-ml Erlenmeyer flask containing a magnetic stirring bar. Set rheostat = 30-35.
7. Tilt flask to allow undigested tissue to settle. Remove cell suspension by gentle suction with a rubber bulb (#R5002-2, American Scientific Products) attached to a sterile Pasteur pipette. Place the cell suspension into a 500-ml Erlenmeyer flask containing 10 ml of "complete" medium (maintain on ice).
8. Repeat tissue digestion (steps 6,7) three times, pooling each subsequent harvest. Add 160 ml ice-cold complete medium (a total volume of about 320 ml) and dispense 40 ml into eight 50-ml sterile plastic centrifuge tubes.
9. Centrifuge 13 sec (full speed setting) and apply brake until rotor stops.
10. Remove supernatant liquid by rapid suction and discard. Wash pellet once with 10 ml of serum-free "basal" medium: blow fluid rapidly against the sidewall to disrupt pellet and then take up the cell suspension and blow it out gently against sidewall. Repeat step 9.
11. Remove by suction and discard clear supernatant fluid; pellets should be "white". Resuspend pellets as in step 10 twice but blow out hard. Transfer to ice-cold sterile bottle, swirl pooled cell suspension (about 80 ml), and resuspend up and down seven times. Titrate a 0.5-ml aliquot.
12. Redistribute in 20 ml portions into four centrifuge tubes and repeat step 9.
13. Aspirate off the supernatants and resuspend (twice, hard) pellets in 10 ml of appropriate plating media (or, pool as needed). Titrate small aliquots in order to determine the appropriate dilutions needed to achieve desired plating densities (usually, about 500,000 cells/ml). Bring up to needed volumes in a large bottle and plate with frequent swirling using a 10-ml plastic pipette.

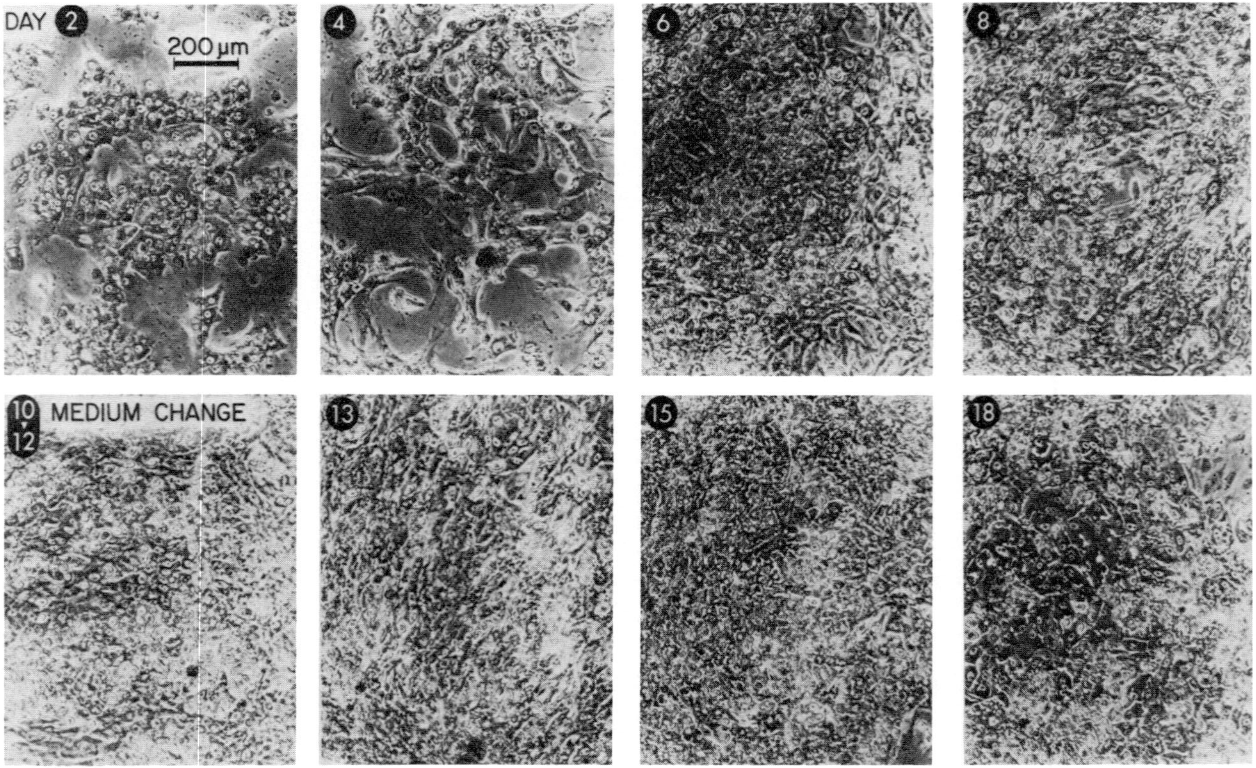

Fig. 1. Morphology of primary cultures of adult rat hepatocytes visualized by phase microscopy. The cells were plated under routine conditions [Leffert et al., 1977, 1978; Koch and Leffert, 1979] and allowed to progress through one growth cycle. On day 12, the cultures were reinitiated to grow, after extensive washing, by shifting the cells into fresh basal medium containing only 50 ng insulin, glucagon, and epidermal growth factor per ml (for growth kinetics, see Koch and Leffert [1979, 1980]). It can be seen that, 6 days after the fluid change into serum-free medium (i.e., 18 days after plating), viable differentiated hepatocytes continue to survive under these culture conditions.

ANIMALS

Fetal livers are obtained from pregnant rats 19-21 days in gestation; fetuses 14-19 days old can be used but hepatocyte yields are reduced. Adult livers are obtained from rats, 150-400 g body weight, fed standard Purina Chow and water ad libitum. Animals are housed at 21°C with alternating 12-h cycles of light and darkness.

SURGERY

Sterile technique is preferred but not necessary; alcohol-washed areas and instruments suffice. Animals are placed onto clean absorbent cotton pads layered over a simple flat cork board and operated on between 0900-1200 h.

BIOLOGICAL REAGENTS

Fetal bovine serum (FBS) is obtained from standard suppliers and is pretested by monitoring growth curves and DNA synthesis reinitiation assays (see below) to ensure its suitability for either fetal or adult cells in culture. Defrost, mix well (avoid air bubbles), and heat the serum at 56°C for 30 min. Dialyze it against 0.15 M NaCl (1:80 [serum:saline], v/v) for 24 h at 5°C, followed by two similar changes, and store aliquots at −20°C for 10-12 months. Dialysis tubing (5/8 inch; Van Waters and Rogers #25225-226) is boiled 5 min in water containing 0.1% ethylenediaminetetracetic acid (EDTA), washed 20 min with deionized water, and stored at 4°C. All serum used is dialyzed (dFBS) unless otherwise noted.

Collagenase (Sigma type I, #C0130) also varies in effectiveness and must be pretested separately for fetal or adult cells (as above, for serum) to ensure its suitability. The enzyme preparation is stored dessicated at −20°C for 5-6 months.

Amino acids, hydrocortisone-succinate, vitamins, inosine, bovine serum albumin (BSA), and sodium pyruvate are from Sigma; penicillin G and streptomycin-sulfate are purchased from Squibb and Pfizer, respectively. Eastman Kodak produces n-butyl-p-hydroxybenzoate (antimycotic). Porcine monocomponent insulin and glucagon are obtained from Dr. William Bromer (Eli Lilly Research Laboratory, Indianapolis, IN). Receptor-grade epidermal growth factor is obtained from Collaborative Research (Waltham, MA).

INORGANIC REAGENTS

Triple-distilled water (Arrowhead Puritas Waters Inc., Los Angeles, CA) is used for serum dialysis and fluid preparations unless otherwise noted. In our hands, further H_2O purification procedures, such as deionization, remove trace impurities that seem essential for long-term survival and growth (this is especially true for the adult system). The nature of the trace material present in Arrowhead H_2O is unknown.

TISSUE CULTURE SUPPLIES

Equipment

Tissue culture incubators supplied with CO_2 and air line filters (to trap particles and remove oil droplets), regulatory valves (settings are 0.5 and 4.0 for CO_2 and air, respectively), and external thermometers are from Napco (#3321) or Shel-Lab (Portland, OR). Water trays are used (gas is bubbled from nontoxic Tygon® tubing through water) to maintain humidity. Calibrate temperature (use internal thermometers at different tray levels) and gas tension (use test media and check pH). Napco also supplies a small air incubator (#310) used to perform tissue digestion and cell "monolayer" trypsinizations at 37°C. Cell numbers are determined electronically with a Coulter Particle Counter (Model Fn or Zf [equipped with a 100 μm aperture tube], Hialeah, FL).

A desk-top centrifuge containing a timer, brake, and rpm indicator is obtained from International Equipment Co. (#HN-S, Needham Heights, MA). A #958 centrifuge head is used, which is fitted with IEC #325 rings and #305 buckets with rubber stoppers.

The electric rheostat-controlled stirrer, from A.H. Thomas (Magne-Stir, #10), is placed into the Napco air incubator.

Aluminum tissue culture trays (12 × 30 cm) with grated open sides (1 cm high) and bottoms are constructed by this Institution's fabrication shop.

Magnetic stirring bars from Van Waters and Rogers (#58948-171; 2 × 5/16 in.) are replaced monthly.

Plasticware and Glassware

Falcon Plastics supplies sterile 50-ml (#2070) and polypropylene test tubes (#2059), petri dishes, 35-mm-diameter tissue culture dishes (#3001), and 1-ml (#7506) and 10-ml (#7530) pipettes. NUNC (Roskilde, Denmark) supplies 5-cm-diameter (#1400) and 9-cm-diameter (#N-1415) tissue culture dishes. Dish lot numbers should be noted in case of toxicity or insterility problems.

Nalgene 0.20- and 0.45-μm filters (#120-0020 and #245-0045) are purchased from Sybron (Rochester, NY) and should be checked for tightness of seal and detergent contamination (removable with warm-water washes).

Glass Erlenmeyer flasks (125 ml; #26500 or #26650) and other routine items are purchased from Kimax. Pasteur Uniform Drop® pipettes are from American Scientific Products.

Cornwall continuous-pipetting outfits (5-cc "flow syringes") are from Becton Dickinson (#3054) and are fitted with autoclavable 1/8-in. ID × 1/4-in. OD silastic tubing from Dow Corning (#601-365).

GROWTH REINITIATION ASSAYS IN SERUM-FREE MEDIUM

Fetal Hepatocytes

Freshly isolated cells are plated (2.5×10^5 cells/35-mm-diameter dish) with 2 ml basal medium supplemented with 10% v/v dFBS. Then 34 h after plating, the medium is aspirated and replaced with 2 ml similar fresh medium; 56 h after plating, the medium is aspirated and 2 ml fresh 50% "conditioned medium" (CM) is added. If dishes are not washed, about 10 μl of old medium are not removed by aspiration and are carried over into the freshly added medium. However, similar results are obtained whether or not the dishes are washed twice with 2 ml serum-free basal medium. 50% CM is formulated by adding 40 ml of a suspension of 10^8 freshly prepared, twice-washed hepatocytes into 215 ml basal medium, and then plating 4×10^6 cells/10 ml basal medium/9-cm-diameter dish. The medium is collected 44 h after plating, centrifuged (IEC Desk-Top, 2,000 rpm, 10 min at 21°C), collected, and sterilized by filtration (Nalgene 0.45-μm filter) to remove debris. CM is stored at 4°C for periods of 6-8 weeks without detectable losses of activity. After the sterilization step at 44 h, the medium is diluted 50% with fresh basal medium. The final amount of residual serum that may be present is equivalent to $\leqslant$0.00005% (v/v). During the next 7-9 days, cell multiplication occurs as indicated by an 80-100% increase in the number of attached cells (day 2: $(0.12\text{-}0.25) \times 10^5$ cells/35-mm dish; day 9: $(0.24\text{-}0.65) \times 10^5$ cells). Between 9 and 11 days after plating, although the cells are not confluent at this time, cell multiplication rates fall, basal rates of DNA synthesis decline to a constant low value (20-100 cpm/dish), and if pulse-labeled as described [Koch and Leffert, 1974], the percentage of DNA-synthesizing cells is less than 2% as determined by autoradiography after a ^{3}H-dT pulse of 12-24 h. The cultures are now ready to be shifted into fresh serum-free media for growth reinitiation assays under chemically defined conditions. Further details and validation of assays are given in Koch and Leffert [1974] and Leffert [1974b] and in a recent review [Leffert and Koch, 1982].

Adult Hepatocytes

Freshly isolated cells (1×10^6 cells/35-mm-diameter dish) are plated and permitted to progress through one growth cycle exactly as described previously [Leffert et al., 1977]. Differentiated properties of this system have been extensively characterized [Koch and Leffert, 1980; Leffert et al., 1978, 1982; Lad et al., 1982]. For unknown reasons ~20% of the platings yield aberrant growth curves (monitored by cell counts [Leffert et al., 1978]) or rapidly dying midphase cultures. Therefore, as plateau levels are reached 8-12 days after plating, the cultures are pretested for their ability to synthesize DNA. First they are washed twice with 37°C fresh serum-free basal medium. Then they receive either 2 ml of 10- to 15-day-old culture fluid (stationary-state conditions) to which is added deionized H_2O to replace evaporative losses (~2-3% v/v day^{-1}), or 2 ml fresh basal medium or 2 ml of medium containing 15% v/v dFBS (= stimulated-state conditions). Finally, they are pulsed for 24 h with ^{3}H-dT (sp. act. 20 Ci/mmol; 1.25 μCi/ml; 3×10^{-6} M TdR). If basal rates are constant after two consecutive days, and if a more than threefold stimulation is obtained, the cultures remaining in spent medium are ready to be shifted into fresh basal media, following washing, for growth reinitiation assays in chemically defined media. Further details and validation of assays are given elsewhere [Koch and Leffert, 1979, 1980].

REFERENCES

Koch K, Leffert HL (1974): Growth control of differentiated fetal rat hepatocytes in primary monolayer culture. VI. Studies with conditioned medium and its functional interactions with serum factors. J Cell Biol 62:780-791.

Koch KS, Leffert HL (1979): Increased sodium ion influx is necessary to initiate rat hepatocyte proliferation. Cell 18:153-163.

Koch KS, Leffert HL (1980): Growth control of differentiated adult rat hepatocytes in primary culture. Ann NY Acad Sci 349:111-127.

Lad PJ, Shier WT, Skelly H, de Hemptinne B, Leffert HL (1982): Adult rat hepatocytes in primary culture. VI. Developmental changes in alcohol dehydrogenase activity and ethanol conversion during the growth cycle. Alcoholism. Clin Exp Res 6:64-71.

Leffert HL (1974a): Growth control of differentiated fetal rat hepatocytes in primary monolayer culture. V. Occurrence in dialyzed fetal bovine serum of macromolecules having both positive and negative growth regulatory functions. J Cell Biol 62:767-779.

Leffert HL (1974b): Growth control of differentiated fetal rat hepatocytes in primary monolayer culture. VII. Hormonal control of DNA synthesis and its possible significance to the problem of liver regeneration. J Cell Biol 62:792-801.

Leffert HL, Koch KS (1982): Hepatocyte growth regulation by hormones in chemically-defined media: A two-signal hypothesis. In "Cell Growth in Hormonally Defined Media." Cold Spring Harbor Symp Cell Prolif 9:597-613.

Leffert HL, Paul D (1972): Studies on primary cultures of differentiated fetal liver cells. J Cell Biol 52:559-568.

Leffert HL, Paul D (1973): Serum-dependent growth of primary cultured differentiated fetal rat hepatocytes in arginine-deficient medium. J Cell Physiol 81:113-124.

Leffert HL, Sell S (1974): Alpha$_1$-fetoprotein biosynthesis during the growth cycle of differentiated fetal rat hepatocytes in primary monolayer culture. J Cell Biol 61:823-829.

Leffert HL, Weinstein DB (1976): Growth control of differentiated fetal rat hepatocytes in primary monolayer culture. IX. Specific inhibition of DNA synthesis initiation by very low density lipoprotein and possible significance to the problem of liver regeneration. J Cell Biol 70:20-32.

Leffert HL, Moran T, Boorstein R, Koch KS (1977): Procarcinogen activation and hormonal control of cell proliferation in differentiated primary adult rat liver cell cultures. Nature 267:58-61.

Leffert H, Moran T, Sell S, Skelly H, Ibsen K, Mueller M, Arias I (1978): Growth state dependent phenotypes of adult hepatocytes in primary monolayer culture. Proc Natl Acad Sci USA 75:1834-1838.

Leffert HL, Koch KS, Lad PJ, Skelly H, de Hemptinne B (1982): Hepatocyte regeneration, replication and differentiation. In Arias I, Popper H, Schacter D, Shafritz DA (eds): "The Liver: Biology and Pathobiology." New York: Raven, pp 601-614.

Methods for Serum-Free Culture of Epithelial and Fibroblastic Cells,
pages 57–71

5
Selective Growth of Human Small Cell Lung Cancer Cell Lines and Clinical Specimens in Serum-Free Medium

Desmond N. Carney, Martin Brower, Virginia Bertness,
and Herbert K. Oie

It has generally been accepted that supplementation of culture medium with serum is essential for the in vitro growth of both normal and malignant cells from humans and animals. However, while many cell lines have been permanently established in culture with the use of serum, there are many problems associated with its use [Barnes and Sato, 1980a,b]. A major disadvantage in using serum-supplemented medium to culture tumor cells directly from biopsies is that such medium is nonselective, and growth of both the malignant cells and of nonmalignant cells (such as fibroblasts, mesothelial cells, or lymphocytes) may occur. In many instances, the nonmalignant cells overgrow the malignant cells, preventing the establishment of permanent malignant cell lines from patients.

The use of serum-free defined media for the growth of cells has many potential advantages over the use of serum-supplemented medium. One advantage is that the growth factor requirements for different cell types are probably distinct, such that in most cases, a defined medium should support the growth of only one particular cell type and not of other cells, either benign or neoplastic. The selectivity of defined medium could be of major benefit in the isolation of tumor cells from fresh biopsy specimens.

Most of the data indicating that growth of cells in serum-free defined media equals or exceeds that found in serum-supplemented media is derived from

NCI-Navy Medical Oncology Branch, Division of Cancer Treatment, National Cancer Institute, National Institutes of Health and the National Naval Medical Center, Bethesda, Maryland 20814 the study of single established cell lines. It is generally not known whether a serum-free medium developed to support one particular cell line supports other cell lines of the same general type. In addition, the hypothesis that tumor cell growth from freshly explanted malignant tissue is as good or better in serum-free medium as in serum-supplemented medium remains to be proven for most human neoplasms.

Small cell lung cancer (SCLC) accounts for 20-25% of all new cases of primary lung cancer. Unlike the other major forms of lung cancer, it is highly responsive to chemotherapy and radiation therapy. With appropriate intensive therapy, the vast majority of patients treated will achieve a clinical remission, and about 10% of all patients will be cured [Carney and Minna, 1982; Minna et al., 1982a]. However, most patients eventually relapse and die, and the median survival of all patients is only 10-11 months.

Recently, we have found that malignant cells from human SCLC can grow in a clonogenic assay, and that in vitro drug and radiation sensitivity in this assay correlates with patients' clinical response to therapy [Carney et al., 1980, 1982, 1983a]. Therefore, the ability to grow human lung cancer for drug and radiation sensitivity testing on a routine basis would be of great clinical value. Toward this end, in addition to growing tumor cells in serum-supplemented medium, in athymic nude mice, and in a soft agarose clonogenic assay [Gazdar et al., 1980, 1981a; Carney et al., 1980, 1981a], we grow both established cell lines of SCLC and fresh clinical biopsy specimens of this tumor in HITES, a selective, serum-free, chemically defined medium [Simms et al., 1980; Carney et al., 1981b]. In this chapter, we report on the development of this medium for SCLC and its application in the selective growth of SCLC cells from clinical specimens. We also describe our initial efforts at growing other types of human lung cancer in serum-free medium.

MATERIALS AND METHODS

Cell Lines

The SCLC cell lines discussed here were started in our laboratory [Gazdar et al., 1980]. All were derived from patients with SCLC, form colonies in agarose, express human isozymes, and form typical SCLC tumors in athymic nude mice. In addition, all of the cell lines express a variety of amine precursor uptake and decarboxylation (APUD) properties including high levels of L-dopa decarboxylase (DDC), bombesin, and neuron-specific enolase, and they also have high levels of the BB isoenzyme of creatine kinase [Gazdar et al., 1980, 1981b; Moody et al., 1981; Marangos et al., 1982]. The adenocarcinoma cell lines also express human isozymes, and clone in soft agarose. They form adenocarcinomas in nude mice and lack APUD characteristics. A549 [Giard et al., 1973] was a gift from Dr. G. Todaro; the other lines were developed in our laboratory.

Clinical Specimens

Clinical specimens were obtained from patients with a confirmed diagnosis of cancer at the time of approved staging or diagnostic procedures, including bone marrow aspirations, thoracenteses, lymph node aspirations or biopsies, and surgical resection of tumor masses. Part of each specimen was processed for culture as previously described [Gazdar et al., 1980] and part was processed for cytologic examination and for DNA content analysis by flow cytometry [Carney et al., 1981b]. We obtained cell suspensions by mincing, passing through a 60-gauge steel mesh, and then gently pipetting in RPMI 1640 culture medium. Specimens containing a large number of red blood cells, including all bone marrow aspirate specimens as well as specimens with extensive debris or nonviable tumor, were clarified by Ficoll-Hypaque centrifugation. Tumors obtained from nude mouse xenografts were handled in an identical fashion.

Growth Factor Studies

The influence of many growth factors on the doubling time of an established SCLC cell line, NCI-H69, was determined with RPMI 1640 medium (GIBCO) as the base medium [Simms et al., 1980]. Stock solutions of hormones and growth factors were prepared at a final concentration of 100–2500 times the final concentration. The procedure for preparing these concentrates varied with each factor. Steroid compounds were dissolved in ethanol and then diluted in 0.9% NaC1 solution. Glucagon, T3, and insulin required pH adjustments to go into solution. The highly concentrated solutions were then filter-sterilized and stored at −20°C in 1- to 3-ml aliquots in glass vials. For study, each factor was thawed and diluted to 100 times the final concentration and was tested in serum-free medium, and then appropriate dilutions were made with the base medium to achieve the desired final concentration.

RESULTS

Cell Line Growth (Table I)

The effects of various growth factors on the in vitro growth of NCI-H69 are shown in Table I. Because prior studies from several other investigators suggested that insulin and transferrin were essential factors for human cell lines, the initial study evaluated the effect of these hormones on the growth of

TABLE I. Effect of Hormones and Growth Factors on the Growth of Human Small Cell Lung Cancer Cell Line NCI-H69 When Added to HITE Medium[a]

Growth conditions	Relative increase in cell number[b] (day 21)
HITE[c]	1.0
HITE + 10% fetal bovine serum	2.4
HITE +	
Selenium (3×10^{-8} M)	1.4
Luteinizing hormone (500 ng/ml)	1.2
Putrescine (10^{-4} M)	1.1
ACTH (5 μU/ml)	0.97
Nerve growth factor (50 ng/ml)	0.89
Calcitonin (0.1 ng/ml)	0.84
Prolactin (10 ng/ml)	0.82
Glucagon (500 ng/ml)	0.82
Growth hormone (5 ng/ml)	0.81
Fibroblast growth factor (50 ng/ml)	0.77
Pitressin (10 ng/ml)	0.76
Multiplication-stimulating activity (50 ng/ml)	0.70
Blood meal (10 ng/ml)	0.64
Epidermal growth factor (1 ng/ml)	0.58
Epidermal growth factor (50 ng/ml)	0.47
Luteinizing hormone-releasing factor (10 ng/ml)	0.45

[a]Adapted from Simms et al. [1980].
[b]Ratio of number of cells in the HITE-supplemented medium plus factor to the number of cells in HITE.
[c]RPMI 1640 medium supplemented with hydrocortisone (10^{-8} M), insulin (5 μg/ml), transferrin (100 μg/ml), and estradiol (10^{-8} M). The dose of transferrin was later reduced to 10 μg/ml.

the SCLC cell line. With the addition of insulin (5.0 μg/ml) and transferrin (10 μg/ml) to the base medium, a significant increase in cell number was observed. When hydrocortisone (10^{-8} M) or 17β-estradiol (10^{-8} M) was added (constituting HITE medium), further stimulation of growth was noted.

The influence of all other hormones was determined with base medium supplemented with HITE. As indicated in Table I, no factor except selenium gave significant growth enhancement, and in some instances growth factors inhibited the proliferation of the SCLC cells.

Based on these data, a defined medium containing hydrocortisone, insulin, transferrin, estradiol, and selenium (HITES medium) was tested for its ability to support the in vitro growth of established cell lines of SCLC. Although the initial studies of the many growth factors were performed on only a single cell line, HITES medium was shown to support the growth of other cell lines of SCLC. When transferred from serum-supplemented medium (SSM; RPMI 1640 medium supplemented with 10% [vol/vol] heat-inactivated fetal bovine serum) to HITES, most established SCLC cell lines proliferate at a rate similar to that observed in SSM after a lag phase.

Growth of Tumor Cells From Fresh Clinical Specimens

We next tested the ability of HITES medium to support the growth of fresh tumor cells from patient biopsy specimens. We obtained 170 specimens by sterile techniques from 117 patients with histologically documented cancer on approved clinical protocols, of which 153 specimens were obtained from 100 patients with SCLC, 7 specimens came from patients with other lung cancers including adenocarcinoma (6 specimens) and mesothelioma (1 specimen), and 10 specimens came from patients with other tumors including malignant melanoma (2 specimens), ovarian carcinoma (2 specimens), Hodgkin's disease, gastric carcinoma, hypernephroma, testicular carcinoma, esophageal carcinoma, and breast carcinoma (one specimen each).

The viability of cell preparations as determined by trypan blue exclusion ranged from 20% to 95%. The lowest viabilities were usually in solid tumor specimens, whereas the highest (often greater than 90%) were in bone marrow aspirates. Approximately 1×10^6 viable mononuclear cells were seeded into 75-cm^2 flasks containing 14 ml of 1) RPMI 1640 medium supplemented with 10–20% (vol/vol) heat-inactivated fetal bovine serum (serum supplemented medium, SSM) and 2) HITES medium.

The nature and origin of cultured cells was determined by cytologic examination, nude mouse tumorigenicity studies, DNA content analysis using flow cytometry, and measurements of cellular L-dopa decarboxylase and bombesin.

Table II shows the results of our growth studies for fresh specimens. Of 53 specimens from 38 patients cytologically positive for SCLC, proliferation of tumor cells was seen in 39 cases in HITES medium (74%), compared with only 26 cases (49%) in serum-supplemented medium. Thirteen specimens from different patients grew in HITES but not in SSM, but only a single specimen grew in SSM but not in HITES.

The SCLC specimens always contained mixtures of malignant and normal cells. Some specimens were composed predominantly of tumor cells, whereas in others, only rare clumps of tumor cells could be found. When plated in SSM, SCLC cells could be distinguished by their typical refractile appearance. The cells grew as floating, tightly packed aggregates, identical to what is observed with SCLC cell lines [Gazdar et al., 1980]. However, concomitant adherent stromal cell proliferation was also observed in the majority of speci-

TABLE II. Growth of Clinical Specimens in Serum-Free HITES Medium

	No. of specimens	Fraction (%) with growth
Histology		
Small cell lung cancer	53[a]	39 (74%)
Other cancers	17[b]	1 (6%)
Nonmalignant	100	0 (0%)
SCLC specimen source		
Bone marrow	25	22 (88%)
Pleural effusion	13	7 (54%)
Lymph node	14	9 (64%)
Brain	1	1 (100%)
SCLC treatment status		
Prior chemotherapy	26	19 (73%)
No prior chemotherapy	27	20 (74%)

[a]In serum-supplemented medium, tumor cell proliferation was observed in 26/53 (49%) SCLC specimens.
[b]In serum-supplemented medium, tumor cell proliferation was observed in 6/17 (33%) non-SCLC tumor specimens.

mens in SSM. In one culture of a bone marrow aspirate, a lymphoblastoid B cell line overgrew the SCLC cells after 6 weeks. Initial tumor cell proliferation in SSM was slow. In contrast, SCLC cells cultured in HITES medium demonstrated early, rapid proliferation. By 24-48 h after plating, light microscope examination of tissue culture flasks clearly revealed large clumps of healthy viable tumor cells. The tumor cells in HITES had a population doubling time of 48-96 h and were passaged (1:2 or 1:3 split) as necessary every 7-10 days. In several of the early cultures,, it was apparent that splitting cultures too early was frequently associated with marked reduction or cessation of growth of tumor cells. This suggested that in some instances, the tumor cells were producing growth factors essential for their own growth and cross-feeding one another.

The most significant difference between HITES and serum-supplemented medium was that HITES absolutely failed to suport the growth of nonmalignant cells. No growth was seen in 100 specimens from 62 patients that were cytologically negative, most of which were bone marrow aspirates or pleural effusions. We did not observe growth of normal bone marrow cells, mesothelial cells from pleural effusions, or fibroblasts or lymphocytes from solid tumor specimens. In cytologically positive specimens, no growth of normal cells was observed even though 1%-99% of each specimen was initially composed of nonmalignant cells. Death and lysis of nontumor cells was observed in HITES medium after 7-14 days in culture, resulting in an almost pure population of rapidly dividing SCLC tumor cells.

As indicated in Table II, SCLC tumor cells were grown successfully from a number of different metastatic sites. The ability to achieve initial growth was very similar with specimens from newly diagnosed, untreated patients and with specimens from patients who had relapsed after intensive combination chemotherapy.

Of the 39 SCLC specimens that demonstrated initial growth in HITES, 7 cultures were eventually lost to microbial contamination. In the remaining 32 specimens, growth of pure populations of tumor cells were maintained for varying periods of time. Continuous cell lines of SCLC were established in 26 cases (66% of specimens showing initial growth), but the tumor cells from 6 patients died after 3-24 weeks in culture. In these six instances, a large number of tumor cells had been grown, permitting in vitro characterization of the cells, but clearly some growth requirement(s) of these cells were lacking in our medium. In most specimens, growth slowed or ceased after 3-15 weeks (approximately 10-40 doublings). In these cases, continuous replication of the tumor cells could be maintained by the addition of 10% (vol/vol) fetal bovine serum.

Characterization of Cells Grown in HITES

In nine instances, continuous cell lines of SCLC were established in HITES medium alone without the addition of serum at the time of "crisis." In six of these instances, parallel continuous cell lines were also established from the same clinical specimens in serum-supplemented medium. These cell lines, established and maintained in serum-free medium, clearly provide a means for determining if different culture conditions (serum-supplemented vs. serum-free) select for the in vitro growth of different subpopulations of tumor cells. Thus far we have found that the cells maintained in HITES retain the typical biochemical and cytologic characteristics of SCLC cultures. All express L-dopa decarboxylase, bombesin, and creatine-kinase BB. There were no significant differences in the expression of these markers in parallel cell lines maintained in HITES and in serum-supplemented medium. The gross morphologic appearance of the proliferating cells in HITES medium was that of SCLC cells (Fig. 1). Cytologic examination of tumor cells from each specimen after several weeks of in vitro growth, whether in serum-supplemented medium or in HITES, always revealed cells with the typical characteristics of SCLC, which were very similar to the tumor cells in the original clinical specimen. No differences were observed cytologically between cells cultured in serum-supplemented medium or in HITES.

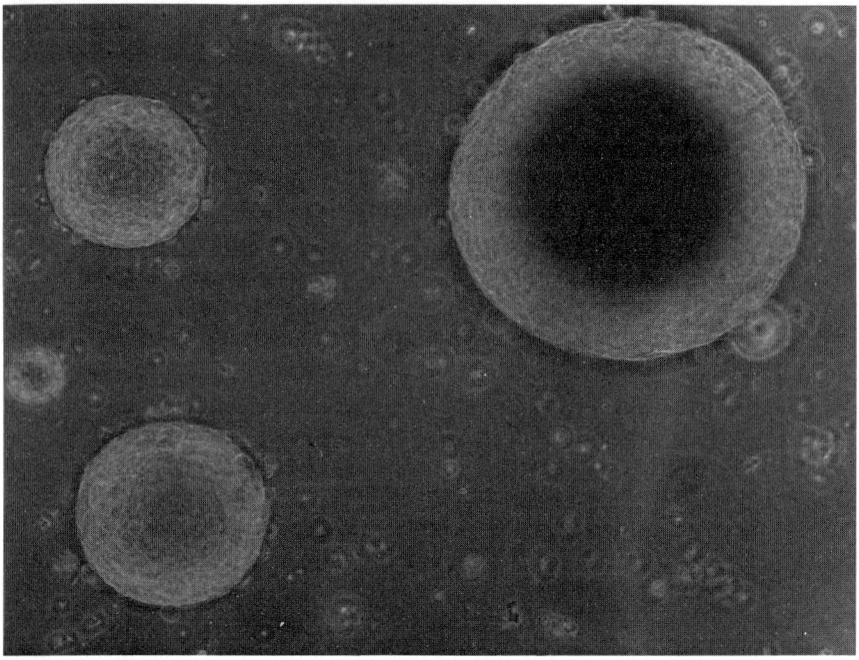

Fig. 1. Morphologic appearance of a clinical specimen of SCLC cultured in serum-free HITES medium. The tumor cells grow as tight floating cell aggregates identical to that previously observed for SCLC cells cultured in serum-supplemented medium.

The DNA content was used as a tumor marker in specimens with aneuploid malignant cells. We found the DNA content of the cultured malignant cells was identical to those of the original clinical specimens in all instances in both culture media. Serial studies of individual specimens in HITES showed that the fraction of malignant (aneuploid) SCLC cells always increased with time whereas that of nonmalignant (diploid) cells decreased with time. In several instances, sequential DNA content analysis of cells in both HITES and serum-supplemented medium clearly demonstrated the selective nature of the HITES medium for the growth of SCLC, in contrast to serum-supplemented medium, in which growth of both malignant and normal cells occurred. In HITES medium, pure populations of rapidly dividing tumor cells were usually seen by 7–14 days after plating, whereas in serum-supplemented medium, in those specimens in which there was no tumor growth, pure populations of tumor cells were frequently not obtained for weeks to months after initial plating.

Cells cultured in HITES medium formed typical SCLC tumors when inoculated into athymic nude mice. When nude mouse SCLC tumors were cultivated in serum-supplemented medium, proliferation of both SCLC tumor cells and supporting rodent stromal cells was observed just as with fresh clinical specimens, whereas in HITES medium, only tumor cell proliferation was observed.

Selectivity of HITES Medium for SCLC

We observed tumor cell proliferation in only one of the 17 non-SCLC tumors cultured in HITES (Table II). This was a pleural effusion cytologically positive for adenocarcinoma of the lung. The tumor cells initially proliferated in HITES, but stopped growing after 4 weeks in culture and could not be successfully passaged. Lung adenocarcinoma cells usually demonstrate substrate adherence and grow as epithelioid monolayers in serum-supplemented medium, but in HITES the adenocarcinoma cells grew as floating cell aggregates of loosely packed cells. Growth of this specimen could not be maintained even by adding serum to the HITES. In contrast, 5 of these 17 non-SCLC specimens (2 ovarian carcinoma, 1 large cell carcinoma, 1 melanoma, 1 hypernephroma) were successfully maintained in serum-supplemented medium and have been in culture for 6–24+ months.

Growth of Non-SCLC in Serum-Free Medium

Because of our success with the in vitro cultivation of SCLC in the serum-free medium HITES, we also attempted to grow non-SCLC in serum-free medium. However, as noted before, we have only once observed proliferation of tumor cells from a non-SCLC specimen in HITES. The established non-SCLC cell line A549 can grow in HITES both in liquid culture and in soft agarose, but other cell lines of non-SCLC cannot grow in HITES alone.

Barnes and his co-workers have described serum-free media for adenocarcinoma and epidermoid lung cancer, called LA and LE, respectively. LA medium, containing insulin, transferrin, epidermal growth factor, selenium, hydrocortisone, and albumin, supported the growth of a human lung adenocarcinoma, T291, that had been maintained in nude mice by successive heterotransplantation. In order for cells to remain permanently adherent, precoating of culture flasks with serum-supplemented medium was required. Barnes reported that new cell lines from fresh clinical specimens could not be established in LA [Barnes et al., 1981].

We tested LA medium and found that it supports the growth of three other established cell lines of lung adenocarcinoma (NCI-H23, NCI-H125, and A549) with doubling times comparable to those in serum-supplemented medium. LA medium was superior to serum-supplemented medium in supporting initial cultures of fresh specimens of adenocarcinoma of the lung, with viable tumor cells present at 1 month in 6 of 9 specimens in LA but in only 2 of 9 specimens in serum-supplemented medium. The serum-free medium did not permit overgrowth of cultures by contaminating stromal cells. However, we were unble to maintain any fresh specimen in LA beyond 5 months, and no permanent cell lines were established with this medium. The established cell lines were maintained in LA for 6+ months [Brower et al., 1982]. Clearly, LA medium needs improvement because of its inability to support indefinite growth of non-SCLC fresh specimens. In addition, the use of serum-precoating of flasks is not desirable in a pure defined system, since recent evidence suggests that various growth factors can adhere to tissue culture plastic and act mitogenically in situ even after brief incubation with serum [Smith et al., 1982].

LE medium contains triiodothyronine, glucagon, selenium, and insulin; it supported the short-term growth of T222 lung epidermoid carcinoma in fibronectin-coated flasks. However, LE did not support the long-term growth of T222. There are no data concerning the ability of this medium to support the growth of fresh human specimens of epidermoid carcinoma of the lung.

Autocrine Growth Factors Are Produced by SCLC Cells in Serum-Free Medium

Although growth of both established cell lines and fresh tumor specimens of SCLC was readily achieved in HITES medium, colony formation of SCLC in agarose was either very poor or did not occur with HITES. When cloned in serum-supplemented medium, SCLC forms colonies with a cloning efficiency of 1–5%. This suggested that some factor(s) stimulatory to SCLC growth are present in serum but missing from HITES.

Conditioned medium from cell lines isolated and permanently maintained in HITES were assayed for their ability to stimulate colony formation of other SCLC cell lines. Serum-free conditioned medium was collected, clarified, filtered, and added to HITES medium (25% vol/vol). The effects of conditioned media from four SCLC cell lines on the soft agarose cloning of cell line NCI-N592 are shown in Table III. Conditioned media from all four lines permitted colony formation in HITES with a cloning efficiency very similar to that observed in serum-supplemented medium. The conditioned media had neither stimulatory nor inhibitory effects on the cloning of normal human bone marrow stem cells, and they did not stimulate colony formation of normal rat kidney fibroblast cells [Carney et al., 1983b].

Further evidence that SCLC growth produces autostimulatory growth factors comes from experiments in which two hormones known to be secreted by SCLC, bombesin and arginine vasopressin, were used to enrich HITES medium. Addition of 10 ng/ml of arginine vasopressin or 10^{-7} M bombesin

TABLE III. Influence of Conditioned Media (25% vol/vol) From Small Cell Lung Carcinoma Cell Lines Cultured in HITES Medium on the Soft-Agarose Cloning of SCLC Cell Line NCI N-592

Growth conditions	Colony formation[a]
HITES + 10% fetal bovine serum	100% (894 colonies)
HITES	1%
HITES + conditioned medium from cell line:	
NCI-H345	106%
NCI-H372	97%
NCI-H462	75%
NCI-N592	71%

[a]Expressed as percentage colony formation compared to that observed in serum-supplemented medium (per 5×10^4 cell plated).

led to 2.4-fold and 1.9-fold increases, respectively, in cell number of the SCLC cell line NCI-H209 in liquid culture over the number of cells in HITES alone. Adding these two hormones to HITES, along with the nutrients ethanolamine, phosphoethanolamine, and albumin, defined a new medium, SCLC-2, that was superior to HITES for short-term culture of established SCLC cell lines [Minna et al., 1982b]. Adding bombesin to HITES also appears to improve the serum-free agarose cloning of SCLC cell lines.

DISCUSSION

We developed a serum-free defined medium containing hydrocortisone, insulin, transferrin, 17β-estradiol, and selenium (HITES) by analyzing the growth requirements of an established SCLC cell line, NCI-H69, and then demonstrating that this and other SCLC cell lines proliferate in HITES at a regular rate similar to that observed in serum-supplemented medium. We then observed tumor cell proliferation in 39 of 53 clinical specimens with SCLC tumor cells when we used HITES, whereas growth of tumor cells in serum-supplemented medium was seen in only 26 of 53 specimens. The proliferation of neoplastic cells from the fresh specimens was monitored by the morphologic appearance of the cultures under phase microscopy and, in the case of aneuploid tumor samples, by DNA content analysis. In HITES medium, but not in serum-supplemented medium, tumor cells were able to replicate so well that by 7–14 days after plating, an almost pure population of rapidly dividing tumor cells was present. Normal cells from SCLC-positive specimens or from cytologically negative specimens, as well as tumor cells from non-SCLC neoplasms, were not able to replicate in HITES. SCLC specimens obtained after growth in athymic nude mice were able to proliferate when explanted into HITES medium, but normal mouse cells could not. The tumor cells proliferating in HITES maintained the typical cytopathology of SCLC and were able to form tumors in athymic nude mice. They expressed the biomarkers previously identified as characteristic of SCLC cell lines isolated and cultured in serum-supplemented medium.

These data suggest that HITES is the medium of choice for the initial isolation and growth of SCLC tumor cells from clinical specimens, regardless of the proportion of tumor cells contained in any specimen. The success of HITES medium in promoting the initial growth and establishing continuous cell lines of SCLC should permit in vitro growth of many new lines of this tumor type, both directly from patient specimens and from nude mouse heterotransplants.

Although a number of studies have reported on the use of defined media to support the growth of both normal and malignant cells in vitro, most of these studies utilized established cell lines. In studies of human cell lines, a nearly universal requirement for insulin and transferrin has been noted. However, other requirements differ among different cell types, suggesting that different cells have different growth requirements. Table IV lists the growth factors used in the serum-free media that have been developed for SCLC, epidermoid lung carcinoma, and lung adenocarcinoma. The factors in each medium are different, suggesting that different types of tumors have different growth requirements.

An approach similar to that used for defining the specific growth factor requirements for SCLC could be used for other human tumors: First define the serum-free requirements of an established cell line of the tumor, then test

TABLE IV. Growth Factor Requirements for Different Types of Lung Cancer in Serum-Free Medium

	Histology		
Growth factors	Small cell	Adenocarcinoma*	Epidermoid*
Hydrocortisone	+	+	−
Insulin	+	+	+
Transferrin	+	+	−
Estradiol	+	−	−
Selenium	+	+	+
Epidermal growth factor	−	+	−
Human fatty acid-free albumin	−	+	−
Glucagon	−	−	+
Triiodothyronine	−	−	+

*To assist attachment of cell to tissue culture plastic, flasks are precoated with serum-supplemented medium or fibronectin [Barnes et al., 1981; Brower et al., 1982].

the formula for its ability to support other cell lines of the same histologic type from the same organ, and then test the formula on biopsy samples containing tumor cells. It is conceivable that in the future, the ability of a tumor to grow in a specific serum-free medium might be of diagnostic value in instances where the histology of the tumor is debatable.

The use of HITES medium has improved our understanding of the biology of SCLC. We observed that in many specimens, growth slowed or stopped after 3–15 weeks in culture after excellent early proliferation of tumor cells. Loss of the cell lines could be avoided by adding fetal bovine serum to the culture medium. This suggested that HITES, while capable of supporting the early growth of SCLC, is not yet a "complete" medium. By using HITES, it is possible to screen and identify other growth factors that may stimulate the growth of SCLC, both in liquid culture and in semisolid medium. We are particularly interested in the concept that SCLC may be under "autocrine" regulation — that one explanation for the uncontrolled growth of this malignancy in the patient is that tumor cells make growth factors that stimulate their own growth [Sporn and Todaro, 1980]. Preliminary data produced with a soft agarose clonogenic assay, for the nine permanent cell lines of SCLC established with HITES medium alone, does indeed suggest that SCLC secretes autocrine growth factors. The isolation and characterization of these factor(s) may greatly help our understanding of the regulation of tumor growth in patients. The ability of conditioned medium from one SCLC cell line to permit serum-free anchorage-independent cloning of other cell lines is similar to recent work with virally transformed 3T3 cells, which also appear to secrete autocrine growth factors in serum-free medium [Kaplan et al., 1982].

An important application of the HITES medium has been its use in the cytogenetic analysis of clinical specimens of SCLC. For most solid-tumor specimens, cytogenetic analysis is limited by the poor proliferation of tumor cells, and thus by the absence of metaphases in the specimens. When HITES is used, cytogenetic analysis of the majority of clinical specimens of SCLC can be obtained after 3 days' culture, when most normal cells are already dying and 20–30% of the tumor cells are in the proliferative phase of the cell cycle [Carney et al., 1981b]. With this technique, the specific cytogenetic abnormality identified in cell lines of SCLC, a deletion on chromosome 3, was identified in all fresh specimens tested [Whang-Peng et al., 1982a,b]. This application of serum-free medium may allow cytogenetic studies to be performed on many other human neoplasms.

Our ability to grow non-SCLC lung tumors in serum-free medium is still limited. LA medium supports the growth of established cell lines of lung adenocarcinoma, and it is superior to serum-supplemented medium for the initial growth of tumor cells from fresh specimens of lung adenocarcinoma. Further work is needed to improve this medium to permit the establishment of permanent cell lines of lung adenocarcinoma in serum-free medium.

SUMMARY

We have shown that the use of serum-free, hormone-supplemented medium, designed on the basis of the study of established cell lines of SCLC, can be used to support the selective outgrowth of human tumor cells from routine clinical biopsy material. Such cells retain all the typical morphologic, biologic, and biochemical characteristics of SCLC, and they do not appear to differ in the phenotypes analyzed in any way from SCLC cells cultured in serum-supplemented medium. The selective nature of this medium for the growth of only SCLC tumor cells may be of diagnostic value in anaplastic tumors, and it provides a means for performing cytogenetic analysis in SCLC. This technology will permit the establishment of more cell lines of this tumor and will aid in the study of the biology of neoplasia.

REFERENCES

Barnes D, Sato G (1980a): Serum-free cell culture: A unifying approach. Cell 22:649-655.

Barnes D, Sato G (1980b): Methods for growth of cultured cells in serum-free medium. Anal Biochem 102:255-270.

Barnes D, van der Bosch J, Masui H, Miyazaki K, Sato G (1981): The culture of human tumor cells in serum-free medium. In Pesika S (ed): "Methods in Enzymology, 79." New York: Raven, pp 369-399.

Brower M, Carney D, Oie H, Matthews M, Minna J (1982): Growth of human adenocarcinoma of the lung (ACL) cell lines and clinical specimens in serum-free defined medium (Abstract). Proc ASCO 1:140.

Carney DN, Minna JD (1982): Small cell lung cancer. Clin Chest Med 3:389-398.

Carney DN, Gazdar AF, Minna JD (1980): Positive correlation between histological tumor involvement and generation of tumor cell colonies in agarose in specimens taken directly from patients with small-cell carcinoma of the lung. Cancer Res 40:1820-1823.

Carney DN, Gazdar AF, Bunn PA, Guccion JG (1981a): Demonstration of the stem cell nature of clonogenic tumor cells from lung cancer patients. Stem Cells 1:149-164.

Carney DN, Bunn PA, Gazdar AF, Pagan JA, Minna JD (1981b): Selective growth in serum-free hormone supplemented medium of tumor cells obtained by biopsy from patients with small cell carcinoma of the lung. Proc Natl Acad Sci USA 78:3185-3189.

Carney DN, Gazdar AF, Minna JD (1982): In vitro chemosensitivity of clinical specimens (CS) and established cell lines (CL) of small cell lung cancer (SCLC) (Abstract). Proc ASCO 1:10.

Carney DN, Mitchell JB, Kinsella TJ (1983a): In vitro radiation and chemosensitivity of established cell lines of human small cell lung cancer and its large cell variants. Cancer Res 43:2806-2811.

Carney DN, Cuttitta FC, Gazdar AF, Minna JD (1983b): Autocrine clonogenic factor(s) (CF) are produced by cell lines (CL) of small cell lung cancer (SCLC) (Abstract). Proc ASCO (in press).

Gazdar AF, Carney DN, Russell EK, Sims HL, Baylin SB, Bunn PA, Guccion JG, Minna JD (1980): Establishment of continuous, clonable cultures of small-cell carcinoma of the lung which have amine precursor uptake and decarboxylation properties. Cancer Res 40:3502-3507.

Gazdar AF, Carney DN, Sims HL, Simmons A (1981a): Heterotransplantation of small-cell carcinoma of the lung into nude mice: Comparison of intracranial and subcutaneous routes. Int J Cancer 28:777-783.

Gazdar AF, Zweig MH, Carney DN, Van Stierteghen AC, Baylin SB, Minna JD (1981b): Levels of creatine kinase and its BB isozyme in lung cancer tumors and cultures. Cancer Res 41:2773-2777.

Giard DJ, Aaronson SA, Todaro GJ, Arnstein P, Kersey JH, Dosik H, Parks WP (1973): In vitro cultivation of human tumors: Establishment of cell lines derived from a series of solid tumors. J Natl Cancer Inst 51:1417-1423.

Kaplan PL, Anderson M, Ozanne B (1982): Transforming growth factor(s) production enables cells to grow in the absence of serum: An autocrine system. Proc Natl Acad Sci USA 79:485-489.

Marangos PJ, Gazdar AF, Carney DN (1982): Neuron specific enolase in human small cell carcinoma cultures. Cancer Lett 15:67-71.

Minna JD, Higgins GA, Glatstein EJ (1982a): Cancer of the lung. In DeVita VT, Hellman S, Rosenberg SA (eds): "Principles and Practice of Oncology." Philadelphia: J.B. Lippincott, pp 396-475.

Minna JD, Carney DN, Oie H, Bunn PA, Gazdar AF (1982b): Growth of human small-cell lung cancer in defined medium. In Pardee A, Sato G (eds): "Growth of Cells in Hormonally Defined Medium," Cold Spring Harbor, New York: Cold Spring Harbor Laboratory, pp 627-639.

Moody TW, Pert CB, Gazdar AF, Carney DN, Minna JD (1981): High levels of intracellular bombesin characterize human small cell lung carcinoma. Science 214:1246-1248.

Simms E, Gazdar AF, Abrams PG, Minna JD (1980): Growth of human cell (oat cell) carcinoma of the lung in serum-free growth factor-supplemented medium. Cancer Res 40:4356-4363.

Smith JC, Singh JP, Lillquist JS, Goon DS, Stiles CD (1982): Growth factors adherent to cell substrate are mitogenically active in situ. Nature 296:154-156.

Sporn MB, Todaro GJ (1980): Autocrine secretion and malignant transformation of cells. N Engl J Med 303:878-880.

Whang-Peng J, Kao-Shan CS, Lee EC, Bunn PA, Carney DN, Gazdar AF, Minna JD (1982a): A specific chromosome defect associated with human small-cell lung cancer: Deletion 3p (14-23). Science 215:181-182.

Whang-Peng J, Bunn PA, Kao-Shan CS, Lee EC, Carney DN, Minna JD, Gazdar AF (1982b): A non-random chromosomal abnormality, del 3p (14-23) in human small cell lung cancer. Cancer Genet Cytogenet 6:119-134.

Methods for Serum-Free Culture of Epithelial and Fibroblastic Cells, pages 73–81

6
Primary Tissue Cultures of Human Colon Carcinomas in Serum-Free Medium: An In Vitro System for Tumor Analysis and Therapy Experiments

Jürgen van der Bosch

The need for in vitro systems of human tumors, which allow the quantitative analysis of tumor responses to defined treatments, has been stressed repeatedly in the literature. Such culture systems are of potential value for clinical drug-sensitivity screenings of individual patients' tumors and for the development of new agents and strategies against cancer.

Until recently, two different systems were available for such purposes: established cell lines derived from human tumors [Fogh and Trempe, 1975; Leibovitz et al., 1976; Murakami and Masui, 1980; Brattain et al., 1982] and the colony-forming assay in soft agar, which uses primary explants of human tumors [Hamburger and Salmon, 1977; Salmon et al., 1978; Rupniak and Hill, 1980; Laboisse et al., 1981; Von Hoff et al., 1981].

The application of these approaches to investigation of human cancer is limited owing to the following difficulties: Tissue culture cell lines often arise from primary explants only after long periods of in vitro adaptation. The processes taking place during this adaptation period include genetic and epigenetic changes as well as selection of subpopulations that constitute only a minor part of, or are not represented at all in, the primary explant. Thus, the advantages of such culture lines for providing reproducibly large amounts

Forschungsinstitut Borstel, 2061 Borstel, Federal Republic of Germany

This work was performed in Dr. Gordon Sato's Laboratory, Cancer Center, University of California, San Diego.

of cells that can be easily passaged (from culture to culture and from culture to host animals) and the environment of which can be exactly manipulated in terms of medium composition, are outweighed, at least in part, by the fact that the results obtained cannot easily be correlated with the properties and behavior of the original tumor. In the colony-forming assay, on the other hand, the goal is to achieve single-cell suspensions of the primary tumor, which are seeded in soft agar gels. This assay seems to work well with some types of human tumors [Salmon et al., 1978; Van Hoff et al., 1981], whereas others, especially some of the solid tumors, have been handled only with limited success [Buick et al., 1980; Rupniak and Hill, 1980]. General disadvantages of this method are 1) very low yields of viable, dividing cells in the cultures and 2) difficulty in controlling the cellular environment in agar gels, in which quantitative medium changes cannot be performed. The populations investigated in this system are probably not representative of the total viable part of the tumor. Furthermore, experiments involving exactly scheduled applications of agents, large populations in defined growth states, transfer of such populations to other cultures or host animals, and quantitative analyses on a per cell basis are very difficult if not impossible to perform.

Recently it has become possible, by the use of defined serum-free media as developed in the laboratories of G. Sato and R. Ham [Bottenstein et al., 1979; Barnes and Sato, 1980; Ham, 1981], to work out procedures and conditions for the preparation and maintenance of primary cultures of various human tumors, which are not subject to the limitations cited above [van der Bosch et al., 1981; Carney et al., 1981; Barnes et al., 1982].

In the following, procedures for setting up primary cultures of human colon carcinomas and a method for quantitative evaluation of growth in such cultures are described as they were developed by working with 11 different human colon carcinomas that were kept as subcutaneous serial transplants in the athymic nude mouse in order to allow repeated experiments with individual tumors [van der Bosch et al., 1981; van der Bosch and Zirvi, 1982]. These procedures also yielded satisfying results when applied to patients' tumor biopsies that had not been transplanted before [van der Bosch, 1983].

PREPARATION OF PRIMARY CULTURES

Cell Preparation

A 2- to 4-ml piece of nonnecrotic tumor tissue is held with forceps over a petri dish and flushed five times with DME/F12, a one-to-one mixture of Dulbecco's modified Eagle's medium and Ham's F12 medium (see below). In a fresh dish the surrounding connective tissue capsule is removed with forceps and scalpel. The remaining tumor piece is flushed with DME/F12 once more and placed in another dish, where it is cut with a scalpel (blade No. 10) into thin slices of about 1-mm thickness. The material is then finely minced with the help of a special tool consisting of a handle with an array of 18 parallel razor blades placed 0.75 mm apart from each other by washers. The mince is suspended in 10 ml of DME/F12 by pipetting several times up and down. This suspension is diluted with DME/F12 to 50 ml and centrifuged for 2.5 min with a clinical centrifuge at 1,200 rpm. The supernatant is discarded and the top layer of the sediment, consisting of small debris and/or mucoid material that is devoid of live cell aggregates, is removed by aspiration. Depending on the individual tumor this procedure must be repeated with the remaining sediment once or twice until a top layer of debris or mucoid substance is no longer detectable after centrifugation, as determined by microscopic examination of the sediment.

Collagenase solution (5 ml/per ml of packed sediment) is then added and the resuspended mince is incubated at 37°C in a closed 25-cm^2 tissue culture flask. After 10 min, 0.1 ml of dispase solution is added per ml of collagenase solution, and after another 10 min, 1 ml of hyaluronidase solution is added per ml of collagenase solution. Incubation at 37°C is continued for 5–10 min more. During the entire enzyme treatment the suspension is pipetted up and down four times every 5 min. Then centrifugation is carried out at 1,200 rpm for 2.5 min. The resulting sediment is carefully freed of overlaying debris by aspiration. It is resuspended in 100 ml of DME/F12 per ml of sediment and filtered through a nylon screen of 100-μm mesh width. The yield of this step is greatly improved by scraping with a spatula back and forth on the screen during filtration. The material remaining on the screen is discarded. Centrifugation and removal of the overlaying debris is carried out another two times, and the final sediment is suspended in an appropriate amount of culture medium, seeded in collagen-coated tissue culture dishes, and incubated at 37°C in a water-saturated atmosphere of 5% CO_2 in air.

The success of this procedure depends critically on the following points.

1. The mechanical mincing should be performed by cutting with sharp blades, not by pressing through screens or shearing in cannulas. By cutting with the bladed tool described above, a mince of rather uniform particle size is produced with a minimum of mechanical stress, thus avoiding unnecessary damage to the tissue and providing a maximum yield of viable material.

2. The repeated careful removal of debris and mucoid material from sediments is important. If not removed, this material will greatly impair the attachment of viable cell aggregates to the culture substrate.

3. During hyaluronidase treatment the progress of tissue disintegration has to be followed closely by microscopic examination, since the sensitivity to this

enzyme varies strongly from tumor to tumor. Centrifugation after hyaluronidase treatment has to be performed well before the major part of the tissue has been disintegrated into single cells: Single-cell suspensions did not give rise to viable cell cultures in the hands of the present author. For the same reason divalent cation chelators could not be applied successfully during the disintegration of colon carcinomas for cell culture purposes.

4. For most colon tumors attachment is greatly improved by use of collagen-coated dishes.

5. The culture medium used is a serum-free medium and completely prevents the growth of fibroblasts.

6. Since the final outcome of this procedure is a suspension of cell aggregates, it is necessary to seed quickly with the help of a dispenser (repipetter) from a stirred suspension in order to obtain replicate cultures with no more than 10% difference in cell number. For the same reason no less than 10^5 cells should be seeded per plate.

For subcultivation, cultures can be detached from the plate by applying the enzymatic procedure described above. Frequently a collagenase treatment alone is sufficient for detachment. The detached material consists primarily of cell aggregates, not of single cells.

Solutions

All solutions used during the disintegration are sterile and prewarmed at 37°C. All instruments and plastic ware used are sterile.

DME/F12 is a 1:1 mixture of Dulbecco's modified Eagle's medium and Ham's F12 medium, containing per liter 1.1 g $NaHCO_3$, 10 ml of 1.5 M HEPES [4-(2-hydroxyethyl)-1-piperazine-ethanesulfonic acid] in 1 M NaOH, 120 mg penicillin G (K^+-salt), 270 mg streptomycin sulfate, and 25 mg ampicillin. For use during the disintegration procedure this medium is adjusted to pH 7.25 with 1 M HCl. For use as culture medium no HCl is added.

Collagenase solution is prepared by adding 0.5 ml of frozen collagenase stock per 5 ml of prewarmed, HCl-adjusted DME/F12. Collagenase stock contains 40 mg collagenase (Worthington type IV) per ml H_2O.

Dispase solution contains 40 mg dispase (Boehringer Mannheim, grade II) per ml H_2O and is stored in frozen 0.5-ml aliquots.

Hyaluronidase solution contains 2 mg hyaluronidase (Sigma, from bovine testes) per ml in 0.9% NaCl, 10 mM KCl, 30 mM MES (2-N-morpholinoethanesulfonic acid, Sigma) at pH 4.0. This low pH is important in order to provide optimum conditions for tissue disintegration following addition of hyaluronidase.

Collagen coating solution is prepared by mixing, at 4°C, 1 ml of Vitrogen 100 (obtained from Collagen Corp., Palo Alto, or Flow Laboratories), containing about 3 mg of calf skin collagen per ml, with 100 ml of DME/F12 (no HCl added). Tissue culture dishes are coated by adding 0.15 ml of this cold solution per cm^2 immediately after its preparation and incubating the dishes for at least 2 h at 37°C in an atmosphere of 5% CO_2 in air. Immediately before plating cells, the collagen solution is removed from the plates by aspiration and plating is performed on still wet plates.

The *culture medium* used is a modification of a serum-free medium developed for a colon tumor cell line [Murakami and Masui, 1980]. It consists of DME/F12 as described above, but additionally containing insulin (2 mg/liter), transferrin (2 mg/liter), epidermal growth factor (4μg/liter), triiodothyronine (2×10^{-10} M), and $NaSeO_3$ (5×10^{-9} M). These five factors are added to the warm medium from concentrated stocks immediately before seeding.

DESCRIPTION OF CULTURES*

The seedings obtained from colon carcinomas by the method described above are suspensions of cell aggregates. The procedure was developed with the aim of seeding aggregates, and not seeding single cells, for the following reasons. Disintegration methods aimed at producing single-cell suspensions regularly led to destruction and loss of the majority of the viable part of the tissue treated. Even the remaining part was not a true single-cell suspension but a mixture of single cells and cell aggregates. The yield of such procedures was so small that only very few cultures could be plated and the majority, if not the total number, of single cells disintegrated during the first few days after seeding. Net growth was not detectable.

In contrast, the procedure given above yields in most cases 5×10^6 to 2×10^8 cells per seeding in the form of cell aggregates, composed as in the original tissue, that attach to the culture plates during the first 10–20 h after seeding. Reproducible net growth in replicate dishes can be detected in these cultures from the first day on after seeding. Dead cells are shed into the supernatant medium and three-dimensional glandular structures are built up, suggesting that cell turnover and morphogenesis take place in these cultures as in normal colonic mucosa and carcinomas of this tissue in vivo. Growth rates and cell density at saturation vary from tumor to tumor, displaying individuality, as is known to occur in vivo in tumors of the same tissue of origin.

*For details see van der Bosch et al. [1981].

EVALUATION OF GROWTH

Counting cells for the purpose of deriving growth curves proved to be impossible, because no method could be found that allowed the reproducible quantitative disintegration of these cultures into single cells. Instead the isolation of cell nuclei by osmotic shock treatment in the presence of a detergent, as used in flow fluorimetry of nuclear DNA contents [Vindelov, 1977], proved to be adaptable for the solution of this problem [van der Bosch et al., 1981]. The procedure outlined below gives reproducible counts of nuclei of the viable cells in replicate cultures. Nuclei of dying or disintegrating cells are not included, since such nuclei do not withstand the osmotic shock treatment.

Preparation of Cell Nuclei

Cultures are washed four times with solution A (see Table I) and four times with solution B, the washes with solution B following each other in rapid succession. Immediately after the last wash, 0.8 ml of solution C is applied per plate. Under this condition cells swell and lyse. In order to facilitate cell lysis and detachment, it is useful to swirl the dishes several times gently. After 5-20 min, the swollen nuclei can be observed under the microscope floating singly and in aggregates. In order to stabilize the nuclei against prolonged action of the detergent and mechanical rupture during pipetting, fixation is effected now by addition of 3.2 ml of solution D. At this stage, the contents of the culture plate must not be pipetted. Addition of solution D is performed by a single stroke from a dispenser with a wide-bore opening, ensuring gentle and thorough mixing with the contents of the culture plate. All solutions are applied at room temperature. Volumes given are for tissue cultures on 35-mm plates.

Counting of Cell Nuclei

After 2-20 h the contents (4 ml) of the culture plates are transferred to counting vessels. In order to adjust to an ionic strength appropriate for counting with the Coulter Counter, 4 ml of solution E is added. Aggregates of nuclei are dispersed into single nuclei at this stage by gentle pipetting. For further dilution, solution F is used. With appropriate settings of current, amplification, and threshold, the nuclei in such a suspension can be counted correctly with a Coulter Counter, and good agreement can be obtained with nuclei counts in a microscopic counting chamber [Barnes et al., 1982]. The stability of such nuclear suspensions is quite satisfactory; counts taken 3 h after the first measurement do not deviate significantly from the original values. If seeding is done properly, differences in counts of replicate dishes do not exceed 10%. If cells do not come off the plate and nuclei are difficult to separate, a higher concentration of Nonidet P-40 should be tried — up to 0.5%. It is important that the pH of solutions A–D be exactly 7.5. At higher values nuclei start disintegrating before fixation. At lower pH values nuclei are difficult to separate. For correct interpretation of the counting results, cultures have to be checked microscopically for cells containing more than one nucleus. This is best done in replicate cultures stained with Giemsa stain.

APPLICATIONS

A major problem in cancer therapy is created by the fact that individual tumors, even if they are derived from the same tissue, display different patterns of sensitivity and of development of resistance against the various treatments available. Another problem is the nonexistence of effective therapies for a large number of tumors (among them colon carcinomas) in general. The availability of in vitro systems that allow a detailed quantitative analysis of tumor responses to defined treatments may contribute to the solution of these problems.

With the culture system presented in this article the following experiments have been performed so far. Drug sensitivities of individual tumors have been analyzed and they showed good agreement with drug sensitivities deduced from drug responses of transplants of the same tumors in the athymic nude mouse. Development of resistance during in vitro drug treatment was investigated and shown to correlate well with development of resistance during in vivo drug treatment [Zirvi et al., 1982]. The responses of a tumor cell population to treatments with interferon and 4′-deoxydoxorubicin (an anthracycline derivative) have been shown to depend qualitatively and quantitatively on the growth state of the population, which can be manipulated by seeding high-density (stationary) and low-density (fast-growing) cultures. In combined applications an unexpected reduction of sensitivity to deoxydoxorubicin was caused by interferon in fast-growing cultures (only) and suggests a rationale for combination treatments with these two agents that was not obvious without

TABLE I. Solutions Used in Preparation and Counting of Cell Nuclei

A.	0.9% NaCl, 10 mM HEPES, pH 7.5.
B.	2 mM EDTA, 1 mM HEPES, pH 7.5.
C.	0.1% Nonidet P-40 (BDH Biochemicals or Sigma), 1 mM HEPES, 2 mM EDTA, pH 7.5.
D.	3 mM HEPES, pH 7.5, 10% formaldehyde solution (37% formaldehyde, 10% methanol in H_2O).
E.	1.8% NaCl, 20 mM HEPES, pH 7.4.
F.	0.9% NaCl, 10 mM HEPES, pH 7.4.

a separate analysis of the two growth states mentioned above [van der Bosch and Zirvi, 1982].

CONCLUSIONS

Since only 9 of 12 colon tumors that have been investigated so far grew continuously for up to 24 days (cultures were terminated at this time), and the remaining three tumors doubled only once and developed large numbers of multinucleated cells in vitro [van der Bosch et al., 1981], the search for a medium supporting the growth of all colon carcinomas, and hopefully also the growth of normal colonic mucosa, has to be continued.

The present in vitro system has the following characteristics that make it a valuable tool, not only for drug sensitivity screening of individual tumors, but also for more sophisticated experiments in tumor biology and medicine. Large numbers of cells, representative of the tumor investigated, can be grown to the exclusion of fibroblasts in a defined culture environment, the composition of which can be controlled by exactly timed quantitative medium changes. Growth can be followed quantitatively. Cells are easily collected in sufficient numbers for subcultures, animal passages, or biochemical and biophysical analysis. Different growth states, as they are found coexisting in fast-growing and stationary (dormant) domains of a single tumor in vivo, can be generated homogeneously in vitro, allowing separate analyses of their individual responses and the reconstruction of the response of a composite tumor. The emergence of subpopulations of cells that are resistant to a treatment to which the overwhelming part of the original tumor was sensitive can be analyzed. In the future, experiments can be done, with expanded resistant subpopulations, that are aimed at the development of strategies for the prevention of the emergence of resistant tumors from previously sensitive ones.

REFERENCES

Barnes D, Sato G (1980): Methods for growth of cultured cells in serum-free medium. Anal Biochem 102:255-270.

Barnes D, van der Bosch J, Masui H, Miyazaki K, Sato G (1982): The culture of human tumor cells in serum-free medium. Methods Enzymol 79:368-391.

Bottenstein J, Hayashi I, Hutchings S, Masui H, Mather J, McClure DB, Ohasa S, Rizzino A, Sato G, Serrero G, Wolfe R, Wu R (1979): The growth of cells in serum-free, hormone-supplemented media. Methods Enzymol 45:94-109.

Brattain MG, Brattain DF, Sarrif AM, McRae CJ, Fine WD, Hawkins JG (1982): Enhancement of growth of human colon tumor cell lines by feeder layers of murine fibroblasts. J Natl Cancer Inst 69:767-771.

Buick RN, Fry SE, Salmon SE (1980): Application of in vitro soft agar techniques for growth of tumor cells to the study of colon cancer. Cancer 45:1238-1242.

Carney DN, Bunn PA, Gazdar AF, Pagan JA, Minna JD (1981): Selective growth in serum-free hormone-supplemented medium of tumor cells obtained by biopsy from patients with small cell carcinoma of the lung. Proc Natl Acad Sci USA 78:3185-3189.

Fogh J, Trempe G (1975): New human tumor cell lines. In Fogh J (ed): "Human Tumor Cells In Vitro." New York: Plenum, pp 115-160.

Ham R (1981): Survival and growth requirements of nontransformed cells. Handb Exp Pharmacol 57:13-88.

Hamburger AW, Salmon SE (1977): Primary bioassay of human tumor stem cells. Science 197:461-463.

Laboisse CL, Augeron C, Potet F (1981): Growth and differentiation of human gastrointestinal adenocarcinoma stem cells in soft agarose. Cancer Res 41:310-315.

Leibovitz A, Stinson JC, McCombs WB, McCoy CE, Mazur KC, Mabry ND (1976): Classification of human colorectal adenocarcinoma cell lines. Cancer Res 36:4562-4569.

Murakami H, Masui H (1980): Hormonal control of human colon carcinoma cell growth in serum-free medium. Proc Natl Acad Sci USA 77:3464-3468.

Rupniak HT, Hill BT (1980): The poor cloning ability in agar of human tumor cells from biopsies of primary tumors. Cell Biol Int Rep 4:479-486.

Salmon SE, Hamburger AW, Soehnlen B, Durie BGM, Alberts DS, Moon TE (1978): Quantitation of differential sensitivity of human tumor stem cells to anticancer drugs. N Engl J Med 298:1321-1327.

van der Bosch J (1983): Tissue culture of human colon carcinomas. Therapy experiments in vitro. In Fischer G, Wieser R (eds): "Hormonally Defined Media. Proceedings of the First European Conference on Serum-Free Cell Culture." Heidelberg: Springer-Verlag (in press).

van der Bosch J, Zirvi K (1982): Growth state specific responsiveness of primary cultures of a nude-mouse-xenografted human colon carcinoma to 4′-deoxydoxorubicin and a crude human α-interferon preparation. Cancer Res 42:3789-3792.

van der Bosch J, Masui H, Sato G (1981): Growth characteristics of primary tissue cultures from heterotransplanted human colorectal carcinomas in serum-free medium. Cancer Res 41:611-618.

Vindelov L (1977): Flow microfluorimetric analysis of nuclear DNA in cells from solid tumors and cell suspensions. Virchows Arch B Cell Pathol 24:227-242.

Von Hoff DD, Casper J, Bradley E, Sandbach J, Jones D, Makuch R (1981): Association between human tumor colony-forming assay results and response of an individual patient's tumor to chemotherapy. Am J Med 70:1027-1032.

Zirvi K, van der Bosch J, Kaplan NO (1982): In vitro responses of nude-mouse-xenografted human colon carcinomas exposed to doxorubicin derivatives in tissue culture and in the animal. Cancer Res 42:3793-3797.

Methods for Serum-Free Culture of Epithelial and Fibroblastic Cells,
pages 83-94

7
Growth and Differentiation of Human Bronchogenic Epidermoid Carcinoma Cells in Serum-Free Media

Kaoru Miyazaki, Hideo Masui, and Gordon H. Sato

In recent decades lung cancers have been rapidly increasing in both incidence and mortality in many countries. Bronchogenic epidermoid carcinoma (or bronchogenic squamous cell carcinoma) represents approximately 70% of lung cancers. This cancer has almost the same morphological and biological properties as other squamous cell carcinomas, which originate in skin, respiratory tract, cervix, esophagus, and many other sites. These squamous carcinoma cells have a variable capacity for differentiation to squamous epithelium. As the result of the terminal differentiation in vivo, they form clusters of keratinized cells surrounded by growing cancer cells, which are often called squamous pearls or keratin pearls. The establishment of culture methods for bronchogenic epidermoid carcinoma cells therefore seems useful for studies on chemotherapy and differentiation of this type and possibly other types of squamous cell carcinomas. For these reasons, we have examined various conditions for primary cultures of human bronchogenic epidermoid carcinoma cells from a transplantable tumor line grown in nude mice, and have established serum-free culture methods that support their growth and differentiation [Miyazaki et al., 1982; Barnes et al., 1981]. These culture methods and some characteristics of the epidermoid carcinoma cells are presented in this paper.

Department of Biology, University of California at San Diego, La Jolla, California 92093

K. Miyazaki was a visiting professor from the Institute for Protein Research, Osaka University, Suita-city, Osaka 565, Japan.

G.H. Sato is now at the W. Alton Jones Cell Science Center, Lake Placid, New York 12946.

MATERIALS AND METHODS

Culture Media

Dulbecco's modified Eagle's medium (DME) and Ham's F12 medium (F12) were obtained in powder form from Gibco and dissolved in triple-distilled water. A 1-to-1 mixture of DME and F12 supplemented with 15 mM HEPES, 1.2 mg/ml $NaHCO_3$, 100 IU/ml penicillin G, and 0.2 mg/ml streptomycin sulfate, which is tentatively called DME:F12, was used as a basal nutrient mixture throughout this study. DME:F12 was stored at 4°C up to 1 month.

For serum-containing medium, DME:F12 was supplemented with 10% newborn calf serum unless otherwise noted. Various serum-free media were prepared by supplementing DME:F12 with a single substance or a mixture of hormones, growth factors, vitamins, and trace elements [Barnes and Sato, 1980]. The stock solutions of main supplements are shown in Table I. Most of these solutions can be either mixed with DME:F12 just before use or directly added into culture vessels containing cells and the basal medium (DME:F12). However, the direct addition of substances dissolved in ethanol, such as retinoids and hydrocortisone, into culture vessels with cells should be avoided, because ethanol is very toxic for cells. For the same reason, the final concentration of ethanol should be kept below 0.05% (vol/vol) in all media to be used. The serum-free medium that was most effective for the growth of bronchogenic epidermoid carcinoma cells in primary cultures was composed of DME:F12, 2.5×10^{-8} M selenium (sodium selenite: Na_2SeO_3), 5 μg/ml insulin, 0.2 μg/ml glucagon, and either 10^{-9} M retinoic acid or 5×10^{-10} M triiodothyronine (T_3).

Cell Preparation

Obviously, basic studies to establish culture methods require a constant supply of the same type of cells. This is sometimes difficult for primary cultures

TABLE I. Concentrated Stock Solutions of Medium Supplements

Substance (species)	Source	Concentration	Solvent	Temp (°C)
Insulin (bovine)	Sigma	1 mg/ml	0.01 N HCl	4
Glucagon	Sigma	0.1 mg/ml	0.01 N HCl	− 20
T_3	Sigma	5×10^{-7} M	0.02 N NaOH	− 20
Hydrocortisone	Sigma	2×10^{-4} M	Ethanol	4
EGF (mouse)	Collab.	2 μg/ml	H_2O	− 20
Fibronectin (human)	Collab.	1 mg/ml	1 M urea-PBS	4
Transferrin (human)	Sigma	1 mg/ml	PBS	− 20
Retinoic acid	Sigma	2×10^{-6} M	Ethanol	− 20
Na_2SeO_3	Difco	10^{-5} M	H_2O	4

Collab.: Collaborative Research.

of human cells. In this study, a human bronchogenic epidermoid carcinoma (T222) obtained from a 70-year-old male patient was maintained by successive transplantations into BALB/c nude mice. Minced tumor tissue (approximately 0.2 cm^3) was subcutaneously injected into the back of mouse. It took about 1 month for the tumor tissue to grow to a size of approximately 5 cm^3.

Immediately after killing of a tumor-bearing mouse by neck dislocation, the tumor was aseptically excised and placed in a petri dish. The tumor tissue (0.5–3 cm^3) was freed of connective tissue and necrotic portions with forceps and scissors, washed with DME:F12, and minced into small pieces around 1 mm in diameter with a pair of fine curved scissors in a culture dish. To wash out a great deal of cell debris released by mincing, the mince was suspended in DME:F12 by pipetting and allowed to stand for several minutes in a test tube. The resulting sediment was collected, washed once more as above, and subjected to treatment with collagenase. Collagenase was purchased from Sigma, and a 1% (wt/vol) solution in PBS (Ca^{2+}-free and Mg^{2+}-free phosphate-buffered saline) was stored in a frozen state after filter sterilization. The washed tissue mince was suspended in DME:F12 supplemented with 1 mg/ml collagenase and transferred into one to three 100-mm culture dishes (10 ml suspension per dish), followed by incubation in a CO_2 incubator at 37°C. After incubation for about 90 min the digested tissue was dispersed by pipetting and transferred into a test tube. Big cell clumps that quickly settled to the bottom were separated from the supernatant and further incubated with a fresh collagenase solution for 1–2 h. The supernatant from which the big cell clumps had been separated was allowed to stand for 10 min to collect as a loose pellet small cell clumps consisting of less than several hundred cells. The resulting second supernatant, which contained mostly single cells, was discarded because these cells exhibit poor growth and contain a high proportion of fibroblastic cells. The loose pellet of small cell clumps was resuspended in serum-containing medium. Small cell clumps were also obtained from the second incubation of the large clumps by the same procedure as above and combined with the first preparation. Before seeding into culture vessels, the combined cell clumps were spun down with a clinical centrifuge and washed again with serum-containing medium.

Primary Cultures and Subcultures

Epidermoid carcinoma cells obtained as above were finally resuspended in serum-containing medium at a concentration of 0.05–0.1 cm^3 packed cell clumps per 200 ml, transferred into culture vessels (4 ml per 60-mm dish or 5 ml per 25-cm^2 flask), and incubated in a humidified atmosphere of 5% CO_2/95% air at 37°C for about 40 h. During the incubation period, cell clumps attached to the plastic substratum, and most of the cells spread well. For the cultivation of the cells in serum-free medium, the serum-containing medium used for the initial incubation was sucked up, the cultures were washed twice with DME:F12, and experimental serum-free media were added. The media were refreshed twice a week in the early stages of cultivation and every 2 days for confluent cultures.

In some cases, cells were directly plated with serum-free medium after being washed once with DME:F12. In the absence of serum, they hardly attached to untreated plastic culture vessels, whereas they did well on vessels precoated with plasma fibronectin (cold-insoluble globulin) or serum [Barnes et al., 1981]. The coated culture vessels were prepared as follows: a solution of 1 mg/ml fibronectin was rapidly mixed with DME:F12 in culture vessels to make a final concentration of 1 $\mu g/cm^2$ of growth area, and then incubated at 37°C for around 1 h. The culture vessels were then washed twice with 2 ml of DME:F12. For serum coating, culture vessels were incubated with DME:F12 containing 10% newborn calf serum at 37°C overnight and then washed as above.

For subcultivation, cultured cells were washed with PBS and then incubated at 37°C with 1 ml of PBS containing 1 mg/ml trypsin and 0.5 mM EDTA (pH 7.2). Most of the cells were detached from the plastic surface by gentle pipetting, and the trypsinization was then terminated by addition of an equal volume of serum-containing medium. The harvested cells were incubated overnight in new culture vessels with serum-containing medium, and then the medium was replaced with serum-free medium or fresh serum-containing medium. Alternatively, the cells could be plated into serum-coated or fibronectin-coated culture vessels with serum-free medium. Since the growth ability of cells was gradually lowered by repeated passages, the harvested cells were plated at a higher density in the following subcultures than in the primary culture.

Quantitative Assay of Growth and Keratinization of Cells

For the determination of cell number, cultured cells were dispersed by trypsinization and pipetting and were counted with either a Coulter counter or a hemocytometer.

Epidermoid carcinoma cells, when cultivated in the proper serum-free medium, underwent a terminal differentiation, forming hard aggregates of keratinized cells. In addition to the marked morphological change, this keratinization was evidenced by the following facts (K. Miyazaki et al., in preparation). First, keratinized cells were stained red with rhodamine B by the method of Liisberg [1968]. For this staining, cells in culture vessels were previously fixed for at least 20 min with 10% (vol/vol) formalin in PBS. The fixed cells were stained with 0.1% (wt/vol) toluidine blue for 10 min, washed briefly with distilled water, and then further stained with 0.1% (wt/vol) rhodamine B in 0.1 M sodium acetate buffer (pH 3.6) for 10 min. The stained cells were washed with distilled water and dried at room temperature. Second, keratinized cells were shown to contain keratins at higher concentrations than nonkeratinized cells by SDS-polyacrylamide gel electrophoresis and immunological methods using rabbit antiserum to human skin keratins. Third, as previously shown by Sun and Green [1976], keratinized cells, but not nonkeratinized cells, were resistant to 1% (wt/vol) SDS though they were mostly dissolved in a mixture of 1% SDS and 1% (vol/vol) 2-mercaptoethanol. This suggested that they contained a large amount of keratins polymerized by disulfide bonds.

The last property of keratinized cells was used for the determination of number of keratinized cells. All cultured cells were detached from the plastic surface by incubation with PBS (pH 7.2) supplemented with 0.5 mM EDTA and 0.5 mM EGTA at 37°C for 10–30 min. The harvested cells were collected by centrifugation, suspended with 10 mM Tris-HCl (pH 7.5) containing 1% SDS, and sonicated at a power of 50 W for 30 sec. After this treatment, only keratinized cells remained unbroken, though their aggregates were dispersed into single cells. These intact keratinized cells were counted with a hemocytometer.

RESULTS AND DISCUSSION

Cell Growth in Primary Cultures

The growth of human bronchogenic epidermoid carcinoma cells in primary cultures has been tested under various conditions [Miyazaki et al., 1982; Barnes et al., 1981]. Unlike the majority of other kinds of cells, the epidermoid carcinoma cells can grow considerably even in serum-free DME:F12 without any supplement if they are plated at a relatively high cell density (Fig. 1). Many substances affect cell growth. Selenium (Na_2SeO_3, 2.5×10^{-8}), insulin (5 μg/ml), glucagon (0.2 μg/ml), retinoic acid (10^{-9} M), and triiodothyronine (T_3, 5×10^{-10} M) stimulate cell growth to varying degrees when added individually to DME:F12 (Table II). Of the five factors selenium is the most important one. Most markedly, when cells are plated at a low cell density in the absence of selenium, they hardly grow, regardless of the presence or absence of other factors. The growth-stimulating effects of retinoic acid and T_3 are not additive: in the presence of one, there is little or no further effect of the other. The best cell growth was obtained by supplementing DME:F12 with a mixture of selenium, insulin, glucagon, and retinoic acid. Serum-free

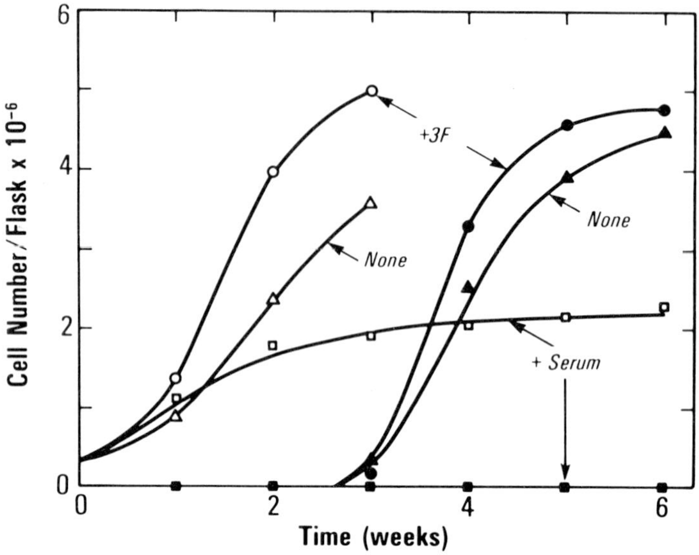

Fig. 1. Time course of growth and keratinization of epidermoid carcinoma cells. Epidermoid carcinoma cells were cultivated in DME:F12 without any supplement (None), in DME:F12 supplemented with 2.5×10^{-8} M Na_2SeO_3, 5 μg/ml insulin, and 0.2 μg/ml glucagon (+3F), and in DME:F12 supplemented with 10% newborn calf serum (+Serum). Open symbols (○, △, □) indicate total cell numbers, and closed symbols (●, ▲, ■) indicate numbers of keratinized cells. Each point represents the average number of cells for triplicate flasks. From Miyazaki et al. [1982], with permission.

medium supplemented with T_3 instead of retinoic acid shows virtually the same effect on cell growth as the above medium.

However, several substances are inhibitory for the growth of epidermoid carcinoma cells. Although epidermal growth factor (EGF) stimulates the expansion of cell colonies as shown in Figure 2, it suppresses the increase in cell number by about 50% at a concentration of 10 ng/ml. Retinoic acid stimulates cell growth at 10^{-10} to 10^{-8} M, but an increase of the concentration to 10^{-6} M completely inhibits cell growth. Hydrocortisone and transferrin are inhibitory at concentrations higher than 10^{-7} M and 5 μg/ml, respectively. In addition, various kinds of sera have been found to be inhibitory at concentrations higher than 1% for cell growth. Sera tested, in order of inhibitory activity, are human serum, horse serum, calf serum, fetal calf serum, and newborn calf serum. Media supplemented with these sera, but not serum-free media, frequently cause the overgrowth of fibroblasts as contaminants in primary cultures of epidermoid carcinoma cells (Fig. 3A). This is the most common problem in

TABLE II. Effect of Medium Supplementation on Cell Growth

Supplemented (concentration)	No. of cells ± SD ($\times 10^{-6}$)	%
None	0.54 ± 0.10	27
Sel (2.5×10^{-8} M)	0.86 ± 0.15	44
Sel, Ins (5 μg/ml)	1.12 ± 0.14	57
Sel, Glu (0.2 μg/ml)	0.94 ± 0.13	48
Sel, RA (10^{-9} M)	1.20 ± 0.18	61
Sel, T_3 (5×10^{-10} M)	1.21 ± 0.16	61
Sel, Ins, Glu	1.39 ± 0.06	71
Ins, Glu, RA	0.44 ± 0.15	22
Sel, Ins, Glu, RA	1.97 ± 0.20	100
Sel, Ins, Glu, T_3	1.93 ± 0.10	98

Sel: Na_2SeO_3; Ins: insulin; Glu: glucagon; RA: retinoic acid. The values for number of cells were obtained from triplicate 60-mm dishes after cultivation for 2 weeks in the specified media.

primary cultures with serum-containing media. Therefore, if serum-containing media are used, the primary cell population should be prepared carefully in order to minimize the contamination of fibroblasts.

Keratinization

In vivo, some epidermoid carcinoma cells undergo a terminal differentiation, forming keratin pearls. We have found that primary cultures of epidermoid carcinoma cells have a strong tendency to differentiate to keratinized cells when serum is deleted from culture medium [Miyazaki et al., 1982].

When cells are kept in serum-containing medium, individual cells in monolayer colonies become extremely flattened, attaching firmly to the plastic substratum (Fig. 3A). This morphology hardly changes upon prolonged cultivation. In contrast, these cells show small polygonal shapes with very distinct intercellular spaces in serum-free medium with or without supplements (Fig. 3B). When cells are incubated in DME:F12 with selenium, insulin, and glucagon for over 2 weeks, they gradually stratify and keratinize. The stratified cells cover whole cell colonies by 4 weeks of cultivation, and after 5-6 weeks the big clusters of keratinized cells begin to slough into the culture medium, leaving only nonkeratinized cells in the bottom layer (Fig. 3C). Thus the majority of cultured cells die upon terminal differentiation. A small number of nonkeratinized, live cells are able to reinitiate cell growth if most keratinized cells in the upper layers are removed by gentle pipetting. The extent of keratinization can be quantitatively assayed by the method described in Materials and Methods.

By this assay we have established that the keratinization of epidermoid carcinoma cells occurs rapidly between 3 weeks and 5 weeks after plating (Fig. 1).

The keratinization described above can be observed even in DME:F12 without any supplement, indicating that selenium, insulin, and glucagon do not affect keratinization, although they stimulate cell growth (Fig. 1). On the other hand, retinoic acid and T_3 have been found to be strong inhibitors of keratinization of epidermoid carcinoma cells. These factors almost completely inhibit stratification and keratinization of the cells at a concentration of 10^{-9} M, approximately the concentration required for both factors to maximally stimulate cell growth (Figs. 3D and 4). Therefore, it is likely that the growth-

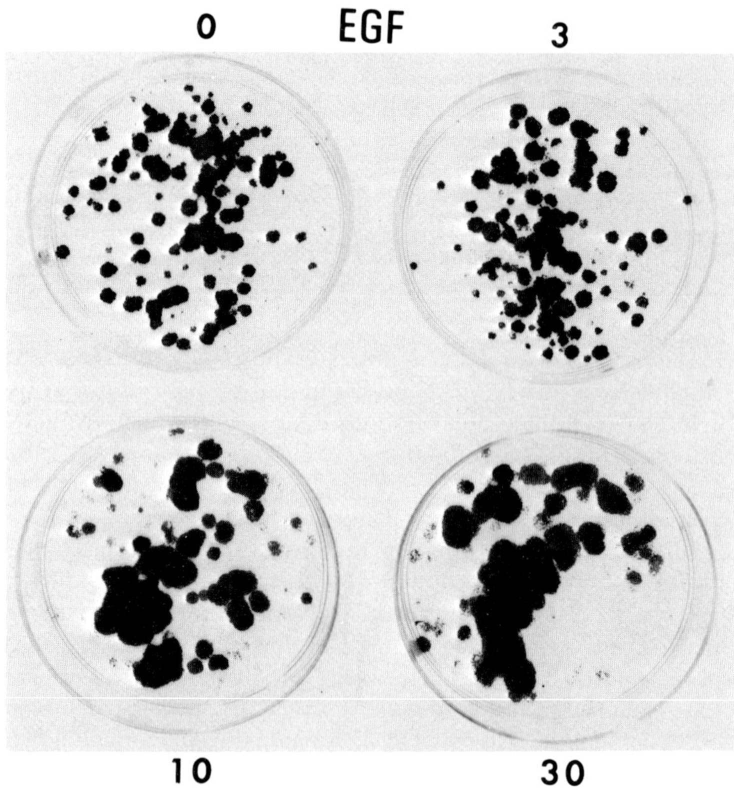

Fig. 2. Effect of EGF on size of cell colonies. Epidermoid carcinoma cells were cultivated for 2 weeks with EGF at the indicated concentrations (ng/ml) and then stained with 0.1% (wt/vol) crystal violet. The basal medium used is DME:F12 supplemented with Na_2SeO_3, insulin, and glucagon (+3F in Fig. 1).

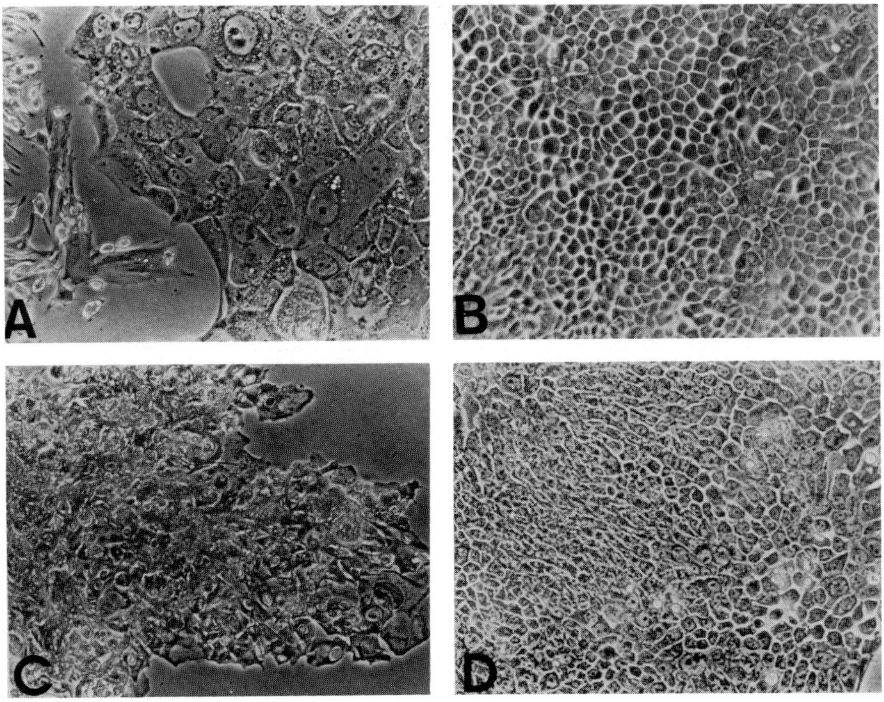

Fig. 3. Photomicrographs (×100) of epidermoid carcinoma cells. A. A 1-week-old culture in serum-containing medium. Fibroblasts are seen at the left of a colony of epidermoid carcinoma cells. B. A 1-week-old culture in DME:F12 supplemented with Na_2SeO_3, insulin, and glucagon. C. A cluster of completely keratinized cells that sloughed into medium after 6 weeks in DME:F12 with Na_2SeO_3, insulin, and glucagon. D. A 5-week-old culture in DME:F12 supplemented with 10^{-9} M retinoic acid in addition to Na_2SeO_3, insulin, and glucagon.

stimulating activities of retinoic acid and T_3 are derived from their antikeratinization activities. Other retinoids such as retinol (vitamin A), retinal, retinyl acetate, and retinyl palmitate, as well as thyroxine (T_4), another thyroid hormone, require concentrations about 10 times higher than retinoic acid and T_3 in order to be as effective in antikeratinization and growth-stimulating activities. Nevertheless, retinol and T_4 may be actual regulators of the keratinization of epidermoid carcinoma cells in vivo, because they are the major forms of retinoids and thyroid hormones in blood, and their effective concentrations (5×10^{-9} to 5×10^{-8}) are still physiologically significant.

As mentioned above, various kinds of sera also inhibit the keratinization of epidermoid carcinoma cells. Newborn calf serum at a final concentration of 0.3% (vol/vol) inhibits keratinization to almost the same extent as 10^{-9} M retinoic acid or 10^{-9} M T_3. The antikeratinization activities of sera can be

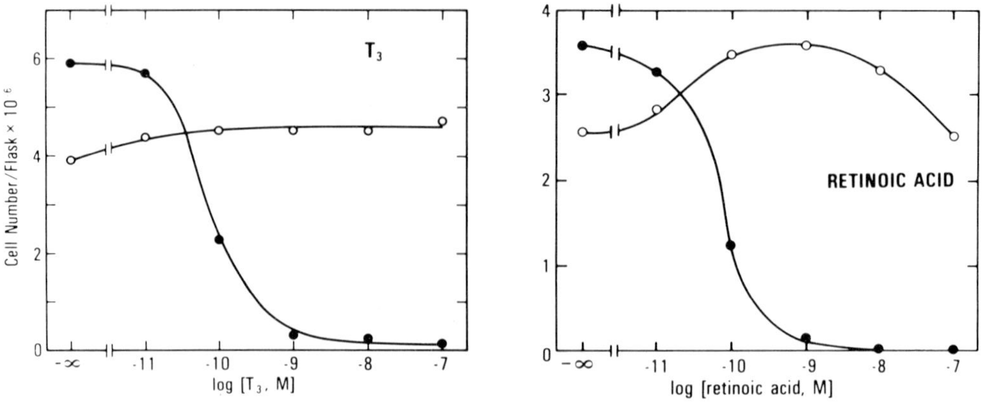

Fig. 4. Dose-response curves of growth and keratinization of epidermoid carcinoma cells to T_3 and retinoic acid. The basal medium is the same as in Figure 2. ○, Total number of cells after 2 weeks; ●, number of keratinized cells after 5 weeks. From Miyazaki et al. [1982], with permission.

removed by extracting sera with cold acetone [Miyazaki et al., 1982]. Furthermore, the addition of a small amount of retinoic acid to the acetone-extracted sera restores their antikeratinization activities. This suggests that the major antikeratinization activities present in sera are due to retinoids.

Serial Passage

In serum-free media without any substances exhibiting antikeratinization activity, epidermoid carcinoma cells gradually lose viability as the keratinization progresses. The carcinoma cells cultivated in these media cannot grow well in subcultures even if they have been harvested before apparent keratinization. Therefore, if subcultivation is needed, they must be cultivated in the presence of one of the antikeratinization substances. For this purpose, retinoic acid (5×10^{-10} to 1×10^{-8}), newborn calf serum (0.3% to 10%), and T_3 (5×10^{-10} to 1×10^{-8} M) can be used as the antikeratinization supplement for primary culture media. For instance, epidermoid carcinoma cells can be subcultured at least several times in DME:F12 supplemented with 10^{-9} M retinoic acid and three other factors (selenium, insulin, and glucagon). Therefore, it seems possible to establish cell lines from primary cultures of epidermoid carcinoma cells with serum-free medium, though our attempts have not yet succeeded. T_3 appears less suitable than retinoic acid and newborn calf serum for the serial passage of epidermoid carcinoma cells.

One cell line has been established with the use of DME:F12 containing 7.5% newborn calf serum (H. Masui and K. Miyazaki, in preparation). In this case, primary epidermoid carcinoma cells were kept in the serum-containing medium for about 2 months without subcultivation. The cells, dividing very slowly, were harvested by trypsinization, and they have been successfully subcultured over 20 times, to date. This established cell line also can grow well in serum-free DME:F12 supplemented with selenium, insulin, and retinoic acid. The growth-stimulating effects of glucagon and T_3 are not apparent for this cell line. Unlike the primary cells, the growth rate of the cell line is higher with serum-containing medium than with serum-free medium. This cell line also keratinizes in vitro when cultivated in DME:F12 without retinoids, thyroid hormones, and sera for over three weeks (Fig. 5). However, these antikeratinization substances are less effective for the cell line than for primary epidermoid carcinoma cells. For example, the minimum concentration of retinoic acid that effectively inhibits keratinization is approximately 10^{-7} M, a concentration about 100 times higher than that necessary for primary epidermoid carcinoma cells.

In vitro keratinization has been reported for several types of cells by other workers. In the cases of a cell line of rat bladder squamous cell carcinoma [Toyoshima and Leighton, 1975; Tchao, 1980] and primary cultures of hu-

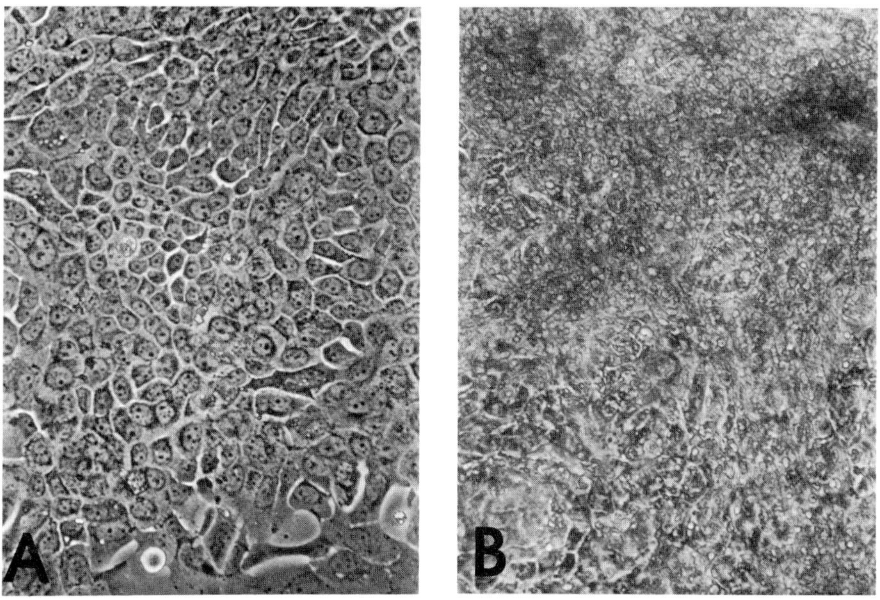

Fig. 5. Photomicrographs (×100) of an established cell line of epidermoid carcinoma cells incubated in DME:F12 with Na_2SeO_3 and insulin. A. A 2-week-old culture. Keratinized cell is not seen. B. A 5-week-old culture. Keratinized cells in upper layers appear much smaller than those of primary epidermoid carcinoma cells (see Fig. 3C).

man skin keratinocytes [Sun et al., 1976; Fuchs et al., 1981], cells can keratinize even in media containing serum at concentrations higher than 10%, and vitamin A at a concentration higher than 10^{-7} M is required to prevent their keratinization. These facts, together with our results, indicate that the effects of retinoids and other factors vary considerably according to the kinds of cells examined. Therefore, some modifications of the serum-free media formulated in this study may be necessary to cultivate, with or without differentiation, squamous carcinoma cells from other sources.

REFERENCES

Barnes D, Sato G (1980): Serum-free cell culture: A unifying approach. Cell 22:649-655.

Barnes D, van der Bosch J, Masui H, Miyazaki K, Sato G (1981): The culture of human tumor cells. Methods Enzymol 79:368-391.

Fuchs E, Green H (1981): Regulation of terminal differentiation of cultured human keratinocytes by vitamin A. Cell 25:617-625.

Liisberg MF (1968): Rhodamine B as an extremely specific stain for cornification. Acta Anat 69:52-57.

Miyazaki K, Masui H, Sato G (1982): Control factors for keratinization of human bronchogenic epidermoid carcinoma cells. In Sato GH, Pardee AB, Sirbasku DA (eds): "Cold Spring Harbor Conferences on Cell Proliferation," Vol. 9: "Growth of Cells in Hormonally Defined Media," Book B. Cold Spring Harbor, NY: Cold Spring Harbor Laboratory, pp 657-661.

Sun T-T, Green H (1976): Differentiation of the epidermal keratinocyte in cell culture: Formation of the cornified envelope. Cell 9:511-521.

Tchao R (1980): Keratinization and the effect of vitamin A in aggregates of a squamous carcinoma cell line NBT II. In Vitro 16:407-414.

Toyoshima K, Leighton J (1975): Vitamin A inhibition of keratinization in rat urinary bladder cancer cell line Nara Bladder Tumor No. 2 in meniscus gradient culture. Cancer Res 35:1873-1879.

Methods for Serum-Free Culture of Epithelial and Fibroblastic Cells,
pages 95-106

8
Serum-Free Cell Culture of A431 Human Epidermoid Carcinoma

David W. Barnes

The A431 cell line was established in continuous culture from a human epidermoid carcinoma [Fabricant et al., 1977]. Cells of this line, compared to most cell types in culture, express an unusually large number of plasma membrane receptors for epidermal growth factor (EGF) [Fabricant et al., 1977], a peptide hormone that is a powerful mitogen for a number of cell types [Carpenter and Cohen, 1979]. A431 cells have proved to be a useful tool for the study of the early biochemical effects of EGF, including internalization of the hormone and its receptor [Cohen et al., 1979; Gross et al., 1983; Haigler et al., 1978, 1979a,b; McKanna et al., 1979], enhancement of phosphorylation by EGF treatment [Chinkers et al., 1981; Carpenter et al., 1978, 1979; Cohen et al., 1980; Cooper and Hunter, 1981; Hunter and Cooper, 1981; King et al., 1980; Ushiro and Cohen, 1980; Gill and Lazar, 1981], effects of EGF on cell morphology and pinocytosis [Chinkers et al., 1979, 1981], and identification and purification of the EGF receptor [Carpenter, 1979; Haigler and Carpenter, 1980; Kamamoto et al., 1983; Wrann and Fox, 1979]. The A431 cell line also has been used as a model for the study of the biology and biochemistry of cell surface receptors for transferrin [Hopkins and Trowbridge, 1983] and low-density lipoprotein [Anderson et al., 1981] and in studies of synthesis and effects of extracellular matrix components [Salomon et al., 1984; Alitalo et al., 1981; Gospodarowicz et al., 1982].

A431 can be grown in continuous, serum-free culture in a rich basal nutrient medium to which is added a hormone, insulin; a binding protein, transferrin; an attachment factor, fibronectin; and a supplementary nutrient,

Department of Biological Sciences, University of Pittsburgh, Pittsburgh, Pennsylvania 15260 ethanolamine [Barnes, 1981, 1982a,b, 1983]. EGF in this medium [Barnes, 1981, 1982a], as well as in serum-supplemented medium [Gill and Lazar, 1981; Buss et al., 1982], inhibits A431 cell growth at concentrations that are mitogenic for other cell types in culture. EGF stimulates growth of A431 at concentrations 10 to 1,000 times lower than those concentrations at which it inhibits their growth [Kamamoto et al., 1983]. The A431 line cultured in the absence of serum has been used for studies of several aspects of the interactions of the cell with its environment [Barnes, 1982b; Kamamoto et al., 1983; Gross et al., 1983; Salomon et al., 1984; Gospodarowicz et al., 1982], and this approach should prove useful in other studies of effects and mechanisms of action of nutrients, hormones, binding proteins, and attachment factors and extracellular matrix components [Barnes, 1984a,b,c; Barnes and Sato, 1980a,b; Barnes et al., 1981; Ham, 1984].

MATERIALS

Bovine insulin, human transferrin, 4-(2-hydroxy-ethyl)-1-piperazine-ethanesulfonic acid (HEPES), soybean trypsin inhibitor, antibiotics, and ethanolamine can be obtained from Sigma Chemical Company, St. Louis. Powdered formulations of Ham's F12 and Dulbecco-modified Eagle's medium (DME) can be obtained from Grand Island Biological Company, Grand Island, NY. Fetal calf serum (FCS) was obtained from Reheis, Irvine, California. Many other serum sources are acceptable. Epidermal growth factor and human plasma fibronectin (cold-insoluble globulin: CIg) can be obtained from Meloy Laboratories, Springfield, VA; Bethesda Research Laboratories, Gaithersburg, MD; or Collaborative Research, Inc., Waltham, MA. Water for culture medium is freshly prepared, either by triple distillation in glass containers [Barnes and Sato, 1980a] or by passage through a Milli-Q water purification system (Millipore).

PREPARATION OF CULTURE MEDIUM AND SUPPLEMENTS

The basal nutrient formulation to which are added supplements that stimulate A431 cell growth is a one-to-one mixture of DME and Ham's F12 [Mather and Sato, 1979] supplemented with 1.2 gm/liter sodium bicarbonate, 15 mM HEPES, pH 7.4, 200 IU/ml penicillin, 200 μg/ml streptomycin, and 25 μg/ml ampicillin (F12:DME). The HEPES and the antibiotics are maintained as 100-fold-concentrated stock solutions in highly purified water. For the preparation of 2 liters of F12:DME, powdered medium in 1-liter preweighed packets (one F12 packet and one DME packet) is added to approximately 500 ml highly purified water and stirred until dissolved. To this is added 20 ml of concentrated HEPES stock, followed by concentrated stock solutions of the antibiotics. The solution is then brought almost to 2,000 ml with additional highly purified water, sodium bicarbonate is added, and the final adjustment to 2,000 ml is made.

The medium is immediately membrane-filtered for sterilization by passage through plastic disposable 0.2-μ filter units and stored frozen in 200-ml aliquots in sterile, disposable tissue culture flasks (75-cm^2 surface area: T75). In general, storage of basal nutrient medium for long periods of time, even in the frozen form, is not recommended. In practice, our laboratory makes medium in 2-liter batches as needed, and the medium is usually used within 2 weeks. While this approach is time-consuming, it minimizes the chances of inadvertent use of F12:DME in which some decomposition of components has occurred during storage. This problem is of little concern when the basal nutrient medium is supplemented with serum, but it may have serious consequences for serum-free cell culture [Ham, 1984].

Growth-stimulating supplements are prepared and maintained as sterile concentrated stock solutions. Insulin is stored at 4°C at 1 mg/ml in 0.05 M HCl. Transferrin is stored frozen in 1-ml aliquots of 1–5 mg/ml in phosphate-buffered saline (PBS). Both of these solutions can be membrane-filtered for sterilization after the commercially available powders are dissolved. EGF and fibronectin can be obtained commercially as sterile lyophylized powders that can be reconstituted with sterile solutions or as frozen sterile solutions. Alternatively, these components can be isolated from primary sources as discussed elsewhere [Yamada and Akiyama, 1984; Rubin and Bradshaw, 1984; Savage and Harper, 1984; Carpenter and Cohen, 1979; Hynes and Yamada, 1982]. Concentrated stocks of these factors can be maintained at 0.5–1.0 mg/ml (fibronectin) and 50–100 μg/ml (EGF). Care must be taken to prevent frequent freezing and thawing of fibronectin solutions in order to avoid aggregation, and it is best to store fibronectin in small aliquots of frozen stocks. General approaches to the use of fibronectin and similar factors in serum-free cell culture are described elsewhere [Barnes, 1984c]. Ethanolamine is prepared as a concentrated (50 mM) stock in highly purified water and is membrane-filtered for sterilization and stored at −20°C.

TRYPSINIZATION AND PLATING OF A431 IN SERUM-FREE MEDIUM

Passage of A431 cells is accomplished first by incubation of culture dishes with 0.1% crude trypsin in PBS with 1.0 mM ethylenediaminetetraacetate (EDTA) at 37°C until the cells are detached. The cell suspension is then

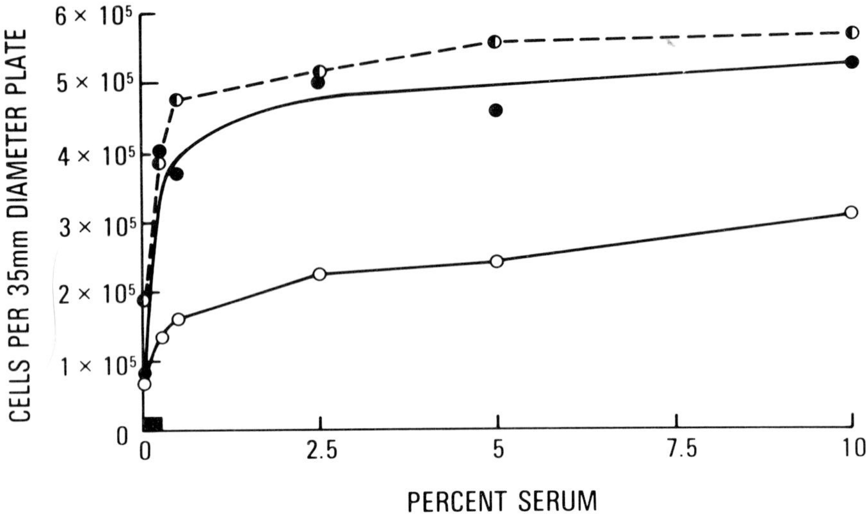

Fig. 1. Growth of A431 cells in F12, DME, and F12:DME supplemented with FCS. Number of cells was determined after incubation for 6 days in plates containing F12 (○——○), DME (●——●), or F12:DME (◐– – –◐) supplemented with the indicated concentrations of FCS. Initial cell inoculum (■) was 2 × 10^5 cells per plate. From Barnes [1983], with permission.

diluted into F12:DME containing 0.1% soybean trypsin inhibitor, and the cells are centrifuged from suspension at low speed and resuspended in fresh F12:DME. For serum-free plating in 35-mm-diameter dishes, cells are diluted in F12:DME to give the final cell density desired and added in 1-ml aliquots to dishes containing 1 ml of F12:DME previously equilibrated for 30 min in a 37° incubator in 5% CO_2-95% air atmosphere. Dishes that are to receive substratum modifications, such as exposure to fibronectin, also should have such factors added to the medium prior to preequilibration and cell seeding. Other growth-stimulating supplements are then added directly to the cells immediately after plating. Supplements are added as small volumes (e.g., 20 μl per plate) of sterile concentrated stocks.

Care should be taken to prevent exposure of the cells to extremes of pH and temperature during the passaging process. Trypsinization, centrifugation, and resuspension should be accomplished as quickly and efficiently as possible so that cells spend as little time exposed to these conditions as can be reasonably achieved. Liberties that can be taken in the passage of these and other cell types when serum-containing media are used may result in greatly reduced cell survival upon passage when the cells are cultured under serum-free conditions. It may be helpful to change the medium the day after the cells are passaged under serum-free conditions, in order to reduce the level in the medium of hydrolase activity released from cells damaged in the passaging

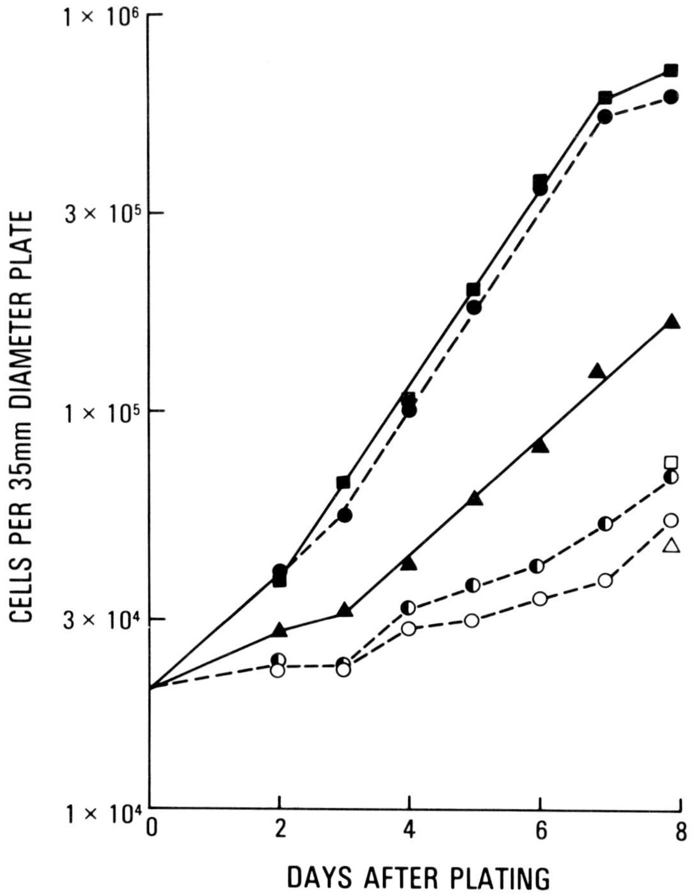

Fig. 2. Growth of A431 cells in serum-containing or serum-free medium in the presence or absence of EGF. F12:DME+FCS (■—■); F12:DME+4F (●– – –●); F12:DME (▲——▲); F12:DME+4F + 10 ng/ml EGF (◐– – –◐); F12:DME+4F + 100 ng/ml EGF (○– – –○); F12:DME+FCS + 100 ng/ml EGF (□); F12:DME + 100 ng/ml EGF (△). From Barnes [1982b], with permission.

procedure. The effects on cell viability of hydrolytic enzyme activities may not be of consequence in serum-containing media because these activities may be inhibited by serum factors, but the presence of these enzymes may represent a problem when cells are passaged under serum-free conditions.

GROWTH CHARACTERISTICS OF A431 CELLS IN VITRO

The A431 cell line is grown in most laboratories in DME supplemented with 10% FCS. Somewhat better growth when fetal calf serum is used as a

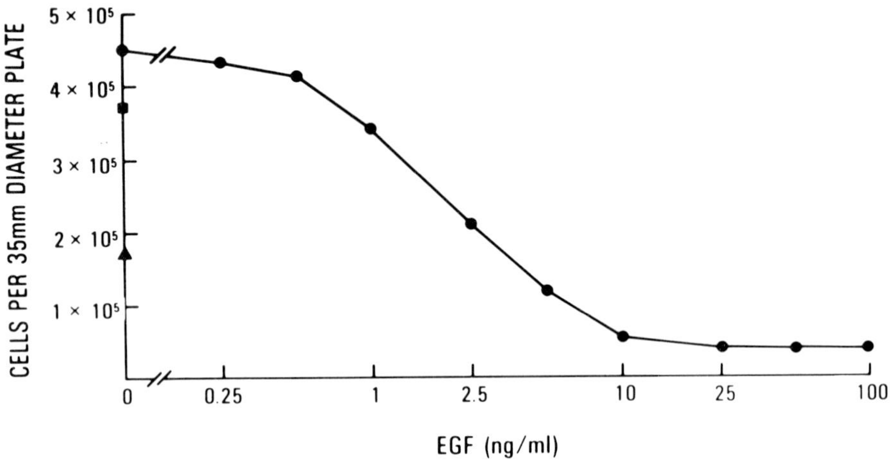

Fig. 3. Concentration dependence of EGF inhibition of A431 cell growth. Cells were counted after incubation for 6 days in plates containing F12:DME:4F with the indicated concentrations of EGF (●——●); F12:DME (▲); F12:DME+FCS (■). From Barnes [1982b], with permission.

supplement is obtained in a basal nutrient mixture consisting of a mixture of Ham's F12 and DME (F12:DME) (Fig. 1), and A431 cells in stock cultures are grown routinely in our laboratory in F12:DME supplemented with 10% FCS (F12:DME + FCS). A431 cells do not grow well in F12 supplemented with FCS. Some cell growth is seen in F12, DME, or F12:DME in the complete absence of FCS, but, of the three media, F12:DME is clearly superior under conditions in which no serum is present.

In a survey of a large number of hormones, binding proteins, attachment factors, and supplementary nutrients previously found to be stimulatory for some cell types in serum-free media tested over a range of concentrations with F12:DME as the basal nutrient medium, supplements found to be the most active in promoting A431 cell growth were insulin at 2-20 μg/ml, transferrin at 2-20 μg/ml, fibronectin at 2-10 μg/ml, and ethanolamine [Kano-Sueoka et al., 1979; Walthall et al., 1980] at 0.1-1.0 mM. A small but consistent effect of glycyl-histidyl-lysine at 2 μg/ml also was observed, and smaller, more variable effects of parathyroid hormone at 5 ng/ml and somatostatin at 100 ng/ml were sometimes observed. Fibronectin could be replaced by another attachment and spreading-promoting protein, serum spreading factor [Silnutzer and Barnes, 1984; Barnes, 1984c].

A431 cells in F12:DME supplemented with four factors—10 μg/ml insulin, 10 μg/ml transferrin, 5 μg/ml fibronectin, and 0.5 mM ethanolamine (F12:DME+4F)—will grow with a doubling time of about 30 h, a growth rate identical to that observed for A431 cells in F12:DME+FCS (Fig. 2). Addition

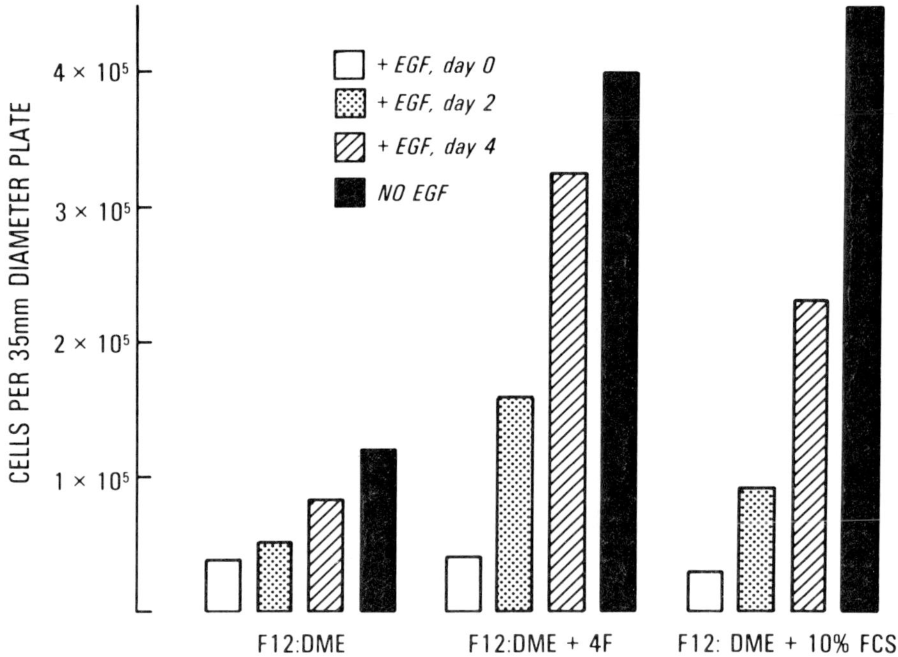

Fig. 4. Inhibition of A431 cell growth by EGF added at various times after plating. Cultures were seeded in the indicated media, either with or without 10 ng/ml EGF. To some cultures not receiving EGF initially, EGF as a small volume of a concentrated stock was added 2 or 4 days after plating to give a final concentration in the culture medium of 10 ng/ml. EGF added at the time of plating (□); EGF added 2 days after plating (▩); EGF added 4 days after plating (▨); no EGF added (■). Number of cells in all plates was determined 6 days after plating. From Barnes [1982a], with permission.

of the four stimulatory factors used in F12:DME+4F to F12:DME+FCS does not improve cell growth beyond that seen in F12:DME+4F or F12:DME+FCS alone. Medium changes every 3 days do not improve the growth rate for the first 6-8 days after seeding of A431 cells in F12:DME+4F, but do increase the cell density at later times. Confluent cultures of A431 cells grown in F12:DME+4F can be passaged repeatedly in the serum-free medium with the use of a trypsin-EDTA solution and a soybean trypsin inhibitor solution. A431 cells are also capable of growth in the absence of serum in unsupplemented F12:DME with a generation time of about 48 h and some time lag before the onset of growth after plating.

INHIBITION OF A431 CELL GROWTH BY EGF

Addition of as little as 1 ng/ml EGF to F12:DME+4F at the time the cells are plated results in inhibition of A431 cell growth (Figs. 2, 3). EGF added to

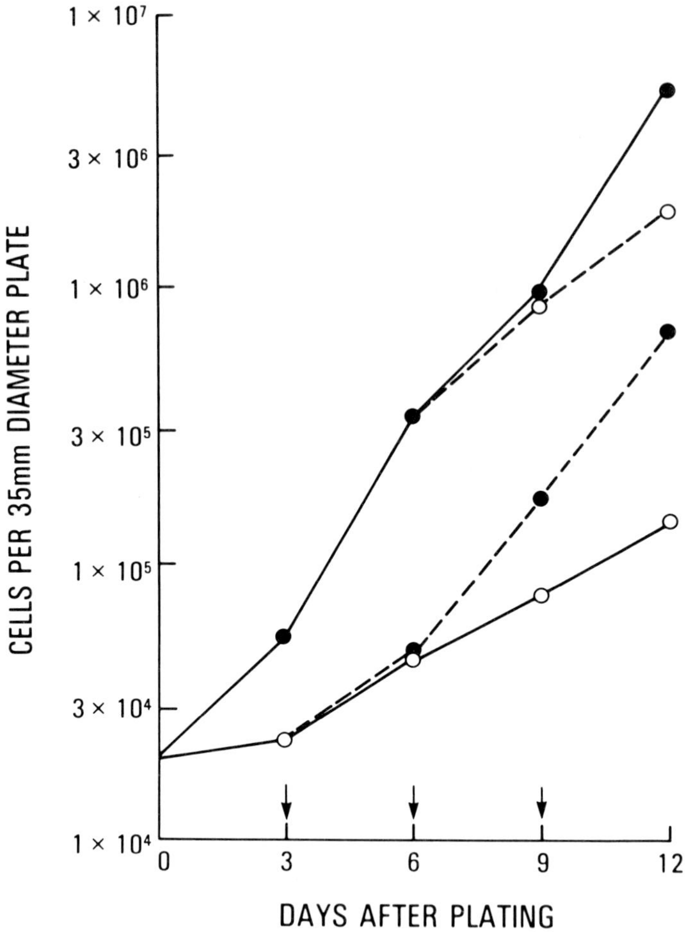

Fig. 5. Reversal of EGF inhibition of cell growth. Arrows indicate days on which medium was changed for all conditions. F12:DME+4F (●——●); F12:DME+4F + 10 ng/ml EGF (○——○); F12:DME+4F changed to F12:DME+4F + 10 ng/ml EGF on day 6 (○———○); F12:DME+4F + 10 ng/ml EGF changed to F12:DME+4F on day 3 (●———●). From Barnes [1982b], with permission.

cells 2 or 4 days after plating also results in a reduced number of cells in those plates compared to controls when the number of cells is determined 6 days after plating (Fig. 4). Inhibition of A431 cell growth by EGF also can be seen with cells in plates containing F12:DME, without the supplements, or in plates containing F12:DME+FCS. The inhibitory effect of EGF on A431 cell growth in F12:DME+4F is reversible; cells incubated for the first 3 days after seeding in F12:DME+4F containing 10 ng/ml EGF and then changed to F12:DME+4F without EGF resume growth after a lag of about 3 days with a generation time near that of cells not exposed to EGF (Fig. 5).

The mechanism by which EGF inhibits the growth of the A431 cell remains to be determined. Inhibitory effects of EGF on the growth of cells in culture that are similar to the effects we have observed on A431 have been reported previously for rat pituitary carcinoma cell lines [Schonbrunn et al., 1980; Johnson et al., 1980]. The availability for both A431 cells and pituitary carcinoma cells [Hayashi, 1984] of a serum-free medium supporting the long-term growth and maintenance of healthy cells in a known extracellular hormonal and nutritional environment allows the design of experiments in which the interesting properties of these cells can be examined in more straightforward ways than previously has been possible in serum-containing media [Barnes, 1984a,b,c; Barnes and Sato, 1980a,b; Barnes et al., 1981].

ACKNOWLEDGMENTS

This work was supported by American Cancer Society grant BC-368 and National Institutes of Health grant CA-35214. The author thanks Dr. Q.A. Wagstaff, Huxley College, for helpful discussions and suggestions.

REFERENCES

Alitalo K, Keski-Oja J, Vaheri A (1981): Extracellular matrix proteins characterize human tumor cell lines. Int J Cancer 27:733–741.

Anderson RGH, Brown MS, Goldstein JL (1981): Inefficient internalization of receptor-bound low density lipoprotein in human carcinoma A431 cells. J Cell Biol 88:441–452.

Barnes D (1981): A serum-free medium for A431 human epidermoid carcinoma cells; inhibition of growth in culture by mouse epidermal growth factor. Endocrinology 108(Suppl):329.

Barnes D (1982a): Growth of A431 human epidermoid carcinoma in serum-free cell culture: Inhibition by epidermal growth factor. Cold Spring Harbor Conf Cell Prolif 9:937–941.

Barnes D (1982b): Epidermal growth factor inhibits growth of A431 human epidermoid carcinoma in serum-free cell culture. J Cell Biol 93:1–4.

Barnes D (1983): Growth characteristics of A431 human epidermoid carcinoma cells in serum-free medium: Inhibition by epidermal growth factor. In Acton RA (ed): "Mammalian Cell Cultures: Basic and Applied." New York: Plenum, pp 49–66.

Barnes DW (1984a): Hormonally-defined serum-free media for epithelial cells in culture. In Taub M (ed): "Tissure Culture in the Study of Epithelial Transport." New York: Plenum (in press).

Barnes D (1984b): Nutritional and hormonal requirements of mammalian cells in culture. World Rev Nutr Diet (in press).

Barnes D (1984c): Attachment factors in cell culture. In Mather J (ed): "The Use of Serum-Free and Hormone-Supplemented Media." New York: Plenum, pp 195-237.

Barnes D, Sato G (1980a): Methods for growth of cultured cells in serum-free medium. Anal Biochem 102:255-270.

Barnes DW, Sato GH (1980b): Serum-free cell culture: A unifying approach. Cell 22:649-655.

Barnes DW, van der Bosch J, Masui H, Miyzaki K, Sato G (1981): The culture of human tumor cells. Methods Enzymol 79:368-391.

Buss JE, Kudlow JE, Lazar CS, Gill GN (1982): Altered epidermal growth factor (EGF)-stimulated protein kinase activity in variant A431 cells with altered growth responses to EGF. Proc Natl Acad Sci USA 79:2574-2578.

Carpenter G (1979): Solubilization of membrane receptor for epidermal growth factor. Life Sci 24:1691-1698.

Carpenter G, Cohen S (1979): Epidermal growth factor. Annu Rev Biochem 48:193-216.

Carpenter G, King L, Cohen S (1978): Epidermal growth factor stimulates phosphorylation in membrane preparations in vitro. Nature (Lond) 276:409-410.

Carpenter G, King L Jr, Cohen S (1979): Rapid enhancement of protein phosphorylation in A431 cell membrane preparations by epidermal growth factor. J Biol Chem 254:4884-4891.

Chinkers M, McKanna JA, Cohen S (1979): Rapid induction of morphological changes in human carcinoma cells A431 by epidermal growth factor. J Cell Biol 83:260-265.

Chinkers M, McKanna JA, Cohen S (1981): Rapid rounding of human epidermoid carcinoma cells A431 induced by epidermal growth factor. J Cell Biol 88:422-429.

Cohen S, Haigler HT, Carpenter G, King L, McKanna JA (1979): Epidermal growth factor: Visualization of the binding and internalization of EGF in cultured cells and enhancement of phosphorylation by EGF in membrane preparations in vitro. Cold Spring Harbor Conf Cell Prolif 6:131-142.

Cohen S, Carpenter G, King L (1980): Epidermal growth factor-receptor-protein kinase interaction. J Biol Chem 255:4834-4842.

Cooper JA, Hunter T (1981): Similarities and differences between the effects of epidermal growth factor and Rous sarcoma virus. J Cell Biol 91:878-883.

Fabricant RN, DeLarco JE, Todaro GJ (1977): Nerve growth factor receptors on human melanoma cells in culture. Proc Natl Acad Sci USA 74:565-569.

Gill GN, Lazar CS (1981): Increased phosphotyrosine content and inhibition of proliferation in EGF-treated A431 cells. Nature (Lond) 293:305-307.

Gospodarowicz D, Lui GM, Gonzales K (1982): High-density lipoprotein and the proliferation of human tumor cells maintained on extracellular matrix-coated dishes and exposed to defined medium. Cancer Res 42:3704-3713.

Gross JL, Krupp MN, Rifkin DB, Lane MD (1983): Down-regulation of epidermal growth factor receptor correlated with plasminogen activator activity in human A431 epidermoid carcinoma cells. Proc Natl Acad Sci USA 80:2276-2280.

Haigler HT, Carpenter G (1980): Production and characterization of antibody affecting epidermal growth factor: Receptor interactions. Biochim Biophys Acta 598:314-325.

Haigler H, Ash JF, Singer SJ, Cohen S (1978): Visualization by fluorescence of the binding and internalization of epidermal growth factor in human carcinoma cells A431. Proc Natl Acad Sci USA 75:3317-3321.

Haigler HT, McKanna JA, Cohen S (1979a): Direct visualization of the binding and internalization of a ferritin conjugate of epidermal growth factor in human carcinoma cells A431. J Cell Biol 81:382-395.

Haigler HT, McKanna JA, Cohen S (1979b): Rapid stimulation of pinocytosis in human carcinoma cells A431 by epidermal growth factor. J Cell Biol 83:82-90.

Ham R (1984): Formulation of basal nutrient media. In Barnes DW, Sirbasku DA, Sato GH (eds): "Cell Culture Methods for Molecular and Cell Biology." New York: Alan R. Liss, Vol 1, pp 3-21.

Hayashi I (1984): Growth of GH_3, a rat pituitary cell line, in serum-free, hormone-supplemented medium. In Barnes DW, Sirbasku DA, Sato GH (eds): "Cell Culture Methods for Molecular and Cell Biology." New York: Alan R. Liss, Vol 2, pp 1-13.

Hopkins CR, Trowbridge IS (1983): Internalization and processing of transferrin and the transferrin receptor in human carcinoma A431 cells. J Cell Biol 97:508-521.

Hunter T, Cooper JA (1981): Epidermal growth factor induces rapid tyrosine phosphorylation of proteins in A431 human tumor cells. Cell 24:741-747.

Hynes RO, Yamada KM (1982): Fibronectins: Multifunctional modular glycoproteins. J Cell Biol 95:369-378.

Johnson LK, Baxter JD, Vlodavsky I, Gospodarowicz D (1980): Epidermal growth factor and expression of specific genes: Effects on cultured rat pituitary cells are dissociable from the mitogenic repsonse. Proc Natl Acad Sci USA 77:394-398.

Kamamoto T, Sato DJ, Le A, Polikoff J, Sato G, Mendelson J (1983): Growth stimulation of A431 cells by epidermal growth factor: Identification of high-affinity receptors for epidermal growth factor by an anti-receptor monoclonal antibody. Proc Natl Acad Sci USA 80:1337-1341.

Kano-Sueoka T, Cohen DM, Yamaizumi Z, Nishimura S, Mori M, Fukiki H (1979): Phosphoethanolamine as a growth factor of a mammary carcinoma cell line of rat. Proc Natl Acad Sci USA 76:5741-5744.

King LE, Carpenter G, Cohen S (1980): Characterization by electrophoresis of epidermal growth factor stimulated phosphorylation using A431 membrane. Biochemistry 19:1524-1528.

Mather JP, Sato GH (1979): The use of hormone-supplemented serum-free media in primary cultures. Exp Cell Res 124:215-221.

McKanna JA, Haigler HT, Cohen S (1979): Hormone receptor topology and dynamics: A morphological analysis using ferritin labelled epidermal growth factor. Proc Natl Acad Sci USA 76:5689-5693.

Rubin JS, Bradshaw RA (1984): Preparation of guinea pig prostate epidermal growth factor. In Barnes DW, Sirbasku DA, Sato GH (eds): "Cell Culture Methods for Molecular Cell Biology." New York: Alan R. Liss, Vol 1, pp 139-145.

Salomon DS, Liotta LA, Panneerselvam M, Terranova VP, Sahai A, Fehnel P (1984): Analysis of basement membrane synthesis and turnover in mouse embryonal and human A431 epidermoid carcinoma cells in serum-free medium. In Barnes DW, Sirbasku DA, Sato GH (eds): "Cell Culture Methods for Molecular and Cell Biology." New York: Alan R. Liss, Vol 1, pp 295-319.

Savage CR, Harper RA (1984): Purification of human epidermal growth factor from urine. In Barnes DW, Sirbasku DA, Sato GH (eds): "Cell Culture Methods for Molecular and Cell Biology." New York: Alan R. Liss, Vol 1, pp 147-158.

Schonbrunn A, Krasnoff M, Westenclorf JM, Tashjian AH Jr (1980): Epidermal growth factor and thyrotropin-releasing hormone act similarly on a clonal pituitary cell strain. J Cell Biol 85:786-797.

Silnutzer J, Barnes DW (1984): Human serum spreading factor (SF): Assay, preparation and use in serum-free cell culture. In Barnes DW, Sirbasku DA, Sato GH, (eds): "Cell Culture Methods for Molecular and Cell Biology." New York: Alan R. Liss, Vol 1, pp 245-268.

Ushiro H, Cohen S (1980): Identification of phosphotyrosine as a product of epidermal growth factor-activated protein kinase in A431 cell membranes. J Biol Chem 255:8363-8365.

Walthall BS, Tsao M, Bettger WJ, Peehl DM, Ham RG (1980): Differences in requirements for growth in defined media of normal human fibroblasts and keratinocytes. J Cell Biol 87:163a.

Wrann MM, Fox CF (1979): Identification of epidermal growth factor receptors in a hyperproducing human epidermoid carcinoma cell line. J Biol Chem 254:8083-8086.

Yamada K, Akiyama S (1984): Preparation of cellular fibronectin. In Barnes DW, Sirbasku DA, Sato GH (eds): "Cell Culture Methods for Molecular and Cell Biology." New York: Alan R. Liss, Vol 1, pp 215-230.

Methods for Serum-Free Culture of Epithelial and Fibroblastic Cells,
pages 107–124

9
Growth and Differentiation of Embryonal Carcinoma Cells in Defined and Serum-Free Media

Angie Rizzino

During the past 15 years, numerous mouse embryonal carcinoma (EC) cell lines have been established from the germ-line tumors known as teratocarcinomas [Solter and Damjanov, 1979]. These cell lines are widely accepted as useful models for studying differentiation in general, and early mammalian development in particular [Martin, 1980]. EC cells are usually referred to as multipotent or nullipotent, depending on their ability to differentiate. As their name suggests, multipotent EC cells are able to differentiate into a wide range of cell types, which are derived from each of the three embryonic germ layers. Nullipotent EC cells, despite their name, are able to differentiate. However, they can only differentiate into a few cell types. It is very likely that these cell lines have undergone determination and consequently are restricted in their developmental potential [Rizzino, 1982a]. Therefore, a more appropriate term for these EC cells would be restricted EC cells, and this is the term that will be used in this text.

In order to understand the molecular mechanisms that regulate differentiation during early mammalian development, many investigators are examining the differentiation of EC cells in culture. Until recently, this work was restricted to serum-containing media. The use of serum poses a variety of problems when studying differentiation, including: 1) concealing the requirements for both growth and differentiation, 2) blocking the expression of differentiated functions [Orly et al., 1980], and 3) obstructing the search for endogenous

Eppley Institute for Research in Cancer and Allied Diseases, University of Nebraska Medical Center, Omaha, Nebraska 68105 growth factors released by EC cells [Rizzino et al., 1983]. To circumvent these and related problems, defined media were developed for EC cells. In addition, it was hoped that appropriately designed defined media would be effective in directing EC cells to differentiate into pure populations of specific cell types.

The main objective of this chapter is to describe the defined and serum-free media that are currently available for studying the growth and differentiation of EC cells. The text has been divided into six sections: 1) development of defined media and preparation of media components, 2) growth and differentiation of restricted EC cells in defined and serum-free media, 3) differentiation of multipotent EC cells in defined and serum-free media, 4) growth of differentiated cell lines derived from EC cells in defined and serum-free media, 5) pitfalls to be avoided, and 6) conclusions.

DEVELOPMENT OF DEFINED MEDIA AND PREPARATION OF MEDIA COMPONENTS

The methodology used to develop the defined media described here will not be discussed in detail, since it has been described elsewhere [Rizzino et al., 1979]. In brief, four basic steps have been followed, which should be generally applicable to developing a defined medium for any cell line. The first step is the selection of the synthetic portion of the medium. It should support clonal growth when supplemented with serum that has been exhaustively dialyzed to remove small molecules. The second step is to select an appropriate attachment factor, since most anchorage-dependent cells will not remain attached to untreated tissue culture plastic in the absence of serum. In most cases, one or more of the known attachment factors will be suitable (fibronectin, laminin, serum-spreading factor, collagen, or basic polymers). Once an appropriate attachment factor has been selected, the third step is to test growth factors, hormones, and serum accessory factors for their ability to promote cell proliferation. Those found to be effective are tested in various combinations in order to achieve continuous growth. When growth at normal cell densities is achieved, the fourth step is to test the medium for its ability to support growth at clonal cell densities. Often cells can be grown at high cell densities in defined medium, but clonal growth requires additional factors, such as trace elements and lipids.

The remainder of this section describes the technical aspects of preparing and handling the components of the various defined and serum-free media used thus far for EC cells. The subsequent sections describe the actual compositions of the media and the behavior of specific cell lines in the absence of serum.

Synthetic Portion of the Medium

The synthetic medium used thus far to culture EC cell lines in the absence of serum is a 1:1 mixture of Dulbecco's modified Eagle's medium (DME) and nutrient mixture F-12 (Ham's F-12). The use of either medium alone does not support satisfactory growth, and most of the cells die when DME is used alone. Although a 1:1 mixture of DME and F-12 has proven to be satisfactory, it is expected that modifications of this synthetic mixture will provide improved growth in defined media.

DME/F-12 medium is prepared fresh each week from powdered DME (Gibco: Cat.# 430-2100) and F-12 (Gibco: Cat.# 413-1700). One liter of DME/F-12 is prepared by mixing the powders for 500 ml of DME and 500 ml of F-12 with 2.4 g of $NaHCO_3$, 10 ml of 5×10^{-6} M selenous acid, and 10 ml of 1.5 M HEPES buffer (pH 7.9) in 900 ml of water. The pH is adjusted to approximately 7.1 with HCl (the pH is usually adjusted by color), and the volume is adjusted to 1 liter. The medium is then filter-sterilized with a Sterivex-GS 0.22-μm filter (Millipore). (If Nalgene filters are used, they should be washed first with water or PBS to minimize contamination of the medium by the detergents with which these filters are impregnated.) To avoid contamination with laboratory reagents, the medium is prepared and stored in glassware used solely for that purpose.

Water quality has been found to be important. In the past, triple-distilled water was used, since water distilled only once did not give consistent results. More recently, satisfactory results have also been obtained with water prepared by Milli Q water systems (Millipore).

Attachment Factors

Fibronectin can be easily and cheaply prepared from Cohn fraction I by the method of Engvall and Ruoslahti [1977], as modified by Ruoslahti et al. [1978]. It can also be purchased from Collaborative Research or from Bethesda Research Laboratories. Most studies have been conducted with lyophilized fibronectin from Collaborative Research. Once the fibronectin is redissolved, it is stored at 4°C for up to 2 weeks. In the defined medium EM-3 (see section on restricted EC cells), fibronectin is bound to the surface of the culture dish. This is accomplished by adding 1 μg of fibronectin per cm^2 to a culture dish containing 1 ml of PBS/10 cm^2. The culture dish is incubated at 37°C for 15-30 min and washed once with PBS, before addition of the DME/F-12 medium.

When laminin is used in place of fibronectin, it is added directly to the DME/F-12 medium (laminin does not bind tightly to the culture dish). The laminin first used for EC cells was prepared at the National Institutes of Health

[Rizzino et al., 1980b]. More recent studies have employed laminin purchased from Bethesda Research Laboratories. Final concentrations of 5–25 μg/ml have been found effective. Stock solutions are stored at −70°C.

When gelatin-coated surfaces are used instead of fibronectin or laminin, the culture dishes are coated (1 ml/10 cm^2) with a 0.1% solution of gelatin (Sigma: type I) for 30 min at room temperature. After coating, the excess gelatin solution is removed by aspiration and DME/F-12 is added. The gelatin solution is prepared by adding the gelatin to water and heating to 70–80°C. Once all of the gelatin has dissolved, the solution is cooled to 40°C and filter-sterilized with a 0.22-μm Nalgene filter. The solution is stored at room temperature until used.

Soluble Factors

Solutions of all the reagents mentioned below, except those prepared in ethanol, are filter-sterilized with 0.22-μm Millex-GS filters (Millipore).

Insulin (Sigma: bovine) is prepared fresh each week as a 1,000 × stock solution. Insulin is first dissolved in 0.1 M HCl at 10 mg/ml. Once the insulin has dissolved, it is diluted with PBS to 1 mg/ml and stored at 4°C. Transferrin (Sigma: human) is prepared as a 100 × stock solution by dissolving it in water to 500 μg/ml. Transferrin solutions are stored at −20°C.

Lipoproteins are prepared by sequential floatation, KBr being used for adjustments of density [Havel et al., 1955; Brown et al., 1974]. The lipoproteins used thus far—very-low-density lipoprotein (VLDL; 1.006–1.019), low-density lipoprotein (LDL; 1.019–1.05), and high-density lipoprotein (HDL; 1.087–1.21)—have been prepared by, and purchased from, Meloy Laboratories. The purified lipoproteins are prepared from fresh human plasma (EDTA). Each lipoprotein fraction is washed once to minimize contamination by plasma proteins, dialyzed (0.15 M NaCl and 0.01% EDTA at pH 7), and concentrated by ultrafiltration (VLDL: 2.8 mg/ml; LDL: 4.6 mg/ml; HDL: 40 mg/ml). The lipoproteins are stored at 4°C and used within 4 weeks.

Fatty acids are used in conjunction with fatty acid-free bovine serum albumin (fafBSA) at a ratio of 1:1. In the case of oleic acid (Sigma), it is prepared as a 100 × stock solution (10^{-3} M) in 100% ethanol and stored at −20°C. fafBSA (Sigma) is prepared as a 100 × stock solution (10^{-3} M) in DME/F-12 medium and stored at −20°C. Ethanolamine (also referred to as 2-aminoethanol; Sigma) is prepared as a 50 × stock solution (10^{-3} M) in DME/F-12 medium and stored at −20°C.

Promoters of Differentiation

Retinoic acid (all-trans-retinoic acid; Eastman Kodak Co.) is prepared as a 100 × stock solution (10^{-6} M) in 100% ethanol and stored at −20°C.

During preparation, storage, and handling, the retinoic acid is protected from light. Hexamethylene-bisacetamide (HMBA; Aldrich) is prepared as a 100 × stock solution (3×10^{-1} M) in DME/F-12 medium, filter-sterilized, and stored at −20°C.

Addition of Factors

All factors are added with an adjustable Pipetman and sterile disposable tips. Prior to dispensing each reagent, the neck of the Pipetman is immersed in 70% ethanol for 10–20 sec to minimize possible contamination by microorganisms. When fatty acids and fafBSA are added to EM-3, they are always added after the culture dishes have been coated with fibronectin, since BSA blocks the binding of fibronectin to the culture dish. Furthermore, since the fatty acids are dissolved in ethanol, they are first diluted with DME/F-12, to avoid denaturing the fibronectin-coated surfaces.

GROWTH AND DIFFERENTIATION OF RESTRICTED EC CELLS IN DEFINED AND SERUM-FREE MEDIA

Growth of F9 EC Cells

The first EC cell line to be grown in a completely defined medium was the restricted EC cell line F9 [Rizzino and Crowley, 1980]. This EC cell line is currently used in many laboratories because it can be easily induced to differentiate in serum-containing media by retinoic acid [Strickland and Mahdavi, 1978]. The defined medium developed for F9 is referred to as EM-3 (Embryonic Medium-3). Its composition is a 1:1 mixture of DME/F-12 plus fibronectin (1 μg/cm^2), insulin (1 μg/ml), and transferrin (5 μg/ml). (Although not employed routinely, laminin can be substituted for fibronectin in EM-3 [Rizzino et al., 1980b].) The F9 cells used to develop EM-3 were from a subclone referred to as F9(H5). Stock cultures of F9(H5) are grown in DME plus 10% fetal bovine serum.

To initiate the growth of F9(H5) in EM-3, the cells from stock cultures are subcultured with PBS-EDTA (1 mM) and replated into the defined medium. Specifically, the serum-containing medium is removed and the cells are washed for 1–2 min with PBS-EDTA. The PBS-EDTA is removed and DME/F-12 is added. The cells are then removed from the culture dish by forcing medium over the cells with a plugged Pasteur pipette that is attached to a 10-ml syringe. (The syringe, which has been modified by adding tygon tubing to accept a Pasteur pipette, allows medium to be briskly forced out over the cells; however, excessive force, as indicated by air bubbles, should be avoided.)

The cells can be grown and passaged in EM-3 for many generations. The doubling time is approximately 30 h as compared to 16 h in serum-containing media [Rizzino et al., 1980b]. F9(H5) cells grown in EM-3 for many generations retain the ability to differentiate and to form tumors. In addition, the cells can be frozen (liquid N_2) in EM-3 containing 10% dimethyl sulfoxide, and growth resumes when the cells are thawed directly into EM-3.

There are some drawbacks to culturing F9(H5) in EM-3. The cell density must be maintained between 2×10^4 and 2×10^5 cells per cm^2, and the medium should be replaced daily with fresh EM-3. In addition, the plating efficiency after several passages in EM-3 is variable. To minimize each of these problems, EM-3 is supplemented with purified lipoproteins. Although low-density lipoprotein (46 μg/ml) and very-low-density lipoprotein (28 μg/ml) are effective, the most consistent results have been obtained with high-density lipoprotein (400 μg/ml) [Rizzino et al., 1980a; Rizzino, 1982a].

In EM-3 plus HDL, the F9(H5) cells can be grown at densities as low as 2×10^2 cells per cm^2. The population doubling time in EM-3 plus HDL is approximately 21 h. Under these conditions, the medium is changed every other day, except when the cell density is above 5×10^4 cells per cm^2, at which point the medium is changed daily.

Until further studies are conducted, EM-3 plus HDL will not be considered a defined medium, since HDL may be contaminated with serum growth factors that stimulate the growth of EC cells. However, it is likely that the primary, if not exclusive, role of HDL is to serve as a source of important lipids. This is supported by the finding that fatty acids or ethanolamine (2×10^{-5} M) also promote the growth of EC cells at low density [Rizzino et al., 1980a]. Several fatty acids (e.g., oleic acid and linoleic acid) have been found effective. As in other systems, the fatty acids must be added in conjunction with fatty acid-free bovine serum albumin, since the fatty acids alone are toxic (fafBSA has no effect on its own). The fafBSA and fatty acid are added separately to the defined medium, and a 1:1 ratio must be maintained. In the case of oleic acid, a final concentration of 10^{-5} M has been found to be the most effective.

Growth of Other Restricted EC Cell Lines

PC-13 EC cells, when originally established in culture, were multipotent [Bernstine et al., 1973]. However, many of the subclones currently in use behave like restricted EC cell lines. Therefore, the subclone PC-13(IA4), used in my laboratory, is tentatively referred to as a restricted EC cell line. Stock cultures are grown on gelatin-coated tissue culture plastic in medium containing 10% fetal bovine serum.

PC-13(IA4) cells will grow in EM-3, but they have been grown in EM-3 for relatively short periods (two passages), and it is not known if they can be grown indefinitely in this medium. However, PC-13(IA4) cells have been grown for over 10 generations in EM-3 plus HDL, and they can also be grown at low cell densities (2×10^2 cells per cm^2) in this medium (Rizzino, unpublished results).

Nulli SSC-1 [Martin and Evans, 1975] is the only other restricted EC cell line that has been examined for growth in the absence of serum (Rizzino, unpublished results). The stock cultures are grown on gelatin-coated tissue culture plastic in medium containing 10% fetal bovine serum. Nulli SSC-1 can be grown for 4-5 doublings in EM-3 plus HDL, but long-term growth does not occur. In addition, Nulli SSC-1 can not be grown in EM-3 plus HDL at densities below 4×10^3 cells per cm^2, and little growth occurs in EM-3 even at high cell densities.

Differentiation of Restricted EC Cells

F9(H5), PC-13(IA4), and Nulli SSC-1 undergo very little differentiation in EM-3, in EM-3 plus HDL, and in serum-containing media unless an inducer is added. In the case of F9(H5), the addition of retinoic acid (10^{-8} M) to EM-3 leads to the formation of cells that exhibit the properties of extraembryonic endoderm [Rizzino and Crowley, 1980]. Based on their morphology and their secretion of plasminogen activator, it is likely that the cells represent parietal extraembryonic endoderm, rather than visceral extraembryonic endoderm. Similar results have been obtained when PC-13(IA4) and Nulli SSC-1 are exposed to retinoic acid in EM-3 plus HDL (Rizzino, unpublished results). In addition, hexamethylene bisacetamide will induce the differentiation of Nulli SSC-1 cultured in EM-3 plus HDL. However, the cells that form when PC-13(IA4) and Nulli SSC-1 are induced to differentiate by retinoic acid or hexamethylene bisacetamide have not yet been identified.

The differentiated cells that form under these conditions differ both morphologically and biochemically (e.g., production of plasminogen activator) from the parental cells. In addition, the differentiated cells respond to a number of growth factors that do not appear to influence the growth of the parental EC cells in either defined media or in serum-containing media. The first evidence of this was the finding that the differentiated cells derived from PC-13 respond to epidermal growth factor (EGF) [Rees et al., 1979]. In most of the experiments cited below, the cells were induced to differentiate with retinoic acid in serum-containing media. The cells were then transferred to serum-free medium and their responses to various growth factors were examined. With this protocol, the growth of the differentiated cells, but not that of the parental PC-13 EC cells, is stimulated by EGF [Rees et al., 1979], insulin [Heath et al., 1981], sarcoma growth factor [Rizzino et al., 1983], and fibroblast growth factor (Rizzino, unpublished results). In the case of EGF and insulin, a large increase in the number of corresponding membrane receptors has been observed when differentiation occurs [Rees et al., 1979; Heath et al., 1981].

Similar results have been observed when F9 EC cells are induced by retinoic acid to differentiate into parietal endoderm-like cells. The major difference is that both F9 and their differentiated cells respond to insulin [Rizzino and Sato, 1978; Rizzino and Crowley, 1980], and F9 EC cells exhibit significant numbers of insulin receptors [Nagarajan and Anderson, 1982]. It is also noteworthy that F9 and their differentiated cells respond by increased growth to multiplication-stimulating activity [Nagarajan et al., 1982; Rizzino, unpublished results].

DIFFERENTIATION OF MULTIPOTENT EC CELLS IN DEFINED AND SERUM-FREE MEDIA

Differentiation of Multipotent EC Cells Into Parietal Endoderm

In direct contrast to F9(H5) and PC-13(IA4), multipotent EC cells do not proliferate as EC cells in EM-3 or in EM-3 plus HDL. Instead, two multipotent EC cell lines, OC-15S1 and 1003, have been found to completely differentiate into one cell type, which has been positively identified as parietal endoderm [Rizzino, 1983a]. (Parietal endoderm is one of the first cell types to appear during the early stages of mammalian development.) This strongly contrasts with the behavior of these multipotent EC cells in serum-containing media, where very little differentiation occurs.

Thus far, the most detailed observations have been made with OC-15S1. The stock cultures are grown in medium containing 10% fetal bovine serum. When OC-15S1 is plated into EM-3 plus HDL (400 μg/ml), over 90% of the cells commit to differentiate within 12 h, and one obtains a pure population of differentiated cells that is completely devoid of EC cells after 72 h [Rizzino, 1983a,b]. The differentiated cells that form exhibit an epithelioid morphology that is characteristic of parietal endoderm, and at least 85% of the differentiated cells secrete plasminogen activator. Furthermore, only one type of plasminogen activator is released, and it has been identified antigenically as the type released by parietal endoderm.

It is important to note that differentiation actually occurs in EM-3 alone and that HDL is required for the survival of the differentiated cells (LDL and VLDL are nearly as effective as HDL). In addition, the same basic results are obtained when laminin is substituted for fibronectin in EM-3 or in EM-3 plus HDL [Rizzino, 1983b].

The differentiation of OC-15S1 and 1003 into parietal endoderm is initiated by plating the cells in EM-3 plus HDL. In these experiments, the cells are removed from stock cultures in serum-containing media by trypsinization. When large aggregates of cells have been broken up and a population composed primarily of single cells has been attained, the action of trypsin is stopped with soybean trypsin inhibitor. Next the cells are spun out and resuspended in DME/F-12 before transfer to EM-3 plus HDL. (If the cells are not transferred in this way, the resulting population of differentiated cells may be composed of several different cell types. This point is discussed below.) In these experiments, the medium is replaced on day 2 with fresh EM-3 plus HDL, from which fibronectin has been omitted. (Fibronectin is not required once the cells have attached.) The cultures of differentiated cells can be maintained for several weeks, and cell death is minimal provided the medium is replaced every 2 days. However, once a monolayer has formed, very little growth occurs.

If the differentiated cells are subcultured, their plating efficiency is very low and the cells have a cloning efficiency of zero at densities below 10^2 cells per cm^2. This is a useful finding that can be used to estimate the time required for the cells to differentiate when plated in EM-3 plus HDL. Since the parental OC-15S1 EC cells have a cloning efficiency of approximately 25% and their differentiated cells have a cloning efficiency of zero, one can estimate the fraction of cells that have become committed to undergo differentiation at any given point by replating the cells into serum-containing medium. The cloning efficiency of cells plated in EM-3 plus HDL divided by the cloning efficiency of cells plated in serum-containing medium is equal to the fraction of uncommitted cells.

Attempts to understand why OC-15S1 and 1003 differentiate in the defined medium led to an important finding: Fibronectin and laminin promote the differentiation of these multipotent EC cells. This is most readily demonstrated by plating the cells at low density onto fibronectin- or laminin-coated culture dishes in serum-containing medium. Under these conditions, 70–80% of the EC cells differentiate into endoderm-like cells, whereas only 5–10% of the EC cells differentiate on untreated culture dishes. The extent of differentiation was estimated by determining the cloning efficiency of the cells on fibronectin- or laminin-coated dishes and comparing it to the cloning efficiency of the cells on untreated culture dishes. (It is important to note that contrasting results are obtained with restricted EC cells. Restricted EC cells do not differentiate when plated on fibronectin- or laminin-coated dishes in serum-containing media. This may explain the failure of restricted EC cells to differentiate in EM-3 or in EM-3 plus HDL.)

At present, the mechanisms by which fibronectin and laminin promote the differentiation of multipotent EC cells are unknown, but it has been suggested that fibronectin, and possibly laminin, promotes differentiation by causing the cells to modify their shape [Rizzino, 1983a]. This possibility is currently under investigation.

The finding that fibronectin promotes the differentiation of these multipotent EC cells strongly suggests that fibronectin is not a major attachment factor in serum, at least for EC cells. This argument is supported by several observations. First, the amount of fibronectin used to promote differentiation (10 μg/35-mm culture dish) is significantly lower than that in culture medium containing 10% fetal bovine serum (30 μg), and this suggests that very little fibronectin binds to the culture dish. In fact, less than 1% of the fibronectin in serum-containing medium binds to the culture dish, as determined by adding ^{125}I-fibronectin directly to the culture medium (Rizzino, unpublished results). Second, serum depleted of fibronectin promotes the attachment and spreading of F9 cells just as well as untreated serum does (Rizzino, unpublished results). Third, serum spreading factor has recently been shown to be a major attachment factor in serum [Hayman et al., 1982], and serum spreading factor is known to promote the attachment of F9 EC cells [Serrero et al., 1980].

It is obvious from the data discussed in this section that a better understanding of how EC cells interact with fibronectin, laminin, and attachment factors such as serum spreading factor, will significantly broaden our understanding of differentiation.

Cultivation of Other Multipotent EC Cells in EM-3 Plus HDL

Thus far, only four other multipotent EC cell lines—P-19 [McBurney and Rogers, 1982], 6050 AJ [Jetten et al., 1979], PCC-3 [Nicholas et al., 1975], and PSA-1 [Martin et al., 1977], have been examined in EM-3 plus HDL (Rizzino, unpublished results), and none have been examined in depth. P-19 and 6050 AJ differentiate when plated in EM-3 plus HDL, but, in contrast to the findings with OC-15S1 and 1003, significant cell death occurs. In addition, it appears that more than one differentiated cell type forms (e.g., nerve-like cells have been observed in the differentiated cultures of 6050 AJ). However, only endoderm-like cells have been positively identified in the differentiated cultures of P-19 and 6050 AJ.

The results with PCC-3 and PSA-1 to date have been discouraging, since most of the cells die within several days of being plated in EM-3 plus HDL. The reason for this is unclear, but it is likely that the differentiated cells that form cannot survive in EM-3 plus HDL. Thus, modifications of the medium will be necessary to proceed further with these EC cell lines in the absence of serum.

Modifications of the Conditions for Differentiation

As mentioned above, when OC-15S1 or 1003 are trypsinized and plated as single cells in EM-3 plus HDL, differentiation occurs and parietal endoderm-like cells appear. However, if OC-15S1 and 1003 are transferred to EM-3 without trypsinization, the results obtained differ significantly. If the cells in serum-containing medium are removed by washing once with DME and then are harvested by forcing DME over the cells, one obtains a population of cells in aggregates of varying size that retain small amounts of serum. When the cells are harvested in this way and transferred to EM-3, cell survival is variable and one obtains a mixed population of differentiated cells that includes endoderm, neuroepithelial and neuronal cells. In the case of 1003, it has been shown that after 6-9 days in EM-3 containing selenous acid (3×10^{-8} M), one observes the appearance of neuronal-like cells with extended processes that will stain with silver [Darmon et al., 1981a]. It has also been shown that the neuronal-like cells in these cultures are positive for several nerve-related antigens, including NG and the nerve-glial antigen NG2 [Darmon et al., 1982]. These results with 1003 are interesting, since 1003 has not been observed to differentiate into nerve-like cells in serum-containing media. In fact, 1003 exhibits very little differentiation in serum-containing media, even though it differentiates into many cell types in vivo [Darmon et al., 1981a]. The fact that a mixed population of differentiated cells forms and the proportion of nerve-like cells varies from one experiment to the next is a drawback to the use of the above conditions. However, such a population may be useful for the isolation of cell lines that represent partially determined stem cells.

GROWTH OF DIFFERENTIATED CELL LINES DERIVED FROM EC CELLS IN DEFINED AND SERUM-FREE MEDIA

This section will discuss five differentiated cell lines derived from EC cells: two endoderm-like cell lines, PSA-5E [Adamson et al., 1977] and PYS-2 [Lehman et al., 1974]; plus a myoblastic cell line (1168), a fibroadipogenic cell line (1246), and a fibroblastic cell line ($1246B_2$) that were isolated from a common precursor [Darmon et al., 1981b]. Stock cultures of these five cell lines are grown in medium containing 10% fetal bovine serum.

PSA-5E, which is a visceral endoderm-like cell line, proliferates in the defined medium EM-3, provided the cell density is maintained above 10^4 cells per cm^2. However, it will clone in EM-3 plus HDL (400 μg/ml) with an efficiency of approximately 20%, which is double the cloning efficiency observed in medium containing 10% fetal bovine serum. PYS-2 is a parietal endoderm-like cell line, and the results obtained with this cell line are similar, though not identical, to those obtained with PSA-5E. The behavior of PYS-2 in the absence of serum depends heavily on the strain of PYS-2 used. Certain strains will proliferate for many generations in EM-3 at densities greater than 10^4 cells per cm^2, whereas others do not. Thus far, the most consistent results have been obtained in EM-3 plus HDL. In this formulation, all the strains of PYS-2 that have been tested proliferate, and at least one, PYS-2(AR), grows at densities as low as 500 cells per cm^2. In the above studies, both PSA-5E and PYS-2 were transferred to EM-3 or EM-3 plus HDL after trypsinization (as described above for the multipotent EC cells).

The other differentiated cell lines that have been studied in media lacking serum were isolated from a myogenic cell line referred to as T984 [Jakob et al., 1978]. T984 was established from a primary culture of a teratocarcinoma that had been induced by the injection of C17S1 EC cells into a syngeneic mouse. In serum-containing medium, at cell densities of 10^4 cells per cm^2, T984 differentiates primarily into embryonic muscle. However, at clonal cell densities, myogenic, fibroadipogenic, and fibroblastic cells are observed, and three corresponding cell lines (1168, 1246, and $1246B_2$) were established from such a population. Each cell line proliferates in serum-free medium containing insulin (5 μg/ml), transferrin (5 μg/ml), fibroblast growth factor (100 mg/ml), and serum spreading factor (5 μg/ml) [Darmon et al., 1981b]. In each case, PBS-EDTA was used to remove the cells from stock culture dishes prior to plating in the serum-free medium.

Both 1246 and $1246B_2$ were grown and replated in the serum-free medium for over 30 generations. These cell lines proliferate in the serum-free medium at rates that are similar to those observed in serum-containing media. However, their saturation densities, especially that of $1246B_2$, were always lower than those observed in serum-containing media. In addition, both 1246 and $1246B_2$ exhibit density-dependent growth inhibition in the serum-free medium, whereas both cell lines overgrow and form multiple cell layers in serum-containing media. Another difference between growth of these cell lines in the serum-free medium and their growth in serum-containing media is the failure of 1246 and $1246B_2$ to clone in the absence of serum. It is not known at this time if this problem can be rectified by supplementing the serum-free medium with purified lipoproteins, such as HDL.

In contrast to 1246 and $1246B_2$, 1168 does not grow continuously in the serum-free medium. Although 1168 initially proliferates, it stops growing after 2-3 generations, and the majority of the cells undergo myogenesis. After 5 days in the serum-free medium, approximately 70% of the nuclei are found in myotubes, and this is accompanied by a large increase (sixfold) in creatine phosphokinase activity.

PITFALLS TO BE AVOIDED

There are three potential problems to be concerned with when working with EC cells in the absence of serum. The first involves the selection of the EC cell line. Thus far, of the more than 20 different EC cell lines that are available, only a small fraction have been tested for growth in serum-free and defined media. Consequently, it is not known whether a particular EC cell line will behave like one of the EC cell lines described above when cultured in the absence of serum. However, it is expected that many EC cell lines will exhibit excessive cell death if cultured in unmodified preparations of the defined and serum-free media described above. Therefore, work with other EC cell lines in the absence of serum will probably require a period of experimentation in order to develop appropriately modified media.

The second problem to be concerned with is the variations that exist within a given EC cell line. The origins of these differences are, more often than not, unknown. However, slight variations in culture conditions from one laboratory to the next may be responsible. Thus, different strains of a given EC cell line may exhibit: 1) different requirements for growth in defined media and 2) different capacities to differentiate. An example of the latter was noted in serum-containing medium after an EC cell line that had been grown on fibroblastic feeder layers was adapted to growth on gelatin-coated surfaces in the absence of a feeder layer [Hogan, 1976]. It is equally evident that the conditions used to culture EC cells in serum-containing media can affect their ability to grow in defined media. Thus, F9(H5), which is grown in serum-containing medium on untreated tissue culture plastic, proliferates actively in either EM-3 or in EM-3 plus HDL, whereas F9 cells that have been grown for many generations on gelatin-coated surfaces do not grow satisfactorily in EM-3 alone, although they grow well in EM-3 plus HDL.

A third complication is that, in light of the first two problems, EC cells grown for prolonged periods in the absence of serum should be monitored for any changes in their properties. In particular, their ability to form tumors and to differentiate in vitro should be checked periodically. In the case of restricted EC cell lines, this can be determined relatively quickly. However, in the case of multipotent EC cells, this will require careful scrutiny to ensure that the number of different cell types formed in vitro and in vivo does not change.

CONCLUSIONS

Several model systems are now available for studying the growth and differentiation of specific restricted and multipotent EC cells in the absence of serum (Table I). These model systems are playing a major role in defining the growth requirements of EC cells and their differentiated cells and in identifying extracellular signals that control differentiation. In addition, defined media are proving to be extremely useful in the development of model systems that facilitate the biochemical investigation of differentiation. For such studies it is highly desirable that an EC cell line differentiate into only one well-characterized cell type. Until recently, this could not be achieved with multipotent EC cells. Today, the use of EM-3 plus HDL enables two different multipotent EC cell lines to rapidly and completely differentiate into one cell type. It is reasonable to expect that, with the development of appropriately designed serum-free/defined media, it will be possible to direct the differentiation of multipotent EC cells into pure populations of other cell types.

TABLE I. Overview of Growth and Differentiation in the Absence of Serum

Cell line	Growth in EM-3 at:		Growth in EM-3 plus HDL at:		Differentiation in EM-3 and EM-3 plus HDL
	High density[a]	Low density[b]	High density[a]	Low density[b]	
Restricted EC cells					
F9(H5)	+	−	+	+	−[d]
PC-13(IA4)	+	−	+	+	−[d]
Nulli SSC-1	−	−	+[c]	−	−[d]
Multipotent EC cells					
OC-15S1	Do not proliferate as EC cells				+[e]
1003					+[e]
P-19					+[f]
6050 AJ					+[f]
PCC-3					+[f]
PSA-1					+[f]
Differentiated cell lines					
PSA-5E	+	−	+	+	NA[g]
PYS-2	+	−	+	+	NA

[a]High density is defined as 10^4 cells per cm^2.
[b]Low density is defined as 2×10^2 cells per cm^2.
[c]Growth is sustained for only 4-5 doublings.
[d]Differentiation requires an inducer, such as retinoic acid.
[e]The cell type formed depends on the culture conditions—see text.
[f]Differentiation occurs but considerable cell death is observed.
[g]NA = not applicable.

Several conclusions drawn from work with EC cells in the absence of serum appear to be relevant to development in vivo. This is particularly true in the area of growth control. It is now well established that extraembryonic endoderm-like cells formed in vitro respond by increased growth to several mitogens that do not affect the growth of their parental EC cells. In addition, differentiation to extraembryonic endoderm is accompanied by a large increase in the number of corresponding membrane receptors (e.g., EGF receptors). These findings raised the possibility that similar events occur in vivo, but this question has not been extensively examined. Thus far, visceral extraembryonic endoderm, which appears very early during development, has been found to have many EGF receptors in vivo [Adamson et al., 1981], and it is likely that EGF, or a related molecule, stimulates its growth in vivo, just as EGF appears to do in vitro.

These findings imply that growth control by extracellular factors begins very early during mammalian development. If this is true, what is the origin of these mitogens, and what is their precise role during development? The answers to these questions are largely unknown, but it is likely that some, and perhaps most, of the mitogens are produced by the embryo itself, in particular by cells derived from the inner cell mass.

This issue obviously deserves further study, but it cannot be examined directly with embryos owing to the size of early embryos. However, given the close relationship between EC cells and cells of the inner cell mass [Martin, 1980], it is possible to address this question using EC cells. With this in mind, it was recently shown that several restricted EC cell lines release endogenous growth factors [Rizzino et al., 1983] and that these factors stimulate the growth of their differentiated cells. Furthermore, by means of an assay developed during work with these EC cells, it was determined that early mouse embryos, in particular the cells derived from the inner cell mass, release similar factors [Rizzino, 1982b]. At present, the role of these factors is not understood, but it is likely that they play an important role in regulating growth and differentiation during early development.

The finding that fibronectin and laminin promote the differentiation of multipotent EC cells may also relate to development in vivo. Of all the molecules currently known to influence EC cell differentiation, only fibronectin, laminin, and retinoic acid are physiological. Moreover, the appearance of both fibronectin [Wartiovaara et al., 1979] and laminin [Leivo et al., 1980] are carefully regulated during the early stages of mammalian development, and both are believed to appear prior to the formation of visceral and parietal extraembryonic endoderm. Thus, the possibility that fibronectin and laminin influence differentiation during early development must be seriously considered.

In conclusion, the use of serum-free media not only offers several important advantages for studying the differentiation of EC cells in culture, but it may also offer a better approach for identifying potential signals that influence differentiation in vivo. The latter is not surprising, since serum-containing media are not representative of the early embryonic environment.

ACKNOWLEDGMENTS

Heather Rizzino is thanked for her excellent editorial assistance, as well as her useful comments and suggestions during the writing of this article. I thank Linda Orme for her excellent technical assistance. The work of this investigator was initially supported by grants from the National Science Foundation (PCM 78-04337) and the National Cancer Institute (5-PO1-CA-23052) and was continued until November 1983 at the Laboratory of Viral Carcinogenesis at the National Cancer Institute.

REFERENCES

Adamson ED, Evans MJ, Magrane GG (1977): Biochemical markers of the progress of differentiation in cloned teratocarcinoma cell lines. Eur J Biochem 79:607-615.

Adamson ED, Deller MJ, Warshaw JB (1981): Functional EGF receptors are present on mouse embryo tissues. Nature 291:656-659.

Bernstine EG, Hooper ML, Grandchamp S, Ephrussi B (1973): Alkaline phosphatase activity in mouse teratoma. Proc Natl Acad Sci USA 70:3899-3903.

Brown MS, Dana SE, Goldstein JL (1974): Regulation of 3-hydroxy-3-methylglutaryl coenzyme A reductase activity in cultured human fibroblasts. J Biol Chem 249:789-796.

Darmon M, Bottenstein J, Sato G (1981a): Neural differentiation following culture of embryonal carcinoma cells in a serum-free defined medium. Dev Biol 85:463-473.

Darmon M, Serrero G, Rizzino A, Sato G (1981b): Isolation of myoblastic, fibroadipogenic, and fibroblastic clonal cell lines from a common precursor and study of their requirements for growth and differentiation. Exp Cell Res 132:313-327.

Darmon M, Stallcup WB, Pittman QJ (1982): Induction of neural differentiation by serum deprivation in cultures of the embryonal carcinoma cell line 1003. Exp Cell Res 138: 73-78.

Engvall E, Ruoslahti E (1977): Binding of soluble form of fibroblast surface protein, fibronectin, to collagen. Int J Cancer 20:1-5.

Havel RJ, Eder HA, Bragdon JH (1955): The distribution and chemical composition of ultracentrifugally separated lipoproteins in human serum. J Clin Invest 34:1345-1353.

Hayman EG, Engvall E, A'Hearn E, Barnes D, Pierschbacher M, Ruoslahti E (1982): Cell attachment on replicas of SDS polyacrylamide gels reveals two adhesive plasma proteins. J Cell Biol 95:20-23.

Heath J, Bell S, Rees AR (1981): Appearance of functional insulin receptors during the differentiation of embryonal carcinoma cells. J Cell Biol 91:293-297.

Hogan BLM (1976): Changes in the behaviour of teratocarcinoma cells cultivated in vitro. Nature 263:136-137.

Jakob H, Buckingham ME, Cohen A, Dupont L, Fiszman M, Jacob F (1978): A skeletal muscle cell line isolated from a mouse teratocarcinoma undergoes apparently normal terminal differentiation in vitro. Exp Cell Res 114:403-408.

Jetten AM, Jetten MER, Sherman MI (1979): Stimulation of differentiation of several murine embryonal carcinoma cell lines by retinoic acid. Exp Cell Res 124:381-391.

Lehman JM, Speers WC, Swartzendruber DE, Pierce GB (1974): Neoplastic differentiation: Characteristics of cell lines derived from a murine teratocarcinoma. J Cell Physiol 84: 13-28.

Leivo I, Vaheri A, Timpl R, Wartiovaara J (1980): Appearance and distribution of collagens and laminin in the early mouse embryo. Dev Biol 76:100-114.

Martin GR (1980): Teratocarcinomas and mammalian embryogenesis. Science 209:768-776.

Martin GR, Evans MJ (1975): Multiple differentiation of clonal teratocarcinoma stem cells following embryoid body formation in vitro. Cell 6:467-474.

Martin GR, Wiley LM, Damjanov I (1977): The development of cystic embryoid bodies in vitro from clonal teratocarcinoma stem cells. Dev Biol 61:230-244.

McBurney MW, Rogers BJ (1982): Isolation of male embryonal carcinoma cells and their chromosome replication patterns. Dev Biol 89:503-508.

Nagarajan L, Anderson WB (1982): Insulin promotes the growth of F9 embryonal carcinoma cells apparently by acting through its own receptor. Biochem Biophys Res Commun 106:974-980.

Nagarajan L, Nissley SP, Rechler MM, Anderson WB (1982): Multiplication-stimulating activity stimulates the multiplication of F9 embryonal carcinoma cells. Endocrinology 110:1231-1237.

Nicholas JF, Dubois P, Jakob H, Gaillard J, Jacob F (1975): Tératocarcinome de la souris: Différençiation en culture d'une lignée de cellules primitives à potentialités multiples. Ann Microbiol (Paris) 126a:3-22.

Orly J, Sato G, Erickson GF (1980): Serum suppresses the expression of hormonally induced functions in cultured granulosa cells. Cell 20:817-827.

Rees AR, Adamson ED, Graham CF (1979): Epidermal growth factor receptors increase during the differentiation of embryonal carcinoma cells. Nature 281:309-311.

Rizzino A (1982a): Growth and differentiation of embryonal carcinoma cells in defined media: The role of fibronectin. In Sato G, Pardee AB, Sirbasku DA (eds): "Cold Spring Harbor Conferences on Cell Proliferation," Volume 9: "Growth of Cells in Hormonally Defined Media." Cold Spring Harbor, NY: Cold Spring Harbor Laboratory, pp 209-218.

Rizzino A (1982b): Embryonal carcinoma cells release factors with transforming growth factor (TGF) activity. J Cell Biol 95:181a.

Rizzino A (1983a): Two multipotent embryonal carcinoma cell lines irreversibly differentiate in defined media. Dev Biol 95:126-136.

Rizzino A (1983b): Fibronectin and laminin promote the differentiation of two multipotent embryonal carcinoma cell lines. In Silver L, Martin G, Strickland S (eds): "Cold Spring Harbor Conferences on Cell Proliferation," Volume 10: "Teratocarcinoma Stem Cells." Cold Spring Harbor, NY: Cold Spring Harbor Laboratory, pp 97-107.

Rizzino A, Crowley C (1980): Growth and differentiation of embryonal carcinoma cell line F9 in defined media. Proc Natl Acad Sci USA 77:457-461.

Rizzino A, Sato G (1978): Growth of embryonal carcinoma cells in serum-free medium. Proc Natl Acad Sci USA 75:1844-1848.

Rizzino A, Rizzino H, Sato G (1979): Defined media and the determination of nutritional and hormonal requirements of mammalian cells in culture. Nutr Rev 37:369-378.

Rizzino A, Crowley C, McClure D (1980a): The growth of F9 embryonal carcinoma cells in defined media at low density. J Cell Biol 87:18a.

Rizzino A, Terranova V, Rohrbach D, Crowley C, Rizzino H (1980b): The effects of laminin on the growth and differentiation of embryonal carcinoma cells in defined media. J Supramol Struct 13:243-253.

Rizzino A, Orme LS, De Larco JE (1983): Embryonal carcinoma cell growth and differentiation: Production of and response to molecules with transforming growth factor activity. Exp Cell Res 143:143-152.

Ruoslahti E, Vuento M, Engvall E (1978): Interaction of fibronectin with antibodies and collagen in radioimmunoassay. Biochim Biophys Acta 534:210-218.

Serrero G, Darmon D, Barnes D, Rizzino A (1980): Effect of a serum factor on attachment and growth of embryonal carcinoma cells (ECC) in culture. In Vitro 16:251.

Solter D, Damjanov I (1979): Teratocarcinoma and the expression of oncodevelopmental genes. In Fishman WH, Busch H (eds): "Methods in Cancer Research, Vol 18." New York: Academic, Chap 7, pp 227-332.

Strickland S, Mahdavi V (1978): The induction of differentiation in teratocarcinoma stem cells by retinoic acid. Cell 15:393-403.

Wartiovaara J, Leivo I, Vaheri A (1979): Expression of the cell surface-associated glycoprotein, fibronectin, in the early mouse embryo. Dev Biol 69:247-257.

Methods for Serum-Free Culture of Epithelial and Fibroblastic Cells,
pages 125–153

10
α_2-Macroglobulin, a Contaminant of Commercially Prepared Pedersen Fetuin: Isolation, Characterization, and Biological Activity

David S. Salomon, Kathryn B. Smith, Ilona Losonczy, Mozeena Bano, William R. Kidwell, Giulio Alessandri, and Pietro M. Gullino

Serum provides a number of low-molecular-weight, nutritional components for the propagation of cells in vitro [Swim, 1967; Bailey and Dunbar, 1973; Higuchi, 1973]. Likewise, the macromolecular species present in serum are also necessary for the attachment, spreading, and subsequent proliferation of cells in culture [Temin et al., 1972; Higuchi, 1973; Barnes and Sato, 1980]. Fibronectin or the serum-associated form, cold-insoluble globulin (CIG), has been identified as one of the serum glycoproteins that functions as a cell attachment factor for fibroblasts or other mesenchymal cells [Kleinman et al., 1981]. More recently laminin, a basement membrane-associated glycoprotein, has been shown to facilitate the attachment and spreading of epithelial cells [Terranova et al., 1980; Kleinman et al., 1981]. The mitogenic activity in serum is provided by the presence of various hormones and growth factors [Temin et al., 1972; Higuchi et al., 1973; Barnes and Sato, 1980]. These can occur in either a free, circulating form or complexed to other carrier proteins [Collipp et al., 1964; Hadden and Prout, 1964; Hoffman et al., 1973; Nevo and Laron, 1979; Carpenter and Cohen, 1979; Ronne et al., 1979; Knauer and Cunningham, 1982]. A major portion of the growth-stimulating activity in serum is found in the α-globulin fraction [Marr et al., 1962; Tozer and Pirt,

Laboratory of Tumor Immunology and Biology (D.S.S., K.B.S., I.L.) and Laboratory of Pathophysiology (M.B., W.R.K., G.A., P.M.G.), National Cancer Institute, National Institutes of Health, Bethesda, MD 20205

1964; Healy and Parker, 1966; Swim, 1967; Temin et al., 1972; Heim et al., 1973; Maeda et al., 1980; Baudysova and Michl, 1981].

In fetal calf serum, fetuin was found to partially account for the growth-promoting activity associated with the α-globulin fraction [Lieberman and Ove, 1958; Fisher et al., 1958, 1959; Lieberman et al., 1959; Bergmann et al., 1962; Puck et al., 1968; Graham, 1972; Temin et al., 1972; Lai et al., 1981]. Crude preparations of fetuin isolated by the method of Pedersen, which utilizes a 45% ammonium sulfate precipitate of fetal calf serum [Pedersen, 1944], were active in promoting the growth of HeLa cells and several other epithelial cell types [Fisher, 1958; Puck et al., 1968]. However, more homogeneous preparations of fetuin as described by Deutsch and Spiro [Deutsch, 1954; Spiro, 1960; Spiro and Spiro, 1962] were biologically inactive in supporting cell proliferation [Temin et al., 1972; Graham et al., 1972; Rizzino and Sato, 1978; Salomon, 1980; Salomon et al., 1981].

A number of other biological activities have been ascribed to Pedersen-prepared fetuin (Table I). These include promotion of cell attachment and spreading, stimulation of anchorage-independent cell growth in soft agar, galactosyltransferase activity, inhibition of trypsin and plasmin activity, and inhibition of collagenase secretion [Galembeck and Cann, 1974; Werb and Aggeler, 1978; Knox et al., 1979; Pierce et al., 1979; Bjornson and Moses, 1981; Rohrlich and Rifkin, 1981]. Hayman et al. [1982] have recently identified a glycoprotein (Mr = 70,000) associated with Pedersen fetuin preparations that promotes the adhesion and spreading of normal rat kidney (NRK) cells and is distinct from fibronectin. This glycoprotein may be related to a gelatin-binding glycoprotein previously described by Vartio and Vaheri [1981] that is synthesized by several different cell types. Likewise, Bjornson and Moses [1981] have isolated a 10,000 molecular weight peptide from Pedersen fetuin preparations that resembles a transforming growth factor [Todaro et al., 1981] in that it induces colony formation in soft agar of NRK and mouse AKR-2B cells.

Attempts to resolve the mitogenic principle(s) in crude preparations of fetuin from fetuin have been relatively unsuccessful [Spiro, 1960; Puck et al., 1968; Temin et al., 1972; Graham, 1972]. This may have been due to the fact that fetuin is an acidic glycoprotein (Mr = 48,500) that tends to aggregate during conventional gel filtration chromatography [Marti et al., 1973; Begbie, 1974]. Moreover, fetuin actually represents a family of polymorphic glycoproteins ranging in molecular weight from 48,000 to 68,000 [Oshiro and Eylar, 1968; Begbie, 1974; Rohrlich and Rifkin, 1981] and possessing isoelectric points from 3.2 to 5.2 [Oshiro and Eylar, 1968; Marti et al., 1973]. These variations probably arise from differences in glycosylation of three oligosac-

TABLE I. Biological Effects of Fetuin and α_2-Macroglobulin

Biological activities associated with Pedersen fetuin
1. Promotes cell attachment and spreading (70,000 Mr glycoprotein)
2. Stimulates cell proliferation (α_2M, unsaturated fatty acids)
3. Possesses serine protease inhibitory activity (e.g., trypsin, plasmin)
4. Inhibits collagenase secretion
5. Possesses galactosyltransferase activity
6. Induces anchorage-independent cell growth in soft agar (10,000 Mr peptide)

Biological effects of α_2-macroglobulin
1. Broad-spectrum irreversible protease inhibitor[a,b]
2. Cleaved following binding to proteases to lower-molecular-weight fragments[c]
3. Stimulates cell proliferation
4. Stimulates collagen production
5. Inhibits neovascularization
6. Carrier protein for hormones and growth factors (i.e., growth hormone, thrombin, insulin, NGF)
7. Binds to cell surface receptors and is internalized

[a]Endopeptidases (trypsin, chymotrypsin, plasmin, thrombin), metalloproteases (collagenases, elastase).
[b]One mole of enzyme bound per protomer of ~ Mr 370,000.
[c]Each 185,000 subunit (4 per molecule) converted to two 85,000 Mr fragments.

charide chains associated with the core protein (Mr = 38,300). Each carbohydrate chain (Mr = 3,400) can possess a maximum of four terminal acid sialic residues or a total of 12 per molecule, resulting in species containing different numbers of terminal sialic acid residues [Oshiro and Eylar, 1968; Graham, 1972].

The microheterogeneity in molecular weight and isolectric point would cause fetuin to behave aberrantly during gel filtration and ion-exchange chromatography. Analysis of fetuin prepared by the method of Pedersen, but not by the method of Spiro, by gel electrophoresis has demonstrated that this preparation is relatively heterogeneous, since it contains several minor high-molecular-weight components other than fetuin (see Fig. 2) [Fisher et al., 1958; Marr et al., 1962; Puck et al., 1968; Temin et al., 1972; Knox et al., 1979]. Attempts to purify this preparation by anion exchange chromatography or sequential ethanol precipitation in the presence of barium and zinc ions yielded a more homogenous preparation of fetuin that was relatively inactive as a growth-promoter. These results suggested that the mitogenic activity resided in one or more of the minor contaminating species or that fetuin per se was the active principle but during purification it was inactivated [Spiro, 1960; Green and Kay, 1963; Graham, 1972; Temin et al., 1972; Rohrlich and Rifkin, 1981].

The present chapter attempts to provide some information with respect to the isolation, purification, and characterization of a growth-promoting activity from crude, commercial Pedersen fetuin preparations. A number of serum-free media have been formulated that require Pedersen-prepared fetuin as a medium supplement [Rizzino and Sato, 1978; Barnes and Sato, 1980; Salomon, 1980; Salomon et al., 1981]. This requirement is also true for the propagation of mouse embryonal carcinoma (EC) and rat mammary epithelial (RME) cells in a serum-free, hormone-supplemented medium [Salomon, 1980; Salomon et al., 1981]. For this reason, these cells were utilized to screen for the growth-promoting activity during purifcation.

A major portion of the growth-promoting activity in the crude fetuin preparation is associated with a high-molecular-weight glycoprotein, embryonin (Mr = 270,000). Three polypeptide chains are observed following denaturation and reduction of the protein, a major polypeptide (Mr = 185,000) and two minor chains (Mr = 116,000 and 68,000). Embryonin differs from fetuin in molecular weight, isoelectric point (pI = 5.4), amino acid composition, and immunological reactivity [Salomon et al., 1982b]. However, embryonin resembles bovine or human α_2-macroglobulin (α_2M) in these physiochemical and immunological properties. Human or bovine α_2M can replace the requirement for embryonin in maintaining EC or RME cell growth in the absence of serum on poly-D-lysine or type I or type IV collagen-coated dishes. Embryonin accounts for approximately 0.1-0.3% of the total protein in the initial Pedersen fetuin preparation. Embryonin and α_2M may function as growth-promoting agents by differentially stimulating the production of extracellular matrix components such as collagen in cell types that respond to these growth factors. In addition, Pedersen fetuin also contains phospholipids (phosphatidylcholine), cholesterol esters, cholesterol, and triglycerides. The phospholipids and neutral lipids in the preparation possess in contrast to Deutsch or Spiro fetuins a high level of the unsaturated fatty acids, oleic, linoleic, and arachidonic acids, which could also contribute to the growth-stimulating activity of this preparation [Ham, 1963; Bailey and Dunbar, 1973; Higuchi, 1973; Kidwell et al., 1978; Wicha et al., 1979].

EXPERIMENTAL PROCEDURES

Materials

Pedersen-prepared bovine fetuin (type III or type IV) was purchased from Sigma Chemical Co., St. Louis. Deutsch and Spiro fetuin preparations were from Grand Island Biological Co., Grand Island, NY (Gibco). Human α_2M and rabbit anti-human α_2M antiserum were obtained from Calbiochem-Behring, Gaithersburg, MD. Bovine α_2M was from Boehringer Mannheim Biochemicals, Indianapolis. Phospholipids, lipids, triglycerides, and fatty acids were from Sigma or Supelco, Bellefonte, PA. Iminodiacetic acid coupled to agarose and Buffalyte (pH 3.0-10.0) were purchased from Pierce Chemical Co., Rockford, IL. Carboxymethylcellulose (CM-52, preswollen) was from Whatman, England. Acrylamide, bisacrylamide, N,N,N′,N′-tetramethylethylenediamine, SDS, ammonium persulfate, and high-molecular-weight standard protein mixture were from Bio-Rad Laboratories, Rockville Centre, NY. LKB capillary gel casting (2117-701) and agarose-EF accessory (1850-100) kits, Isogel agarose, and ampholine carriers (pH 3.5-9.5) were obtained from LKB, Inc., Hicksville, NY. Ouchterlony immunodiffusion agarose plates were from Miles Laboratories, Elkhart, IN. Pevicon C-870 was from Accurate Chemical and Scientific Corp., Hicksville, NY. Thin-layer chromatography plates (TLC, Silica gel IB2) were obtained from J.T. Baker Chemical Co., Phillipsburg, NJ.

Cell Cultures

Mouse OTT-6050 EC cells were originally obtained from Dr. John Lehman, University of Colorado Medical Center, Denver. NRK cells (clone 49F) were generously supplied by Dr. Joseph DeLarco, Laboratory of Viral Carcinogenesis, NCI, Frederick, MD. EC cells were propagated in Eagle's minimal essential medium supplemented with 20 mM HEPES (pH 7.4), antibiotics, and 10% fetal calf serum as previously described [Salomon, 1980; Salomon et al., 1982a]. NRK cells (clone 49F) were maintained in HEPES-buffered Dulbecco's modified Eagle's medium (DMEM) supplemented with antibiotics and 10% fetal calf serum as previously described [DeLarco and Todaro, 1978]. Primary cultures of RME cells were obtained from isolated ductal and alveolar structures [Wicha et al., 1979; Salomon et al., 1981].

Assay for Mitogenic Activity

Growth-promoting activity was monitored by determining the changes in cell number of EC or RME cells after 2-3 days in culture in the presence of various fractions obtained during the purification steps. Briefly, EC cells (1×10^5) were seeded into 35-mm tissue culture dishes (3506, Costar, Cambridge, MA) previously coated with poly-D-lysine (high molecular weight, Sigma) or 10 μg per dish of type I or type IV collagens [Salomon, 1980; Salomon et al., 1981] in 2 ml of improved Eagle's minimal essential medium-zinc option (IMEM) supplemented with 4 mM L-glutamine, 20 mM HEPES (pH 7.4), 100 μg/ml streptomycin, 100 units/ml penicillin, epidermal growth factor (EGF, receptor grade, 10 ng/ml, Collaborative Research, Waltham, MA), bovine pancreatic insulin (Ins, 5 μg/ml, Sigma), human transferrin (Tf, 5 μg/ml, Sigma), and aliquots of the test material, which had been filter-sterilized through 0.22-μm Millex-GV (Millipore Corp., Bedford, MA) filter units. RME cells (1 × 10^5) were plated into 35-mm tissue culture dishes coated with 10 μg per dish of type IV collagen in 1.5 ml HEPES-buffered IMEM supplemented with Deutsch fetuin (1 mg/ml), EGF (10 ng/ml), Ins (5 μg/ml), Tf (5 μg/ml), and filter-sterilized aliquots of the test fractions. Duplicate cultures of EC and RME cells were removed from the dishes with 0.25% trypsin and counted with a ZBI Coulter particle counter.

Assay for Collagen-Stimulating Activity

Collagen production by EC, RME, or NRK cells was measured by the method of Peterkofsky and Diegelmann [1971], which monitors the amount of newly synthesized protein that is sensitive to digestion with pure, protease-free bacterial collagenase (type III, Advanced Biofractures, Lynbrook, NY). EC and RME cells synthesize type IV (basement membrane) collagen [Salomon et al., 1981, 1982a]. In the case of EC cells, 90% of the collagen is secreted into the medium. In contrast, NRK cells predominantly synthesize type I (stromal) collagen. Duplicate cultures of cells were grown for 2–4 days in IMEM supplemented with Deutsch fetuin (1 mg/ml), EGF (10 ng/ml), Ins (5 μg/ml), Tf (5 g/ml), and various test fractions. Cells were labeled with 2–5 μCi/ml of ^{14}C-proline (275 mCi/mmole, Amersham Corp., Arlington Heights, IL) or ^{3}H-lysine (48 Ci/mmole, Amersham) for the last 18–72 h of the culture period. The amount of labeled collagen in trichloroacetic acid-precipitable (TCA-precipitable) protein associated with the cells or medium was determined following digestion of the TCA-solubilized precipitate with bacterial collagenase. Incubations performed in the absence of collagenase served as controls. The radioactivity that was solubilized by collagenase was normalized against the radioactivity incorporated into total protein and expressed as collagen percentage of total protein (See Salomon et al. and Kidwell et al., this series).

Extraction and Thin-Layer Chromatography (TLC) Separation of Lipids and Phospholipids

Neutral lipids and phospholipids were extracted from 50 gm of the various fetuin preparations, which were dissolved in 1 ml of phosphate-buffered saline (PBS) (pH 7.4), with 20 ml of chloroform:methanol (2:1, vol./vol.) containing the antioxidant 0.005% butylated hydroxytoluene (BHT). The lipid extracts were mixed with 0.2 volumes of 0.74% KCl to separate the lipids from nonlipid material. The mixtures were centrifuged for 20 min at 2,400 rpm to achieve separation of the organic and aqueous phases. The upper, aqueous phase in each case was discarded and the lower, organic phase was evaporated to dryness under N_2. The dried extracts were reconstituted in 0.5 ml of chloroform:methanol (2:1, vol./vol.) containing known amounts of either 14C-labeled free fatty acid, triglyceride, or phospholipid to correct for recovery. Standard neutral lipids, free fatty acids, triglycerides, and phospholipids were cochromatographed with the samples on 20-cm × 5-cm silica gel IB2 TLC plates. The TLC plates were first developed in chloroform: methanol:water (65:30:5, vol./vol.) up to 4 cm from the bottom of the plates. The plates were then dried and developed in a second solvent system consisting of hexane:ethyl ether:acetic acid (80:20:1, vol./vol.). The first system separates phospholipids and the second resolves neutral lipids. The phospholipids, free fatty acids, and triglycerides were located on the TLC plates following fluorography. The appropriate spots were scraped from the plates and eluted from the silica gel with 10 ml chloroform:methanol (2:1, vol./vol.) containing 1% acetic acid in screw-capped centrifuge tubes. The tubes were shaken for 10 min and centrifuged for 15 min at 1,500 rpm. The solvent solutions, organic phase, were filtered through a fine-porosity sintered-glass funnel to remove any silica gel particles and subsequently were evaporated to dryness under N_2.

Fatty Acid (FA) Analysis by Gas-Liquid Chromatography (GLC)

The individual lipids and phospholipids were saponified and esterified with Meth-Prep II reagent (Applied Science Corp., State College, PA) for GLC analysis. Briefly, the lipids were dissolved in 50 μl of benzene in reacti-vials (Supelco) with 20 μl of Meth-Prep II reagent and reacted for 30 min at room temperature before injection into the GLC system. A Shimadzu GC-Minil gas chromatograph equipped with a hydrogen flame ionization detector coupled to a 50-m Silar 10 C-coated capillary column (Quadrex Corp., New Haven, CT) was utilized to resolve the fatty acids. The initial temperature of the column was 100 °C and the final temperature was 200°C, with a rate of temperature change of 8°C per minute and a flow rate of H_2 of 120 ml per minute. The area under each FA peak was determined with a Hewlett Packard integrator, model 3380A. Fatty acid mixtures (Applied Science) were used to standardize the column. TLC plates that were not scraped for FA analysis were sprayed with a solution of phosphoric acid (80%) and cupric acetate (20 ml:80 ml water containing 3 g of cupric acetate) and heated for 10 min at 130°C to visualize the lipid, phospholipid, and fatty acid bands.

Analytical Polyacrylamide Gel Electrophoresis (PAGE)

PAGE was performed according to the method of Laemmli [1970]. Protein samples (25–50 μg) were denatured at 100°C for 10 min in 25 mM Tris–190 mM glycine (pH 8.3) containing 1% SDS, 1 mM EDTA, and bromophenol blue tracking dye with or without 40 mM dithiothreitol (DTT). Cylindrical disk gels (60 mm) of either 5% or 7.5% acrylamide containing 0.1% SDS and a constant ratio of 3.3% bisacrylamide/total acrylamide monomer concentration were run at 3 mA per gel. The gels were stained for proteins with 0.05% R250 Coomassie brilliant blue in 25% isopropanol for 2 h and destained for 24 h in 7% acetic acid (Figs. 2 and 8).

Analytical and Preparative Isoelectric Focusing

Analytical isoelectric focusing (AIEF). Ampholytes (Ampholine, 1818, preblended) covering the pH range of 3.5 to 9.5 were utilized in horizontal slab gels of 1.8% IsoGel agarose-EF to determine the isoelectric points (pI). Protein samples (40 μg) were applied in 10-μl aliquots on Whatman filter paper strips to the agarose gels and focused at a constant voltage of 500 V for 30 min at 10°C in a water-cooled LKB Multiphor unit using acetic acid (0.5 M) and NaOH (0.5 M) at the anode and cathode, respectively. Isoelectric points were determined from the migration of a mixture of proteins (Isoelectric focusing kit, pH range 3.0–10.0, Pharmacia Fine Chemicals, Piscataway, NJ) of known pIs. Following the focusing, the gels were fixed in a TCA-sulfosalicylic acid mixture (10% TCA–1% sulfosalicylic acid) for 10 min, washed in 95% ethanol, stained with 0.5% R250 Coomassie brilliant blue for 5 min, and destained for 20 min in 10% acetic acid–30% ethanol.

Preparative isoelectric focusing (PIEF). PIEF was performed according to the method of Harpel and Kueppers [1980] with Pevikon C-870 as the support matrix. Briefly, protein samples (0.5–2 g) were dissolved in 58 ml of 10 mM Buffalyte (pH 3.0–10.0) and mixed with 111 g of Pevikon C-870. The horizontal gels were casted and focused at 1,200 V, 50 mA, and 10 W for 4 h at 10°C in an LKB Multiphor flat-bed template with 0.1 M acetic acid and 0.1 M NaOH at the anode and cathode, respectively. At the termination of the run, the Pevikon bed was fractionated with an LKB fractionating grid into thirty 1-cm sections, and the pH of each fraction was determined with an LKB surface-contact electrode. Each gel section was scraped and eluted with 2 ml of 0.2 M ammonium formate (pH 5.2) for 12 h at 0–4°C, lyophilized, and reconstituted in 1 ml of distilled water (Fig. 3).

Amino Acid Analysis

Amino acid (AA) composition was determined with a Beckman model 121 MB analyzer (AA-10 resin) equipped with a model 126 data system integrator. Protein samples (40–60 μg) were hydrolyzed under N_2 in 6 N HCl for 24 h at 110°C. Amino acids were resolved to baseline with the following program:

buffer 1, 0.2 M sodium citrate, pH 3.1, 9 min; buffer 2, 0.2 M sodium citrate, pH 3.5, 15.5 min; buffer 3, 0.2 M sodium citrate, pH 3.9, 12 min; buffer 4, 0.2 M sodium citrate, pH 4.95, 36 min [Salomon et al., 1982b].

Ouchterlony Double Immunodiffusion

Antisera prepared against Pedersen-, Deutsch-, or Spiro-prepared fetuins were obtained from immunized rabbits as previously described [Salomon et al., 1982b]. Immunological reactivity of these antisera against the fetuin preparations, bovine embryonin, and α_2M was measured by double immunodiffusion in 0.9% agarose plates after 48 h.

Isolation and Purification of Embryonin From Pedersen Fetuin

Carboxymethylcellulose chromatography (CMC). Four grams of Pedersen (type III) fetuin was dissolved in 40 ml of 10 mM ammonium formate (pH 5.2) and dialyzed against 3 liters of the same buffer for 24 h at 0–4°C. The dialyzed material was then centrifuged at 10,000 rpm at 4°C for 30 min to remove any insoluble protein. The clarified supernatant was then applied to a CM-52 column (4 × 20 cm) previously equilibrated at 0–4°C in 10 mM ammonium formate (pH 5.2). The column was washed extensively with 150 ml of 10 mM ammonium formate (pH 5.2) at 0–4°C until the absorbance at 280 nm of the eluent was less than 0.05 absorbance units. At this point, a 200-ml linear gradient of 0.01–1.0 M ammonium formate (pH 5.2) was initiated and 4-ml fractions were collected. Protein that eluted between a concentration of 0.05 and 0.1 M ammonium formate from the CMC column (CMC peak I, Fig. 4) was pooled and dialyzed against 3 liters of triple-distilled water for 24 h at 0–4°C. The dialyzed material was lyophilized and reconstituted in 0.2 M ammonium formate (pH 5.2) to a protein concentration of 20–30 mg/ml.

High-performance gel filtration chromatography (HPLC). Fifteen milligrams of the concentrated CMC-eluted protein was chromatographed on a 60-cm TSK-3000 SW HPLC gel filtration column (Altex, Beckman Instruments) that had previously been equilibrated in 0.2 M ammonium formate (pH 5.2) at 23°C. The column was eluted at a flow rate of 0.5 ml/min at 600 psi with a Waters HPLC chromatography pump equipped with a model 440 absorbance detector. The biologically active high-molecular-weight material (HPLC peak I, see Salomon et al. [1982b]) that eluted from the HPLC column was dialyzed against distilled water, lyophilized, and stored at −20°C in distilled water at a protein concentration of 2–4 mg/ml.

Metal (Zn) chelate affinity chromatography. One gram of Pedersen fetuin (type III) was dissolved in 10 ml of triple-distilled water and centrifuged at 10,000 rpm for 30 min and filtered through a 0.45-μm Millipore filter. Ten milliliters of agarose-iminodiacetic acid II was coupled to zinc by reaction of the resin with 100 ml of 0.2 M $ZnSO_4$ for 30 min at 23°C. The resin was subsequently washed with 300 ml of distilled water five times and equilibrated with 30 ml of 20 mM $NaPO_4$/0.5 M NaCl (pH 6.4) [Porath et al., 1975; Kurecki et al., 1979; Sottrup-Jensen et al., 1980]. The crude fetuin solution was applied to the column and the column was washed with the equilibration buffer until the absorbance at 280 nm was less than 0.05 absorbance units. Bound material was eluted with 50 ml of 0.2 M NaEDTA (pH 7.0) (Fig. 7).

Affinity-dye binding chromatography. Blue Sepharose CL-6B (Pharmacia) contains the dye Cibacron blue F3G-A, cross-linked to Sepharose 6B. Virca et al. [1978] have utilized this affinity resin to purify α_2M from human plasma in 50 mM Tris-HCl (pH 8.0). Under these conditions, most serum proteins including albumin are strongly absorbed to the resin, whereas α_2M remains unbound and elutes in the first several fractions and lipoproteins and immunoglobulins are retarded in their elution. This procedure, like zinc-chelate chromatography, can be utilized to isolate bovine α_2M (embryonin) from commercial Pedersen fetuin preparations in a single step.

ANALYSIS OF FETUIN PREPARATIONS AND ISOLATION OF EMBRYONIN (α_2M)

Pedersen fetuin (type III or IV), but not Deutsch or Spiro fetuin preparations, is active in maintaining EC or RME cell growth in serum-free, hormone-supplemented media [Salomon, 1980; Salomon et al., 1981]. In the presence of EGF, insulin, and transferrin, maximum stimulation of EC (Fig. 1) or RME cell growth (data not shown) occurs at a concentration of 0.5–1.0 mg/ml of the crude Pedersen fetuin preparation. Analysis of these commercially available preparations by SDS-PAGE following denaturation and reduction (Fig. 2) demonstrates the heterogeneity of Pedersen fetuin (Fig. 2a) compared to Spiro fetuin (Fig. 2c). Two major doublet bands of approximate Mr of 60,000 and 66,000 are present in both preparations. In addition, the Pedersen preparation possesses three minor, high-molecular-weight bands at 185,000, 116,000, and 85,000. These three minor protein components account for less than 5% of the total protein. To determine if these species could be resolved from fetuin because of differences in their isoelectric points, Pedersen fetuin was fractionated by PIEF (Fig. 3). Analysis of the three protein peaks by SDS-PAGE (data not shown) demonstrated that peak A represents fetuin (pI 3.5–4.0, Mr 60,000 and 66,000) whereas peak B (pI 4.5–5.5) contains a mixture of fetuin and the 185,000 Mr component. Peak C (pI 6.8–7.8)

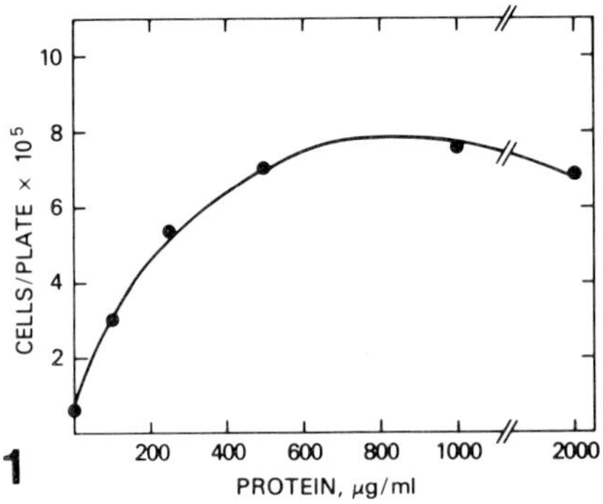

1

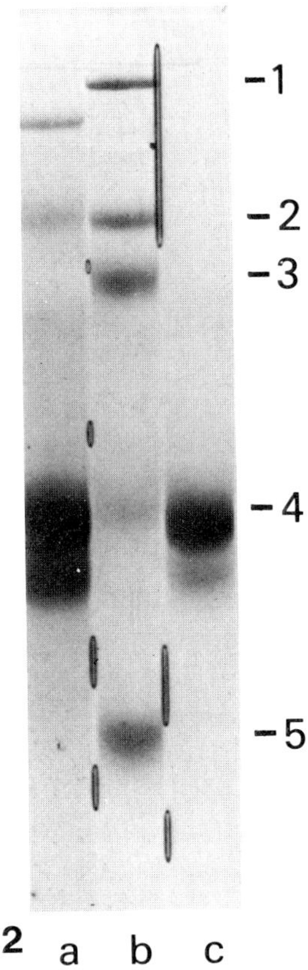

2

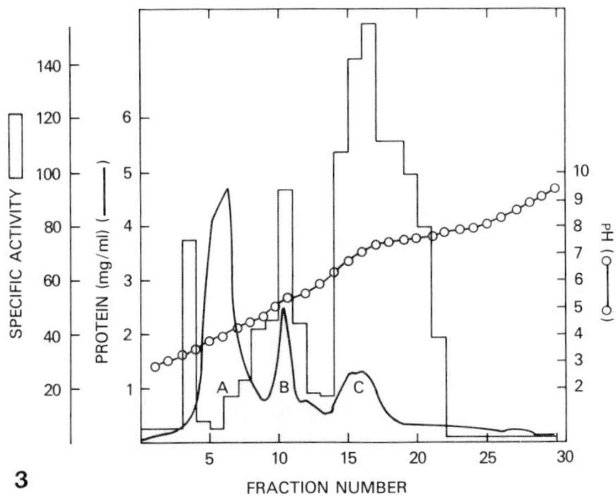

3

Fig. 1. Effect of Pedersen fetuin on EC cell growth. EC cells (2 × 10^5) were plated on polylysine-coated dishes in IMEM supplemented with EGF (10 ng/ml), Ins (5 µg/ml), Tf (5 µg/ml), and different concentrations of fetuin. Cells were grown for 2 days, trypsinized, and counted. All points represent the average of duplicate determinations. Data reproduced from Salomon [1980].

Fig. 2. SDS-PAGE (7.5% gel) of commercial fetuin preparations. Pedersen fetuin (50 µg, gel a), Spiro fetuin (25 µg, gel c), and high-molecular-weight protein standards (25 µg, gel b) were reduced with 40 mM DTT and denatured at 100°C for 10 min. Bands 1-5 represent myosin (1, Mr = 200,000), β-galactosidase (2, Mr = 116,250), phosphorylase B (3, Mr = 92,500), bovine serum ablumin (4, Mr = 66,200), and ovalbumin (5, Mr = 45,000). Data reproduced from Salomon et al. [1982b].

Fig. 3. PIEF profile of Pedersen fetuin. Pedersen fetuin (0.5 g) was dissolved in 10 mM Buffalyte (pH 3-10) and focused for 4 h at 10°C. The Pevicon C-870 gel was then fractionated, the pH (○—○) determined, the gel eluted, and the protein (———) concentration determined in each fraction. The specific activity (gray area) for growth stimulation of EC cells represents the multiple by which growth was stimulated by each fraction over the growth produced by crude Pedersen fetuin (1 mg/ml) normalized against the protein concentration in each fraction. Data reproduced from Salomon et al. [1982b].

contains a mixture of the 185,000, 116,000, and 85,000 species. The bulk of the growth-promoting activity is associated with peaks B and C in pH regions above 4.5.

Although the biological activity could be separated from fetuin by PIEF, the recovery was variable and ranged between 40% and 50%. Moreover, the resolution capabilities of PIEF are generally limited to samples containing less than 1 g of protein. For these reasons, preparative cation exchange chromatography with a CMC resin was selected as a potential method for the separation of gram quantities of the Pedersen preparation into its respective components (Fig. 4). CMC chromatography was selected in preference to DEAE chromatography because at pH 5.2 and under low-salt conditions fetuin does not absorb to the CMC resin whereas other contaminants are bound [Marti et al., 1973]. These components were sequentially eluted from the resin with a linear ammonium formate gradient at pH 5.2 and were arbitrarily subdivided into four regions. Regions I and II stimulated the growth of EC cells in a dose-dependent manner (data not shown; see Salomon et al. [1982b]) with region I possessing greater activity than region II. Maximal growth stimulation by region I occurred at a concentration of protein between 100 and 200 μg/ml in serum-free, hormone-supplemented medium on poly-D-lysine-coated dishes. The flow-through fractions and regions III and IV were biologically inactive. Region I contained equal amounts of fetuin and the 185,000 Mr protein following reduction and SDS-PAGE analysis with minor quantities of the 116,000 Mr species. The total protein recovered in region I starting from a total of 4 g of the initial Pedersen fetuin preparation ranged

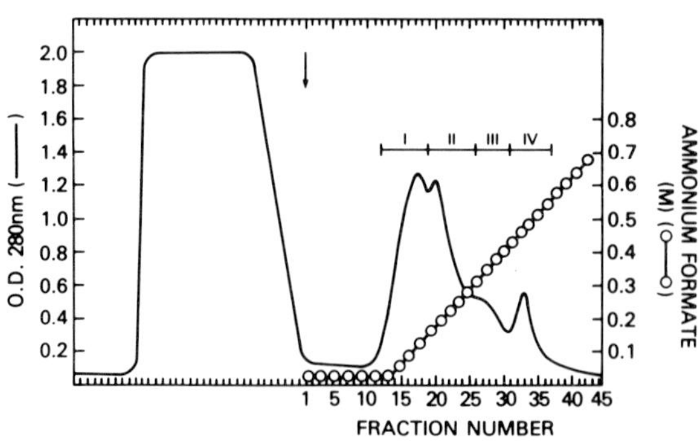

Fig. 4. CMC chromatography profile of Pedersen fetuin. Fetuin (4 g) was applied to a CMC-52 column in 10 mM ammonium formate (pH 5.2). The column was washed and eluted with a linear gradient (O—O) of ammonium formate (pH 5.2). Data reproduced from Salomon et al. [1982b].

from 40 to 50 mg. This material was dialyzed against distilled water, lyophilized, and reconstituted in 1 ml of 0.2 M ammonium formate (pH 5.2). Region I from the CMC column was further purified by gel filtration on an HPLC system using a 60-cm TSK-3000 SW ultrafiltration column that has an exclusion limit of 300,000 (data not shown; see Salomon et al. [1982b]). The majority of the growth-promoting activity was associated with a high-molecular-weight fraction (by SDS-PAGE analysis 270,000 and following reduction, 185,000; see Fig. 8) that eluted (21–24 min retention time) just after the void volume (18 min retention time). An additional peak (II) also possessed activity (Fig. 5). Maximal stimulation of EC cell growth occurred at a concentration of 2-4 μg/ml of protein on either poly-D-lysine-coated or type IV collagen-coated dishes (Fig. 5). The 185,000 Mr protein was tentatively classified as embryonin but, as will be demonstrated, this protein physiochemically and immunologically resembles α_2-macroglobulin of either bovine or human origin. Approximately 4–12 mg of embryonin are recovered at this stage of purification, or 0.1% to 0.3% of the total protein in the initial preparation assuming complete recovery, which is probably not the case. This would be equivalent to a 100- to 200-fold purification (compare Fig. 1 to Fig. 5).

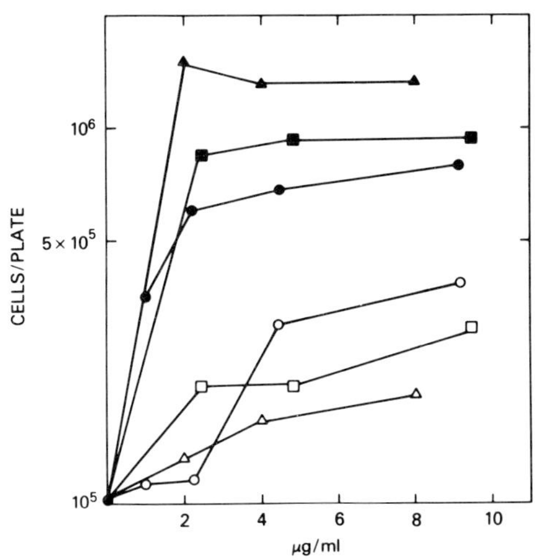

Fig. 5. Dose-response curves for HPLC peaks I (embryonin) and II and human α_2M on EC cell growth. Various amounts of protein from HPLC peak I (▲, △), peak II (■, □), or α_2M (●, ○) were added to EC cell cultures (10^5 cells per dish). Cells were grown for 3 days on polylysine-coated (closed symbols) or type IV collagen-coated (open symbols) dishes before determination of cell numbers. Data reproduced from Salomon et al. [1982b].

Alternative Fractionation Procedures for the Purification of Embryonin From Pedersen Fetuin

Gel filtration HPLC. Embryonin can be directly isolated from the crude commercial Pedersen fetuin preparations by direct chromatography on the HPLC ultrafiltration column (Fig. 6). However, the yields tend to be low and the amount of material that can be applied to the column is limited to approximately 20 mg of protein. Figure 6 demonstrates the elution profiles for Pedersen (A, 2.5 mg), Deutsch (B, 5 mg), and Spiro (C, 5 mg) fetuin preparations. The presence of peak I (embryonin, 21–24 min retention time) in the Pedersen preparation and its apparent absence in the Deutsch and

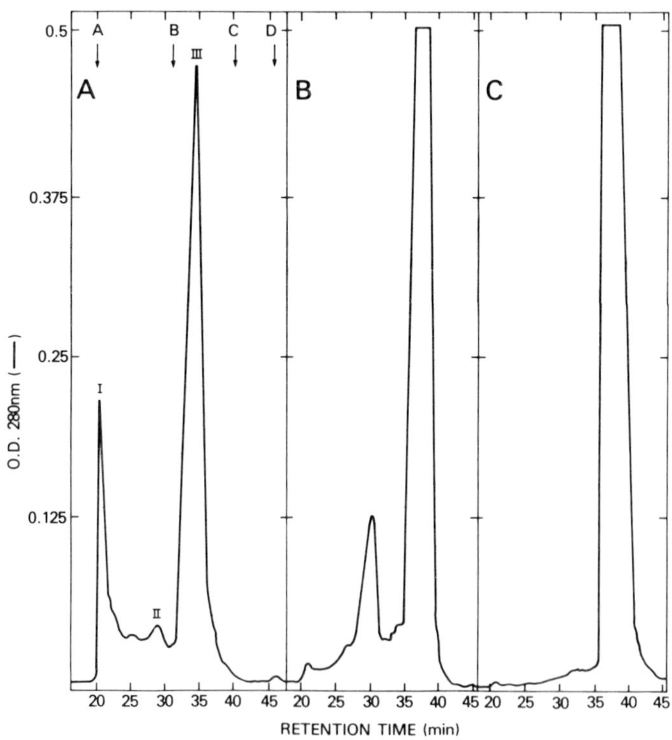

Fig. 6. Gel filtration, HPLC profiles of commercial fetuin preparations. Pedersen (2.5 mg, A), Deutsch (5 mg, B), or Spiro (5 mg, C) fetuins were dissolved in 0.2 M ammonium formate (pH 5.2) and chromatographed on a TSK-3000 SW ultrafiltration HPLC column (60 cm) at a flow rate of 0.5 ml/min. Elution positions of ferritin (A, Mr = 450,000), aldolase (B, Mr = 158,000), albumin monomer (C, Mr = 45,000), and cytochrome C (D, Mr = 12,500).

Spiro fractions lends further support to the notion that the biological activity resides in this species, as other regions of the profiles between these preparations are similar. Material eluting between 32 and 40 min represents fetuin in all the preparations as determined by SDS-PAGE.

Zinc-chelate affinity chromatography. Since embryonin probably represents bovine α_2M (see below), metal-chelate affinity chromatography was tested as a potential method for separating the growth-promoting activity (embryonin) directly from Pedersen fetuin. This approach was selected because α_2M binds zinc [Starkey and Barret, 1977] and can be purified directly from human plasma in large amounts by zinc-chelate affinity chromatography [Kurecki et al., 1979; Sottrup-Jensen et al., 1980]. Figure 7 demonstrates that fetuin fails to bind to a Zn-agarose column, whereas the α_2M (embryonin) contaminant is absorbed and can be eluted with EDTA. From 1 g of crude fetuin, approximately 1 mg of α_2M can be recovered by this procedure.

PHYSIOCHEMICAL, IMMUNOLOGICAL, AND BIOLOGICAL PROPERTIES OF FETUIN, EMBRYONIN, AND α_2M

Physiochemical Properties of Fetuin, Embryonin, and α_2M (Table II)

The high-molecular-weight fraction that was collected from the HPLC column was hydrolyzed with 6N HCl and the total amino acid composition was determined assuming a subunit molecular weight of 185,000 daltons. The amino acid composition of embryonin was compared to purified Spiro-prepared fetuin [Spiro and Spiro, 1962] and bovine [Nagasawa et al., 1970] and human α_2M [Swenson and Howard, 1979]. The compositions of embryonin and human α_2M are very similar (data not shown; see Salomon et al. [1982b]) with respect to the amount of glycine, histidine, lysine, methionine, serine, threonine, and valine. In contrast, fetuin lacks methionine and it has a

TABLE II. Physiochemical Properties of Fetuin and α_2-Macroglobulin

	Fetuin	α_2-Macroglobulin
$S°_{20,w}$	3.4	20.6
Carbohydrate content	28%	8.6%
Molecular weight	48,500[a]	740,000
Isoelectric point (pI)	3.2-5.2	5.4
$E^{1\%}_{1cm}$	4.10	9.15
Concentration in serum	10-20 mg/ml (fetal)[b]	
	0.4 mg/ml (adult)	2 mg/ml (adult)

[a]Core protein 38,300. Three carbohydrate chains each of ~ 3,400 with a maximum of four sialic acid residues/chain.
[b]Bergmann et al. [1962] and Lai et al. [1981].

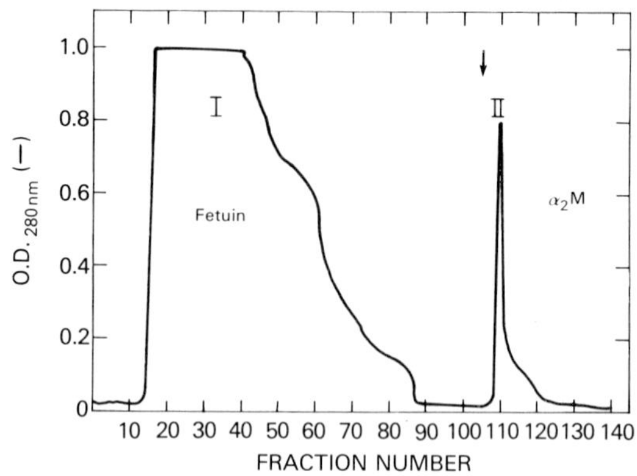

Fig. 7. Zinc-chelate affinity chromatography profile of Pedersen fetuin. Fetuin (1 g) was dissolved in 1 ml distilled water and applied to a Zn-agarose column. The column was washed with 20 mM $NaPO_4$/0.5 M NaCl (pH 6.4) until the absorbance was reduced and bound protein was eluted (arrow) with 0.2 M NaEDTA (pH 7.0).

higher content of aspartic acid and proline and a lower content of tyrosine. Moreover, following 5% SDS-PAGE analysis without reduction (Fig. 8), embryonin and human α_2M both have a molecular weight of approximately 270,000-300,000 (Fig. 8, a and c) and both are glycosylated (data not shown). Following reduction with DTT (Fig. 8, b and d), a major species of 185,000 Mr is present in both samples, with two minor species of apparent molecular weights of 116,000 and 68,000. These latter two peptides are probably generated by cleavage of the 185,000 Mr subunits owing to heat denaturation in the presence of SDS and DTT [Howard et al., 1980]. Native α_2M has a molecular weight of 620,000-820,000 and can be denatured by acid or urea into two protomers having a molecular weight of 310,000-370,000 (Fig. 9). Each protomer in turn consists of two 185,000 subunits linked by disulfide bridges [Dunn and Spiro, 1967; Jones et al., 1972; Harpel, 1973; Starkey and Barret, 1977; Swenson and Howard, 1979]. Interaction of α_2M with protease leads to the cleavage of the 185,000 subunits to 85,000 Mr peptides [Harpel, 1973; Werb et al., 1974; Howard et al., 1980; Howard, 1981], the predominant species in region III of the CMC eluate [Salomon et al., 1982b], suggesting that the α_2M detected in commercially prepared fetuin preparations may be partially denatured owing to the presence of enzymes, such as thrombin or plasmin, that are activated during the preparation of serum from plasma [Harpel, 1973].

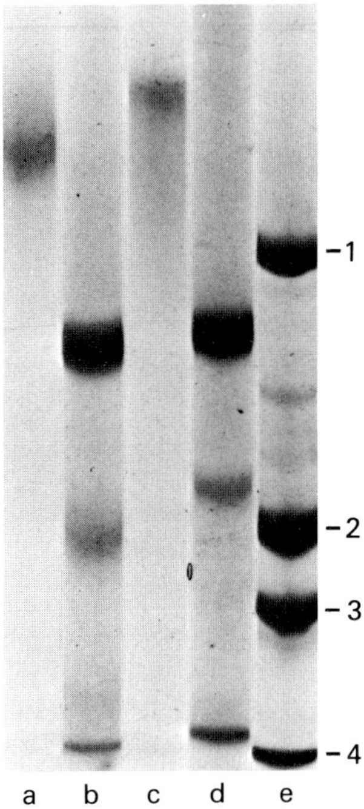

8

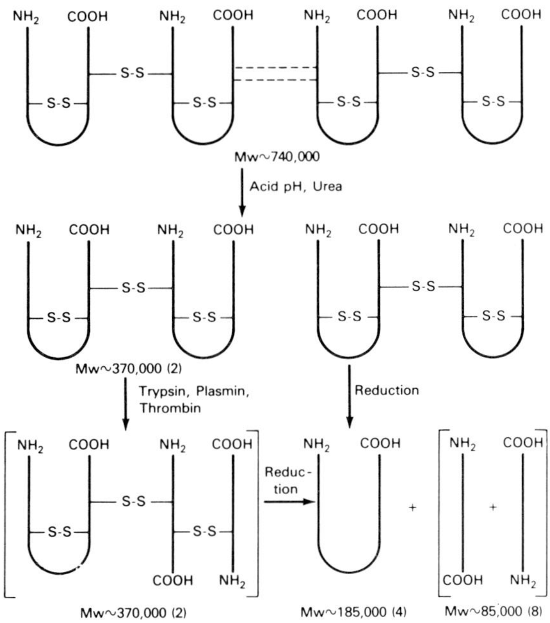

9

Fig. 8. SDS-PAGE (5.0% gels) of embryonin (HPLC peak I) and human α_2M. Protein samples (50 μg) were heat-denatured in the absence (gels a and c) or presence (gels b and d) of 40 mM DTT for 10 min at 100°C. Gels a and b represent embryonin and gels c and d represent human α_2M. Gel e, bands 1-4, represents myosin, β-galactosidase, phosphorylase B, and bovine serum albumin, respectively. Data reproduced from Salomon et al. [1982b].

Fig. 9. Proposed subunit structure of human α_2M. Native molecule (Mr = 740,000) consists of two protomers (Mr = 370,000) that are noncovalently linked. Following reduction of the inter- and intrachain disulfide bridges, each protomer consists in turn of two polypeptide subunits (Mr = 185,000). Interaction of proteases with α_2M leads to the formation of eight cleavage fragments (Mr = 85,000), two fragments being derived from each 185,000 subunit. Model reproduced from Harpel [1973], with permission.

The similarities in the electrophoretic patterns of embryonin and human α_2M and their amino acid composition suggest that embryonin may actually represent bovine α_2M [Nagasawa et al., 1970]. Neither protein resembles fetuin in these and other physiochemical properties (Table II). Moreover, analytical isoelectric focusing demonstrated that both embryonin and human α_2M have a similar pI of approximately 5.4. Rotary shadowing of commercially prepared human α_2M and embryonin (L.A. Liotta, personal communication) demonstrated that both molecules have the following shape: ш , probably representing the protomeric subunits. Native undenatured α_2M resembles the cyrillic letter Ж [Bloth et al., 1968; Starkey and Barret, 1977].

Immunological Properties of Fetuin, Embryonin, and α_2M

Embryonin and human α_2M are further related in their antigenic properties (data not shown; see Salomon et al. [1982b]) and are distinct from purified fetuin. Polyclonal antibodies raised against Pedersen fetuin produce a strong precipitin reaction with either the crude fetuin preparation or purified embryonin, and a weak precipitin band against human α_2M. Antiserum raised against human α_2M fails to react with the crude fetuin preparation but exhibits a weak reaction with embryonin. In contrast, antibodies raised against Spiro fetuin, which lacks any embryonin, show only a single diffuse precipitin band with the Pedersen preparation and no reaction with either embryonin or human α_2M.

Biological Properties of Fetuin, Embryonin, and α_2M (Table I)

Cell growth. Both embryonin and α_2M stimulate EC cell growth on either polylysine-, type I collagen- (data not shown), or type IV collagen-coated dishes (Fig. 5). In both instances, growth was more pronounced on polylysine-coated dishes. α_2M stimulated EC cell growth to the same extent and over the same concentration range as embryonin. RME cell growth was also stimulated (25–50% increase over controls) by either embryonin or α_2M in serum-free, hormone-supplemented medium (data not shown). The magnitude of growth promotion by embryonin or α_2M at 2–5 μg/ml is equivalent to that observed with 0.5–1.0 mg/ml of the crude Pedersen preparation. Pedersen fetuin has been routinely utilized as a medium supplement to propogate cells in serum-free, hormone-supplemented media [Rizzino and Sato, 1978; Barnes and Sato, 1980; Salomon, 1980; Salomon et al., 1981]. Since embryonin accounts for approximately 0.1–0.3% of the total protein in the initial fetuin prepration, the concentration present in 0.5–1 mg/ml of the Pedersen preparation is within the concentration range needed to stimulate EC cell growth [Rizzino and Sato, 1978; Salomon, 1980].

α_2M from human, horse, and bovine sera have been demonstrated to possess growth-promoting activity either alone or in combination with other proteins associated with the α-globulin fraction of serum [Marr et al., 1962; Tozer and Pirt, 1964; Healy and Parker, 1966; Temin et al., 1972; Baudysova and Michl, 1981]. Marr et al. [1962] demonstrated that two glycoproteins could be isolated from fetal calf serum, fetuin, and an α-macroglobulin fraction, which were active together in stimulating the growth of HeLa cells. Subsequently, Healy and Parker [1966] demonstrated that horse α_2M in combination with a low-molecular-weight α_1-acid glycoprotein from calf serum (Mr = 40,000-50,000) could synergistically enhance the growth of mouse embryonic fibroblasts. Since α_2M is known to bind to thrombin, insulin, growth hormone, and nerve growth factor (NGF), it is still unclear whether the growth-promoting properties are due to α_2M per se or some other smaller growth factor or hormone that is associated with this macromolecule [Collipp et al., 1964; Hadden and Prout, 1964; Hoffmann et al., 1973; Ronne et al., 1979; Maeda et al., 1980]. Nevertheless, the demonstration that a variety of fibroblast cell lines possess cell surface receptors for α_2M (approximately 600,000 sites per cell), and that the α_2M-receptor complex is internalized and processed in a manner analogous to other growth factor- and hormone-cell surface receptor complexes, suggests that α_2M or some component bound to α_2M may function to regulate cell proliferation [Dickson et al., 1981; Van Leuven et al., 1979, 1981]. Moreover, α_2M is a broad-spectrum protease inhibitor (see below) [Harpel, 1973; Werb et al., 1974; Starkey and Barret, 1977]. In fact, α_2M that is complexed to proteases such as thrombin, plasmin, trypsin, or collagenase is bound to and internalized by cells at a significantly higher rate than uncomplexed, native α_2M [Van Leuven et al., 1979, 1981].

Collagen production and protease inhibition. The mechanism(s) by which embryonin and α_2M function as mitogens is unknown. However, several growth factors including EGF, platelet-derived growth factor, mammary tumor factor, α-fetoprotein (AFP), and prostaglandin E_1 (PGE_1) differentially stimulate collagen production in cells that mitogenically respond to these agents [Salomon et al., 1981; Stavenow et al., 1981; Bano et al., 1982]. Figure 10 demonstrates that both embryonin and α_2M differentially stimulate collagen production in mouse EC cells in a dose-dependent manner and over a concentration range that stimulates cell growth. Embryonin or α_2M also produce a twofold to ninefold increase in the percentage of total labeled collagen in either RME or NRK cells [Salomon et al., 1982b]. This differential increase in collagen production is not selective for enhancing the production of a particular collagen, since NRK cells synthesize type I (stromal) collagen, whereas EC and RME synthesize type IV (basement membrane) collagen

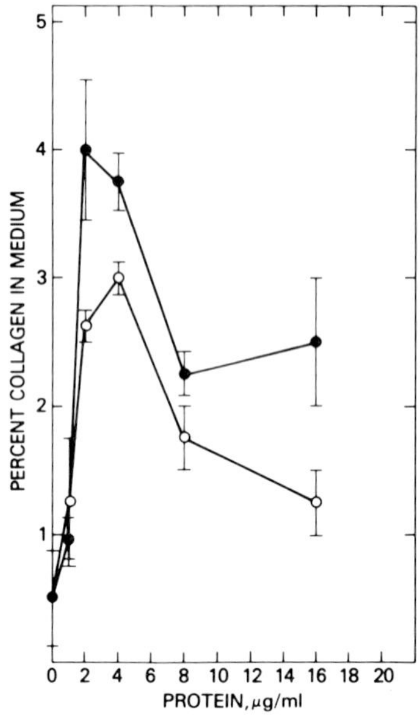

Fig. 10. Dose-response curve for type IV collagen production by embryonin and human α_2M in EC cells. EC cells were cultured in serum-free medium for 48 h in the absence or presence of either embryonin (●—●) or α_2M (○—○). The cells were labeled with ^{14}C-proline (10 μCi/ml) for 18 h, and the percentage collagen in the medium was determined. Data reproduced from Salomon et al. [1982b].

[Salomon et al., 1981; Kidwell et al., 1982; Salomon et al., 1982a]. The mitogenic response to several different growth factors in various cell types is related to the ability of the cells to interact with and deposit an appropriate extracellular matrix consisting of collagen, laminin, entactin, fibronectin, hyaluronic acid, and other proteoglycans [Salomon et al., 1981; Gospodarowicz and Tauber, 1980; Salomon et al., 1982a] (also see Salomon et al., this series). Factors that accentuate or attenuate the synthesis and/or turnover of these matrix components may serve to modulate the proliferative response of cells to various hormones or growth factors [Gospodarowicz and Tauber, 1980]. For example, EGF and glucocorticoids function as mitogens in RME cells by either enhancing type IV collagen synthesis or suppressing the appearance of a specific type IV collagenase, respectively [Salomon et al., 1981]. α_2M functions as an irreversible, broad-spectrum protease inhibitor for such diverse enzymes as trypsin, chymotrypsin, plasmin, thrombin, elastase, and collagenases [Harpel, 1973; Werb et al., 1974, 1977; Starkey and Barret, 1977]. Likewise, fetuin has been demonstrated to be a reversible inhibitor of serine proteases such as trypsin and to inhibit the secretion of collagenase by synovial fibroblasts [Galembeck and Cann, 1974; Werb and Aggeler, 1978]. Moreover, fetuin, like α_2M can inhibit plasmin acitivity [Rohrlich and Rifkin, 1981]. This latter effect by either protein may be significant in regulating collagen turnover, since various mammalian collagenases exist in a latent inactive form requiring proteolytic cleavage via plasmin or trypsin for conversion to an active form [Werb et al., 1974, 1977; Liotta et al., 1981b; Salomon et al., 1981; Liotta et al., 1982; Salomon et al., 1982a]. Likewise, plasmin and thrombin have been demonstrated to be involved in the degradation of laminin [Liotta et al., 1981a]. Therefore, both α_2M and fetuin may control the turnover of various matrix components via their ability to function as protease inhibitors.

Modulation of capillary endothelial cell migration. Neovascularization occurs during several normal and pathophysiological processes such as osteogenesis, luteinization, chronic inflammation, and tumor development [Zetter, 1980]. The growth of solid tumors is dependent upon the acquisition of nutrients from the host by connection of the host to the tumor through a capillary network of blood vessels [Folkman, 1974]. One of the earliest events during neovascularization is the migration and proliferation of capillary endothelial cells [Ausprunk and Folkman, 1977]. Endothelial cells within the capillary bed proliferate and form new branches or sprouts. These branches eventually are directed toward the tumor presumably by the secretion of diffusible, angiogenic factors by the tumor [Folkman, 1974]. The directional migration of capillary endothelial cells in response to factors derived from a variety of tumor (chemotaxis) has been demonstrated in vivo and in vitro [Schor et al., 1980; Zetter, 1980; Gullino, 1981; Ziche et al., 1982]. Factors isolated from nonneoplastic tissues have also been demonstrated to be chemoattractants for capillary endothelial cells [Folkman, 1974; Gullino, 1981]. The directional migration of endothelial cells in response to these various agents occurs along a concentration gradient [Ausprunk and Folkman, 1977]. The extracellular matrix functions as an important determinant during neovascularization and subsequent tumor growth [Gospodarowicz and Tauber, 1980; Schor et al., 1980]. In fact, degradation of the endothelial basement membrane occurs prior to and during migration of the capillary endothelial cells [Ausprunk and Folkman, 1977]. Since α_2M is an effective collagenase inhibitor and since various synthetic protease inhibitors block tumor cell migration in vitro [Thorgeirsson et al., 1982], α_2M either alone or in combination with a capillary endothelial cell chemoattractant was tested on the chemokinesis and chemotaxis of bovine adrenal capillary endothelial cells (BACE) in vitro by the

Boyden chamber migration assay. This assay has been demonstrated to be a reliable indicator for monitoring capillary endothelial cell migration in response to chemoattractants [Alessandri et al., 1983]. α_2M inhibited in a dose-dependent fashion the random migration of cells (chemokinesis) across the filter in the absence of any chemoattractant (Table III). Likewise, when chemoattraction of endothelial cells was induced with a PGE_1-primed rabbit corneal extract [Ziche et al., 1982], α_2M attenuated the chemotaxis of BACE cells. A 50% inhibition of chemokinesis and chemotaxis by α_2M occurred at a dose of 12 μg per assay. Capillary endothelial cells secrete collagenases in response to chemoattractants [Kalebic et al., 1983]. These collagenases degrade type IV (basement membrane) and type V collagens, which are constituents of the vascular basement membrane. If one accepts the hypothesis that mobilization

TABLE III. Effect of α_2-Macroglobulin on the Migration of Capillary Endothelial Cells

Treatment	Dose (μg/assay)	Cells/field	% of control
A. Chemokinesis			
DMEM + 1% CS	—	23 ± 3	100
α_2M	4	31 ± 1	134
	20	14 ± 5	60
	40	9 ± 3	39
B. Chemotaxis			
PGE_2 corneal extract	0.6	15 ± 4	100
(E_2)	0.6	75 ± 12	500
PGE_1 corneal extract	4	69 ± 8	460
(E_1)	12	33 ± 7	220
E_1 extract + α_2M	40	10 ± 5	66

Boyden chamber assay is based upon the migration of cells across a filter against a concentration gradient of a putative chemoattractant. The upper chamber has a 200-μl capacity, the lower chamber a capacity of 40 μl. The two wells are separated by a 5-μ PVP Nucleopore filter, 13 mm in diameter and precoated with gelatin. For measurement of chemokinesis, the upper wells received 10^5 BACE cells in 200 μl of DMEM + 1% calf serum (CS); the lower wells received 40 μl of DMEM + 1% CS and different concentrations of α_2M. For chemotaxis, the lower wells also received a fixed concentration of a chemoattractant (PGE_1 rabbit corneal extracts) in the absence or presence of different concentrations of α_2M. After a 3-4 h incubation at 37°C in 5% CO_2-95% air, the cells on the upper surface of the filters were removed with a cotton swab. The cells that had migrated to the lower surface were fixed in 80% methanol and stained with Wright's stain and counted in five different fields for each filter. Extracts of PGE_2 corneas served as controls. Each determination is the average of triplicate samples in two separate experiments $\pm$ SD.

of the capillary endothelium in vitro by chemoattractants is an accurate reflection of their angiogenic activity in vivo, then α_2M may attenuate angiogenesis by its ability to inhibit the activity of type IV and type V collagenases and thereby perturb endothelial degradation of the capillary basement membrane and subsequent migration.

Fatty acid carrier function. Serum components such as albumin and AFP can function as carrier or transport proteins for fatty acids [Ham, 1963; Ashbrook et al., 1975; Parmelee et al., 1978], which may be relevant to the ability of these two proteins to affect cell growth and other physiological responses such as lymphocyte blastogenesis [Kelly and Parker, 1979]. In this respect, unsaturated but not saturated fatty acids can stimulate cell proliferation [Ham, 1963; Swim, 1967; Bailey and Dunbar, 1973; Higuchi, 1973; Kidwell et al., 1978; Wicha et al., 1979]. This is particularly true for linoleic, oleic, and arachidonic acids [Ham, 1963; Higuchi, 1973; Kidwell et al., 1978; Kelly and Parker, 1979; Wicha et al., 1979]. For these reasons, the various commercial fetuin preparations, embryonin, and α_2M were analyzed for the presence of neutral lipids and phospholipids by TLC and the FA composition was determined by GLC (Table IV). Pedersen fetuin has quantities of free cholesterol, cholesterol esters, triglycerides, free fatty acids, and phospholipids that can be detected as discrete bands following extraction with chloroform:methanol and analysis by TLC. In contrast, the more homogeneous Deutsch and Spiro fetuin preparations lack any neutral lipids and phospholipids but do contain detectable amounts of free fatty acids. Quantitatively, the total amount of fatty acids in the Pedersen fetuin preparations is generally higher than in the Deutsch and Spiro fractions. Moreover, the free fatty acids present in these latter two preparations are predominantly the saturated fatty acids, palmitic and stearic acids, whereas the Pedersen preparations contain lesser amounts of palmitic acid but contain in addition the unsaturated fatty acids, palmitoleic, linoelic, and arachidonic acids. These latter three fatty acids are probably derived from cholesterol esters or phospholipids, which include predominantly phosphatidylcholine (PC), phosphatidylethanolamine (PE), and sphingomyelin, as determined by TLC analysis. Purified human α_2M contains no detectable lipids, whereas bovine embryonin contains PC, PE, and cholesterol esters. It is therefore possible that some of the growth-promoting acitivity associated with the Pedersen fetuin preparations and bound bovine α_2M (embryonin) in this fraction may relate to the presence of these unsaturated fatty acids and cholesterol.

CONCLUSIONS

It was the purpose of this chapter to describe the isolation of one of the growth-promoting activities that is associated with commercially prepared

TABLE IV. Total Fatty Acid (FA) and Cholesterol Composition of Fetuin Preparations

Preparation	Pedersen	Deutsch	Spiro
μg of FA or cholesterol/mg protein			
FA[a]	3.4	0.65	2.36
Cholesterol	3.7	0	0
FA analysis of fetuin preparations (%)[b]			
Palmitic (16:0)	34.1	60.0	60.0
Palmitoleic (16:1)	7.9	0	0
Stearic (18:0)	21.5	20.0	20.0
Oleic (18:1)	20.6	20.0	20.0
Linoleic (18:2)	12.6	0	0
Arachidonic (20:4)	3.3	0	0
Neutral lipid, triglyceride, free fatty acid, and phospholipid composition[c]			
Cholesterol	+	Trace	−
Cholesterol esters	+	−	−
Triglycerides	+	−	−
Free fatty acids	+	+	+
Phospholipids[d]	+	−	−

[a]Total fatty acids (FA) determined from neutral lipid, triglyceride, free fatty acid, and phospholipid fractions.
[b]Individual fatty acids (total) determined following separation by GLC and expressed as a percentage of the total amount of FA in each preparation.
[c]Presence (+) or absence (−) of visible spots in each fraction following TLC in the two-solvent systems as described in text.
[d]Phospholipids consisted primarily of phosphatidylcholine, phosphatidylethanolamine, or sphingomyelin.

Pedersen fetuin. The physiochemical, immunological, and biological characteristics of this glycoprotein, which we have tentatively named embryonin, suggest that it strongly resembles bovine and human α_2M. The ability of α_2M to function as a broad-spectrum protease inhibitor may be relevant to some of the other biological properties ascribed to this protein. Both α_2M and possibly fetuin may control the level of extracellular proteases by sequestering them for cellular uptake [Van Leuven et al., 1979; Baker et al., 1980; Low et al., 1981; Rohrlich and Rifkin, 1981]. Likewise, the generation of specific cleavage peptides from either fetuin or α_2M following their interaction with extracellular or ectopic cell surface proteases could serve to generate localized growth factors [Hubbard, 1978; Low et al., 1981]. It is not known whether the growth-promoting or collagen-stimulating activities are associated with the native molecules of embryonin or α_2M or with smaller polypeptides or growth factors that are present in, bound to, or derived from, embryonin and α_2M. In vivo α_2M is known to regulate the activity of extracellular and plasma-associated proteases during clotting and inflammation [Harpel, 1973; Starkey and Barret, 1977]. In a variety of pathophysiological situations such as rheumatoid arthritis, hepatic cirrhosis, diabetes, and nephrotic syndrome the plasma levels of α_2M are elevated, suggesting that this protein may also subserve some physiological function in these states. Moreover, the ability of α_2M to inhibit neovascularization may be important in regulating angiogenic response during tumor formation. It should be stressed that although α_2M is a contaminant of Pedersen fetuin, other proteins are also present in this crude preparation, suggesting that this fraction actually represents a composite of fatty acids, phospholipids, and several different proteins, each possessing discrete biological activities (Table I).

REFERENCES

Alessandri G, Raju K, Gullino PM (1983): Mobilization of capillary endothelium in vitro induced by effectors of angiogenesis in vivo. Cancer Res 43:1790-1797.

Ashbrook JJ, Spector AA, Santos EC, Fletcher JE (1975): Long chain fatty acid binding to human plasma albumin. J Biol Chem 250:2333-2338.

Ausprunk DH, Folkman J (1977): Migration and proliferation of endothelial cells in preformed and newly formed blood vessels during tumor angiogenesis. Microvasc Res 14:53-65.

Bailey JM, Dunbar LM (1973): Essential fatty acid requirements of cells in tissue culture: A review. Exp Mol Pathol 18:142-161.

Baker JB, Low DA, Simmer RL, Cunningham DD (1980): Protease-nexin: A cellular component that links thrombin and plasminogen activator and mediates their binding to cells. Cell 21:37-45.

Bano M, Zweibel JA, Kidwell WR (1982): Growth factor effects on collagen synthesis by mouse mammary cells. (Abstract 5007.) J Cell Biol 95:114a.

Barnes D, Sato G (1980): Methods for growth of cultured cells in serum-free medium. Anal Biochem 102:255-270.

Baudysova M, Michl J (1981): Complete replacement of serum in primary cultures of chick embryo brain cells by growth-promoting alpha-globulin. Experientia 37:826-828.

Begbie R (1974): Studies on fetuin from fetal bovine serum: The composition and amino acid sequences of glycopeptides from fetuin. Biochim Biophys Acta 371:549-567.

Bergmann FH, Levine L, Spiro RG (1962): Fetuin: Immunochemistry and quantitative estimation in serum. Biochim Biophys Acta 58:41-51.

Bjornson CL, Moses HL (1981): Fetal bovine serum contains "transforming" growth factor. (Abstract 648.) J Supramol Struct 186:235.

Bloth B, Chesebro B, Svehag S-E (1968): Ultrastructural studies of human and rabbit αM-globulins. J Exp Med 127:749-756.

Carpenter G, Cohen S (1979): Epidermal growth factor. Annu Rev Biochem 48:193-216.

Collipp PJ, Kaplan SA, Boyle DC, Shimizu CSN (1964): Protein bound human growth hormone. Metabolism 13:532-538.

DeLarco JE, Todaro GJ (1978): Epithelioid and fibroblastic rat kidney cell clones: Epidermal growth factor (EGF) receptors and the effect of mouse sarcoma virus transformation. J Cell Physiol 94:335-342.

Deutsch HF (1954): Fetuin: The mucoprotein of fetal calf serum. J Biol Chem 208:669-678.

Dickson RB, Willingham MC, Pastan I (1981): Binding and internalization of $125_{I-\alpha_2}$ macroglobulin by cultured fibroblasts. J Biol Chem 256:3454-3459.

Dunn JT, Spiro RG (1967): The α_2-macroglobulin of human plasma: Isolation and composition. J Biol Chem 242:549-555.

Fisher HW, Puck TT, Sato G (1958): Molecular growth requirements of single mammalian cells: The action of fetuin in promoting cell attachment to glass. Proc Natl Acad Sci USA 44: 4-10.

Fisher HW, Puck TT, Sato G (1959): Molecular growth requirements of single mammalian cells: Quantitative clonal growth of single S3 cells in a medium containing synthetic small molecular constitutents and two purified protein fractions. J Exp Med 109:649-660.

Folkman J (1974): Tumor angiogenesis. Adv Can Res 19:331-358.

Galembeck F, Cann JR (1974): Fetuin as a trypsin inhibitor. Arch Biochem Biophys 164:326-331.

Gospodarowicz D, Tauber J-P (1980): Growth factors and the extracellular matrix. Endocr Rev 1:201-227.

Graham ERB (1972): Fetuin. In Gottschalk A (ed): "Glycoproteins: Their Composition, Structure and Function." Amsterdam: Elsevier, Vol 5, pp 717-731.

Green WA, Kay CM (1963): The influence of organic solvents and enzymatic modification on the secondary structure of fetuin. J Biol Chem 238:3640-3644.

Gullino PM (1981): Angiogenesis factor(s). In Baserga R (ed): "Handbook of Experimental Pharmacology." New York: Springer-Verlag, Vol 57, pp 427-449.

Hadden DR, Prout TE (1964): A growth hormone binding protein in normal human serum. Nature 202:1342-1343.

Ham RG (1963): Albumin replacement by fatty acids in clonal growth of mammalian cells. Science 140:802-803.

Harpel PC (1973): Studies on human plasma α_2-macroglobulin-enzyme interactions. J Exp Med 138:508-521.

Harpel BM, Kuepper F (1980): Preparative isoelectric focusing with pevicon as a supporting medium. Anal Biochem 104:173-174.

Hayman EG, Barnes D, Pierschacher M, Engvall E, Ruoslahti E (1982): Serum spreading factor is a major attachment protein in fetal bovine serum and the active principle in attachment-promoting fetuin. (Abstract 5038.) J Cell Biol 95:121a.

Healy GM, Parker RC (1966): Cultivation of mammalian cells in defined media with protein and nonprotein supplements. J Cell Biol 30:539-553.

Heim WG, Morrison AC, Shoemaker JR (1973): Effects of rat alpha-2 macroglobulin on growth in vitro. Differentiation 1:205-218.

Higuchi K (1973): Cultivation of animal cells in chemically defined media, a review. Adv Appl Microbiol 16:111-136.

Hoffman R, Ristow H-J, Veser J, Frank W (1973): Properties of two growth-stimulating proteins isolated from fetal calf serum. Exp Cell Res 85:275-280.

Howard JB (1981): Reactive site in human α_2-macroglobulin: Circumstantial evidence for a thiolester. Proc Natl Acad Sci USA 78:2235-2239.

Howard JB, Vermeulen M, Swenson RP (1980): The temperature-sensitive bond in human α_2-macroglobulin is the alkylamine-reactive site. J Biol Chem 255:3820-3823.

Hubbard WJ (1978): Alpha-2-macroglobulin-enzyme complexes as suppressors of cellular activity. Cell Immunol 39:388-394.

Jones JM, Creeth JM, Kekwick RA (1972): Thiol reduction of human α_2-macroglobulin: The subunit structure. Biochem J 127:187-197.

Kalebic T, Garbisa S, Glaser B, Liotta LA (1983): Migrating endothelial cells degrade basement membrane collagen. Science 221:281-283.

Kelly JP, Parker CW (1979): Effects of arachidonic acid and other unsaturated fatty acids on mitogenesis in human lymphocytes. J Immunol 122:1556-1562.

Kidwell WR, Monaco ME, Wicha MS, Smith GS (1978): Unsaturated fatty acid requirements for growth and survival of a rat mammary tumor cell line. Cancer Res 38:4091-4100.

Kidwell WR, Salomon DS, Liotta LA, Zwiebel JA, Bano M (1982): Growth factor effects on mammary epithelial cell proliferation and basement membrane synthesis. In Sato G (ed): "Cold Spring Harbor Symposium on Cell Proliferation," Vol 9: "Growth of Cells in Hormonally Defined Media." Cold Spring Harbor, NY: Cold Spring Harbor Press, pp 807-818.

Kleinman HK, Klebe RJ, Martin GR (1981): Role of collagenous matrices in the adhesion and growth of cells. J Cell Biol 88:473-485.

Knauer DJ, Cunningham DD (1982): Epidermal growth factor carrier protein binds to cells via a complex with released carrier protein nexin. Proc Natl Acad Sci USA 79:2310-2314.

Knox P, Griffiths S, Whatelely JG (1979): Fetal calf serum and the adhesion of cells in culture. Biochem Soc Trans 7:1011-1012.

Kurecki T, Kress LF, Laskowski M (1979): Purification of human plasma α_2-macroglobulin and $\alpha 1$ proteinase inhibitor using zinc chelate chromatography. Anal Biochem 99:415-420.

Laemmli UK (1970): Cleavage of structural proteins during the assembly of the head of bacteriophage T-4. Nature 227:680-685.

Lai PCW, Huang LL, Panrucker DE, Church RB, Lorscheider FL (1981): Distribution of bovine fetuin and albumin in plasma, allantoic and amniotic fluids during development. J Reprod Fertil 63:53-60.

Lieberman I, Ove P (1958): A protein growth factor for mammalian cells in culture. J Biol Chem 233:637-642.

Lieberman I, Lamy F, Ove P (1959): Nonidentity of fetuin and protein growth (flattening) factor. Science 129:43-44.

Liotta LA, Goldfarb RH, Terranova UP (1981a): Cleavage of laminin by thrombin and plasmin: Alpha thrombin selectively cleaves the beta chain of laminin. Thrombosis Res 21: 663-673.

Liotta LA, Tryggvason K, Garbisa S, Robey PG, Abe S (1981b): Partial purification and characterization of a neutral protease which cleaves type IV collagen. Biochemistry 20:100-104.

Liotta LA, Thorgeirsson UP, Garbisa S (1982): Role of collagenases in tumor cell invasion. Can Metastasis Rev 1:277-288.

Low DA, Baker JB, Koonce WC, Cunningham DD (1981): Released protease-nexin regulates cellular binding, internalization, and degradation of serine proteases. Proc Natl Acad Sci USA 78:2340-2344.

Maeda H, Murakami O, Kann M, Yamane I (1980): The growth-stimulating effect of α_1-acid glycoprotein in cells in culture. Proc Soc Exp Biol Med 163:223-227.

Marr AGM, Owen JA, Wilson GS (1962): Studies on the growth-promoting glycoprotein fraction of fetal calf serum. Biochim Biophys Acta 63:276-285.

Marti J, Aliau S, Bonfils C, Vigne C, Moretti J (1973): Preparation, physical constants and chemical composition of lamb fetuin. Biochim Biophys Acta 303:348-359.

Nagasawa S, Sugihara H, Han BH, Suzuki T (1970): Studies on α_2-macroglobulin in bovine plasma: Purification, physical properties and chemical compositions. J Biochem 67: 809-819.

Nevo Z, Laron Z (1979): Growth Factors. Am J Dis Child 133:419-428.

Oshiro Y, Eylar EH (1968): Physical and chemical studies on glycoproteins. III. The microheterogeneity of fetuin, a fetal calf serum glycoprotein. Arch Biochem Biophys 127:476-489.

Parmelee DC, Evenson MA, Deutsch HF (1978): The presence of fatty acids in human α-fetoprotein. J Biol Chem 253:2114-2119.

Pedersen KO (1944): Fetuin, a new globulin isolated from serum. Nature 154:575-576.

Peterkofsky B, Diegelmann R (1971): Use of a mixture of proteinase-free collagenases for the specific assay of radioactive collagen in the presence of other proteins. Biochemistry 10:988-994.

Pierce M, Cummings RD, Cebula TA, Roth S (1979): Galactosyltransferase from commercial preparations of fetuin. Biochim Biophys Acta 571:166-170.

Porath J, Carlsson J, Olsson I, Belfrage G (1975): Metal chelate affinity chromatography, a new approach to protein fractionation. Nature 258:598-599.

Puck TT, Waldren CA, Jones C (1968): Mammalian cell growth proteins. I. Growth stimulation by fetuin. Proc Natl Acad Sci USA 59:192-199.

Rizzino A, Sato G (1978): Growth of embryonal carcinoma cells in serum-free medium. Proc Natl Acad Sci USA 75:1844-1848.

Rohrlich ST, Rifkin DB (1981): Isolation of the major serine protease inhibitor from 5-day serum-free conditioned medium of human embryonic lung cells and demonstration that it is fetuin. J Cell Physiol 109:1-15.

Ronne H, Anundi H, Rask L, Peterson PA (1979): Nerve growth factor binds to serum alpha-2-macroglobulin. Biochem Biophys Res Commun 87:330-336.

Salomon DS (1980): Growth of mouse embryonal carcinoma cells in serum-free, hormone-supplemented medium correlated with receptors for growth factors. Exp Cell Res 128: 311-327.

Salomon DS, Liotta LA, Kidwell WR (1981): Differential response to growth factor by rat mammary epithelium plated on different collagen substrata in serum-free medium. Proc Natl Acad Sci USA 78:382-386.

Salomon DS, Liotta LA, Rennard SI, Terranova VP, Foidart J-M, Yaar M (1982a): Stimulation by retinoic acid of synthesis and turnover of basement membrane in mouse embryonal carcinoma-derived endoderm cells. Collagen Related Res 2:93-110.

Salomon DS, Bano M, Smith KB, Kidwell WR (1982b): Isolation and characterization of a growth factor (embryonin) from bovine fetuin and its resemblance to α_2-macroglobulin. J Biol Chem 257:14093-14101.

Schor AM, Schor SL, Weiss JB, Brown RA, Kumar S, Phillips P (1980): Stimulation by a low-molecular-weight angiogenic factor of capillary endothelial cells in culture. Br J Cancer 41:790-799.

Sottrup-Jensen L, Petersen TE, Magnusson S (1980): A thiol-ester in α_2-macroglobulin cleaved during proteinase complex formation. FEBS Lett 121:275-279.

Spiro RG (1960): Studies on fetuin, a glycoprotein of fetal serum. I. Isolation, chemical composition and physiochemical properties. J Biol Chem 235:2860-2869.

Spiro MJ, Spiro RG (1962): Composition of peptide portion of fetuin. J Biol Chem 237:1507-1510.

Starkey PM, Barret AJ (1977): α_2-Macroglobulin, a physiological regulator of proteinase activity. In Barret AJ (ed): "Proteinases in Mammalian Cells and Tissues." Amsterdam: Elsevier/North-Holland, pp 663-696.

Stavenow L, Kjellstrom T, Malniquist J (1981): Stimulation of collagen production in growth-arrested myocytes and fibroblasts in culture by growth factor(s) from platelets. Exp Cell Res 136:321-325.

Swenson RP, Howard JB (1979): Structural characterization of human α_2-macroglobulin subunits. J Biol Chem 254:4452-4456.

Swim HE (1967): Nutrition of cells in culture—A review. In Rothblat GH, Kritchevsky D (eds): "Lipid Metabolism in Tissue Culture Cells." Philadelphia: Wistar Institute Press, pp 1-14.

Temin HM, Pierson RW, Dulak NC (1972): The role of serum in the control of multiplication of avian and mammalian cells in culture. In Rothblat GH, Cristotolo VJ (eds): "Growth, Nutrition and Metabolism of Cells in Culture." New York: Academic, Vol 1, pp 50-81.

Terranova VP, Rohrbach DH, Martin GR (1980): Role of laminin in the attachment of PAM 212 (epithelial) cells to basement membrane collagen. Cell 22:719-726.

Thorgeirsson UP, Liotta LA, Kalebic T, Marguilies IM, Thomas K, Rios-Candelore M, Russo RG (1982): Effect of natural protease inhibitors and a chemoattractant on tumor cell invasion in vitro. J Natl Cancer Inst 69:1049-1054.

Todaro GJ, DeLarco JE, Fryling C, Johnson PA, Sporn MB (1981): Transforming growth factors (TGFs): Properties and possible mechanisms of action. J Supramol Struct 15:287-301.

Tozer BT, Pirt SJ (1964): Suspension culture of mammalian cells and macromolecular growth-promoting fractions of calf serum. Nature 201:375-378.

Van Leuven F, Cassiman J-J, Van Den Berghe H (1979): Demonstration of an α_2-macroglobulin receptor in human fibroblasts, absent in tumor-derived cell lines. J Biol Chem 254: 5155-5160.

Van Leuven F, Cassiman J-J, Van Den Berghe H (1981): Functional modifications of α_2-macroglobulin by primary amines. I. Characterization of α_2M after derivatization by methylamine and factor XIII. J Biol Chem 256:9016-9022.

Vartio T, Vaheri A (1981): A gelatin-binding 70,000-dalton glycoprotein synthesized distinctly from fibronectin by normal and malignant adherent cells. J Biol Chem 256:13085-13090.

Virca GD, Travis J, Hall PK, Roberts RC (1978): Purification of α_2-macroglobulin by chromatography on Cibacron blue sepharose. Anal Biochem 89:274-278.

Werb Z, Aggeler J (1978): Proteases induce secretion of collagenase and plasminogen activator by fibroblasts. Proc Natl Acad Sci USA 75:1839-1843.

Werb Z, Burleigh MC, Barrett AJ, Starkey PM (1974): The interaction of α_2-macroglobulin with proteinases. Binding and inhibition of mammalian collagenases and other metal proteinases. Biochem J 139:359-368.

Werb Z, Mainardi C, Vater CA, Harris ED (1977): Endogenous activation of latent collagenases by rheumatoid synovial cells. N Engl J Med 296:1017-1023.

Wicha MS, Liotta LA, Kidwell WR (1979): Effects of free fatty acids on the growth of normal and neoplastic rat mammary epithelial cells. Can Res 39:426-435.

Zetter BR (1980): Migration of capillary endothelial cells is stimulated by tumor-derived factor. Nature 285:41-43.

Ziche M, Jones J, Gullino PM (1982): Role of prostaglandin E_1 and copper in angiogenesis. J Natl Cancer Inst 69:475-482.

Serum-Free Culture of Fibroblastic Cells

Methods for Serum-Free Culture of Epithelial and Fibroblastic Cells,
pages 157–166

11
On Deciding Which Factors Regulate Cell Growth

Arthur B. Pardee, Paul V. Cherington, and Estela E. Medrano

Cells in culture require many soluble factors to permit their rapid growth. These factors include ions, sugars, amino acids, vitamins, and other low molecular weight compounds generally provided in tissue culture media. Also needed are hormones, peptides, and proteins that are usually provided by serum. Absence of any of these factors prevents rapid growth. Cells deprived of certain factors go into a quiescent state from which they may emerge if they are provided with complete media [1]. The factors necessary for growth may be divided between those that are required as nutrients for example, and those that somehow regulate growth, e.g., that under physiological conditions determine continued growth vs. quiescence. Deciding which factors are nutrients and which are regulatory is not a simple problem. Thus, essential amino acids such as isoleucine are required for growth but they also have been suggested to be growth regulatory [2].

The problem addressed in this article is "What criteria can we at present use to decide whether a given substance is or is not a regulatory factor?" Our approach to this question is similar to one suggested by E.A. Adelberg about a quarter century ago. He asked how one decides which enzymes are involved in a given metabolic pathway as contrasted to those enzymes which only seem to be involved because they act on potential intermediates [3]. Adelberg proposed that no one criterion was adequate at the present state of knowledge, and that one had to apply several criteria. If all were satisfied, one might

Laboratory of Tumor Biology, Sidney Farber Cancer Institute and Department of Pharmacology, Harvard Medical School, Boston, Massachusetts 02115

Reprinted from "Control of Cellular Division and Development, Part A," D. Cunningham, E. Goldwasser, J. Watson, and C.F. Fox, editors. New York: Alan R. Liss, pp 495–502, 1981.

conclude that the enzyme in question was likely to be involved. Similarly, for growth control we suggest that three criteria can be used. We illustrate this proposal with results from our recent studies.

FIRST CRITERION: THE "ARREST TEST"

Regulatory factors are mainly functional in and required for traverse of the G_1 phase of the cell cycle (Fig. 1). In general, cells that are at rest in vivo have a 2N DNA content and are described as being in G_1 or G_0 [1]. Similarly, in culture mammalian cells come to rest with a 2N DNA content under most physiological conditions that are suboptimal for growth, e.g., at high density, in the absence of a substratum, or with insufficient serum factors (Fig. 2). We have proposed earlier that cells stop in G_1 because they are unable to complete some crucial biochemical events under adverse conditions. These events are required for cells to initiate DNA synthesis [4]. Thus, in the absence of a truly regulatory factor, cells should arrest in G_0/G_1, but removal of a nutrient could cause arrest in other parts of the cycle.

As an illustration of this test we refer to cytofluorometric patterns of CHEF/18 [5] cells (Fig. 3). These cells grow in a defined medium lacking serum and containing epidermal growth factor (EGF), fibroblast growth factor (FGF), insulin, and transferrin [6]. In this medium the population has a typical distribution of cells in the G_1, S, and G_2 +M phases of the cycle. If these cells are placed for 12 hours in medium lacking EGF and insulin, their DNA

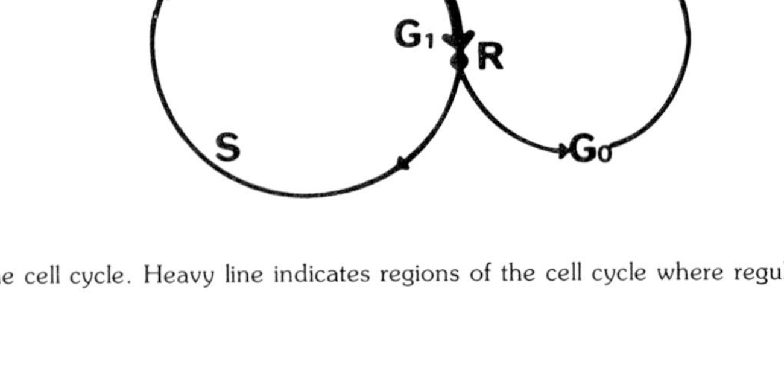

Fig. 1. The cell cycle. Heavy line indicates regions of the cell cycle where regulatory events may occur.

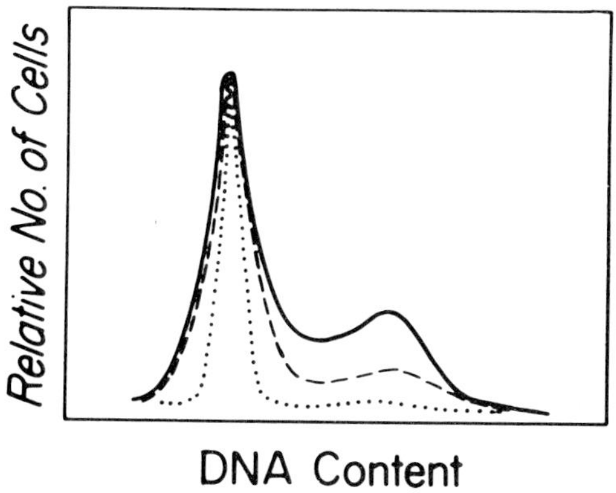

Fig. 2. Cell cycle distributions of Balb/c 3T3 cells. Balb/c 3T3 fibroblasts were plated into 60-mm dishes in medium containing 10% fetal calf serum at an initial density of 3 × 10^3 cells/cm^2. (—) Cells processed for cytofluorometric analysis 24 h after plating (exponentially growing). (----) Cells processed after 72 h (at high density). (····) Cells shifted to 0.5% calf serum and processed for cytofluorometric analysis after 48 h (serum-starved cells).

Fig. 3. Cell cycle distributions of CHEF/18 deprived of various growth factors. CHEF/18 cells were plated into 60-mm dishes in medium containing 10% fetal calf serum at an initial density of 3 × 10^3 cells/cm^2. After one day the cells were extensively washed and fed with media containing EGF, insulin, FGF, and transferrin (4F) or with media lacking one or more of these factors. After 12 h, plates were sampled for cytofluorometric analysis.

histograms reveal arrest in the G_1 part of the cycle. This specific arrest suggests that these two factors are involved in growth regulation. By contrast, after transferrin is removed, the cell number no longer increases but the population remains distributed around the cycle. Thus, according to this test, transferrin is not growth regulatory although it is essential for growth.

That deletion of an individual factor preferentially arrests cells in G_0/G_1 does not prove or disprove that this factor also limits growth under other conditions, for example at confluence or after serum starvation. Addition of EGF and insulin, but not nutrients, stimulates the growth of both serum starved or confluent 3T3 cells [7,8] indicating that the peptide growth factors, but not nutrients, are regulatory for both conditions.

Nutrients have been proposed to be growth regulatory [2] because their rates of transport are reduced when cells become quiescent [9]. For instance, lower transport of phosphate or uridine by untransformed cells was correlated with G_0/G_1 arrest [10]. However, these observations do not rule out an alternative explanation: that cells shut down their underutilized active transport processes as a consequence of growth arrest. Indeed, a reduced rate of uridine transport, which is not essential for growth, is observed when cells come to confluence. Transport of glucose and of phosphate are also not regulatory [11,12].

SECOND CRITERION: THE TRANSFORMATION TEST

This test is essentially based on mutation. We propose that when a mutation causes growth control to become relaxed or lost so that cells do not arrest as they did previously, a true regulatory factor should no longer have the same role. If removal of the factor does have the same effect, still stopping the mutated cells in G_0, this factor is either not regulatory or its function is not modified by this mutation (though it might be modified by another mutation).

We already have a plentiful supply of cells whose growth control is relaxed or eliminated; they are transformed cells [1]. These cells generally have a diminished requirement for serum: They continue to grow at much lower serum concentrations than normal cells require. Some, but not others, depending on the specific transformation, can be arrested with a 2N DNA content under conditions more stringent than needed to arrest untransformed cells [6,13]. These cells generally become arrested at high density owing to their using up nutrients rather than from lack of serum factors [14].

As an example of selective mutational loss of a growth factor requirement, we can consider effects of transformation on the CHEF/18 line [6,26]. In totally defined medium two transformed derivatives of CHEF/18, CHEF/16, and T30-4 still required insulin and transferrin, but both lost their EGF requirements, as did some transformed BHK lines [6]. By this test, EGF is a regulatory factor. Insulin is not eliminated as a regulatory factor; the test shows only that these particular transformations did not diminish its role.

Transferrin is an example of a nonregulatory factor. This was indicated by examining the growth requirements of the polyoma-virus-transformed BHK-21 cells, which by all criteria seem to be totally transformed. These cells continued to grow slowly in the absence of serum, but cells did not accumulate at G_1 (Table I). They still required transferrin but not EGF, FGF, or insulin for rapid growth. Similarly, we have observed that SV40-transformed Swiss 3T3 cell (SV3T3) growth in the presence of 0.5% serum is greatly enhanced by transferrin (results not shown), confirming a previously reported result [15] that SV3T3 cells require transferrin, as well as biotin, for growth in serum-free medium. Thus transferrin is needed, but according to these tests it is not regulatory.

THIRD CRITERION: AVAILABILITY TEST

A true regulatory factor should be present under physiological conditions at a reasonable concentration. This criterion rules out some substances that affect growth and fulfill the first two criteria, but which are not found in vivo in the animal in question. Examples that come to mind are caffeine [16], cycloheximide (CHM) [17], lectins [18], or phorbol myristate acetate [19].

Such compounds may not be natural regulators, but they can provide clues as to mechanisms of regulation. For example, we consider the effects of CHM [17] which satisfies the first two criteria. At low CHM concentrations normal cells are arrested with a G_0/G_1 DNA content. Transformed cells are not

TABLE I. Transferrin is Nonregulatory, but Essential for Rapid Growth of Polyoma-Transformed BHK Cells

Condition	Cells/cm^2 ($\times 10^{-3}$ $\pm$SEM)	%G_1
4F	99 $\pm$ 1	45
4F minus transferrin	58 $\pm$ 7	47
4F minus EGF, FGF, insulin	116 $\pm$ 7	47
No factors or serum added	38 $\pm$ 6	50

Polyoma-virus-transformed BHK cells were plated at 3×10^3 cells/cm^2 into 60-mm dishes in medium containing 10% serum. After one day the medium in each plate was replaced with serum-free media containing EGF, insulin, FGF, and transferrin (4F) or 4F with the indicated factors omitted. After one day, samples were prepared for cytofluorometric analysis. Analysis of pyBHK cells after two days under similar conditions gives the same result. Duplicate cell counts were made at the time of the shift and two days later. Cell density at the shift to experimental conditions was $12 \pm 0.5 \times 10^3$ cells/cm^2.

arrested in G_1 by CHM as readily as are normal cells [20]. The small effects of CHM on four differently transformed 3T3 lines are shown in Figure 4. CHM diminishes the growth rates of all these lines similarly, but cells remain distributed through the cell cycle [20]. These results with transformed cells strongly support our original conclusion that CHM specifically affects the major growth regulatory event that is important in transformation [17]. We have proposed that CHM's specific effect is to limit the accumulation of a labile protein that has a half life of a few hours. This protein is proposed to be required for transit past the restriction point in G_1 in preparation for DNA synthesis.

Limiting the concentrations of amino acids [21] might cause cells to be arrested in G_1 by the same mechanism as CHM, i.e., by limiting the rate of protein synthesis. Although growth of yeast is not dependent on intercellular communication and hormone-like factors (except for mating hormone) [22], limiting an amino acid essential for an auxotrophic yeast does restrict its growth [23], as CHM also does [24]. This basic mechanism might persist in animal

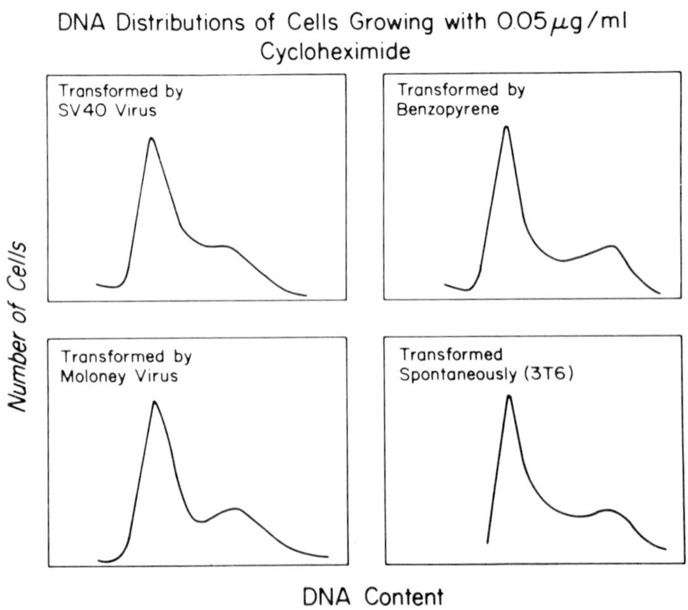

Fig. 4. Effects of CHM on various transformed 3T3 cells. Cells were plated into 60-mm dishes in medium containing 10% calf serum at an initial density of 3×10^3 cells/cm^2. After 24 h, 0.05 μg/ml CHM was added and cells processed for cytofluorometric analysis 48 or 72 h later.

cells and have superimposed on it the effects of growth factors such as EGF and insulin.

THE CHOICE BETWEEN POSSIBLE REGULATORY FACTORS

More than one factor sometimes fulfills the three criteria just proposed. For example, somatomedins appear to replace insulin for permitting growth of Balb/c 3T3 cells [25]. Inability of serum from hypophysectomized animals to support cell growth unless somatomedin C is added suggests that somatomedin C may be a physiologically important factor. The low concentration at which it functions in culture also indicates that somatomedin C is physiologically important. Indeed, the fact that insulin is needed in culture at concentrations of micrograms/ml as compared to its much lower (nanogram/ml) concentrations in vivo, and the nanogram/ml concentrations required for other factors, has suggested that insulin might substitute for the actual growth factor for these cells.

We further illustrate the dilemma of choice between factors with some of our recent observations. Thrombin supports growth in defined medium of CHEF/18 cells even better than does EGF. Other proteases, including trypsin and urokinase, did not stimulate growth of these cells. Thrombin almost completely eliminated the cells' need for EGF, and EGF could not support as rapid growth as did thrombin [26]. Both proteins functioned optimally at similar concentrations (Fig. 5). Transformation eliminated the requirements for both factors (Table II). Although both thrombin and EGF are present in vivo, the role of thrombin in wound repair (blood clotting) makes it a reasonable candidate for regulating fibroblast growth, because fibroblasts fill a wound following clotting. Which is the true regulatory factor?

Among all the compounds of nature some could accidentally interact with a cellular process unrelated to its native action. For example, factors involved with other aspects of growth or development, such as EGF or insulin, can stimulate the growth of cells in culture that are otherwise stimulated by thrombin or somatomedin C. Compounds derived from unrelated organisms such as plants, e.g., lectins or phorbol esters, can also interact with mammalian cells and perhaps mimic endogenous factors (like opiates with enkephalins). These compounds that mimic regulatory factors could replace the endogenous factor by acting at the cell surface. Thus, the presence of receptors is not proof that the compound has a physiological role. Accidentally a compound of similar structure may be involved. These factors could act intracellularly on events normally stimulated as a secondary response to the endogenous factor.

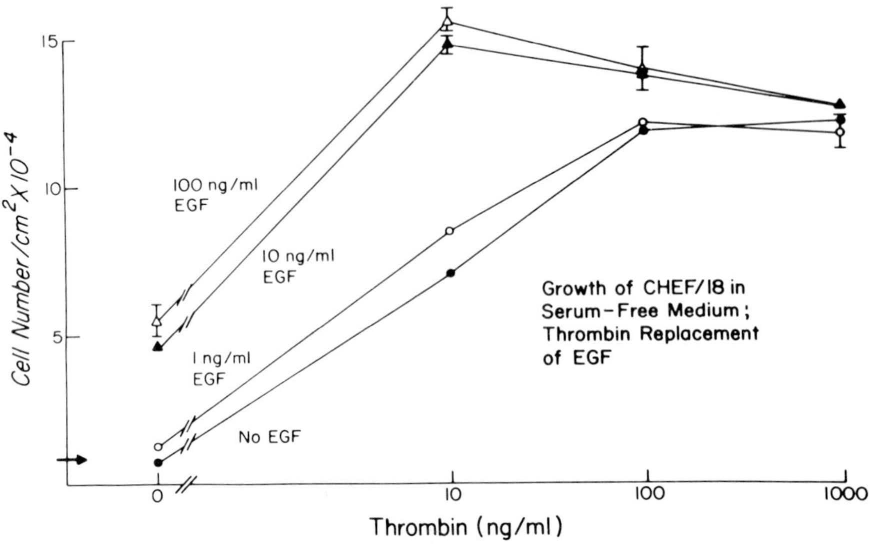

Fig. 5. Growth of CHEF/18 in serum-free medium: thrombin replacement of EGF. CHEF/18 cells were plated at 3×10^3 cells/cm^2 into 35-mm dishes in medium containing 10% serum and one day later shifted after extensive washing into media containing FGF, insulin, and transferrin with no EGF (●), 1 ng/ml EGF (○), 10 ng/ml EGF (▲), or 100 ng/ml EGF (△). Thrombin concentrations were varied as indicated. Duplicate cell counts were made at the time of the shift (arrow) and again three days later. SEM (bar) is noted only where the SEM is larger than the symbol.

Clearly, these are still some dilemmas regarding the nature of the true growth factors. We need other criteria. As possible tests genetic defects can be indicative: Defective platelet formation has consequences that can be attributed to deficiencies in the platelet-derived growth factor [27]. The in vivo function of all hormones was pinpointed by organ ablation surgically or loss owing to genetic diseases [28]. Surgical removal of the source of a growth factor, such as hypophysectomy to eliminate somatomedin C production, can be used together with supplementation of a test system with the factor in question [25]. The possibility also exists of using specific antibodies against a growth factor for in vivo testing of a factor's role, as reported some time ago for nerve growth factor [29].

ACKNOWLEDGMENTS

We are grateful to Gail Morreo for technical assistance and to Eileen Fingerman for typing. This work was supported by U.S. Public Health Service

TABLE II. Transformed CHEF Cells Have Lost Requirements for EGF and Thrombin, but Not for Insulin or Transferrin

	Population doublings in three days (±SEM)		
Condition	CHEF/18	CHEF/16	T30-4
4F + thrombin (10 ng/ml)	3.3 ± 0.1	4.0 ± 0.1	3.9 ± 0.1
Minus EGF	$1.8 \pm < 0.1$	3.9 ± 0.1	4.0 ± 0.1
Minus thrombin	$2.4 \pm < 0.1$	3.7 ± 0.2	3.9 ± 0.1
Minus insulin	$1.3 \pm < 0.1$	$1.9 \pm < 0.1$	2.9 ± 0.1
Minus transferrin	1.1 ± 0.1	2.1 ± 0.3	(decreased cell number)

Each cell line was plated at 3×10^3 cells/cm^2 into 35-mm dishes in medium containing 10% serum. After one day, the medium in each plate was replaced with serum-free medium containing 4F (10ng/ml EGF, 10 μg/ml insulin, 10 ng/ml FGF, 5 μg/ml transferrin) and thrombin (10 ng/ml) or with media lacking one of the above factors. Duplicate cell counts were made at the time of the shift into experimental conditions and three days later. The number of population doublings under each condition was calculated as follows:

$$\text{Number of doublings} = \frac{\log \dfrac{\text{cell number 3 days after shift}}{\text{cell number at shift}}}{\log 2}.$$

grant GM 24571 to A.B.P., a U.S. Public Health Service International Research Fellowship to E.E.M., and a National Research Service Award Predoctoral Training grant 5 T32 GM 07306 to P.V.C.

REFERENCES

1. Baserga R: "Multiplication and Division in Mammalian Cells." New York: Marcel Dekker, 1976.
2. Holley R: Proc Natl Acad Sci USA 69:2840, 1972.
3. Adelberg EA: Bacteriol Rev 17:253, 1953.
4. Pardee AB: Proc Natl Acad Sci USA 71:1286. 1974.
5. Sager R, Kovac PE: Somatic Cell Genet 4:375, 1978.
6. Cherington PV, Smith BL, Pardee AB: Proc Natl Acad Sci USA 76:3937, 1979.
7. Rossow PW, Pardee AB: Unpublished data.
8. Mierzejewski K, Rozengurt E: Biochem Biophys Res Commun 83:874, 1978.
9. Cunningham DD, Pardee AB: Proc Natl Acad Sci USA 64:1049, 1970.
10. Dubrow R, Pardee AB, Pollack R: J Cell Physiol 95:203, 1978.
11. Barsh GS, Greenberg DB, Cunningham DD: J Cell Physiol 92:115, 1977.
12. Naiditch WP, Cunningham DD: J Cell Physiol 92:319, 1977.
13. Dubrow R, Riddle VGH, Pardee AB: Cancer Res 39:2718, 1979.
14. Moses HL, Proper JA, Volkenant ME, Wells DJ, Getz MJ: Cancer Res 38:2807, 1978.
15. Young DV, Cox FW III, Chipman S, Hartman SC: Exp Cell Res 118:410, 1979.
16. Pardee AB, James LJ: Proc Natl Acad Sci USA 72:4994, 1975.

17. Rossow PW, Riddle VGH, Pardee AB: Proc Natl Acad Sci USA 76:4446, 1979.
18. Gunther GR, Wang JL, Edelman GM: Exp Cell Res 98:15, 1976.
19. Slaga TJ, Sivak A, Boutwell RK (eds): "Mechanism of Tumor Promotion and Cocarcinogenesis." New York: Raven Press, 1978.
20. Medrano EE, Pardee AB: Proc Natl Acad Sci USA 77:4123, 1980.
21. Ley KD, Tobey RA: J Cell Biol 47:453, 1970.
22. Hartwell LH: Bacteriol Rev 38:164, 1974.
23. Shilo B, Simchen G, Pardee AB: J Cell Physiol 97:177, 1978.
24. Shilo B, Riddle VGH, Pardee AB: Exp Cell Res 123:221, 1979.
25. Stiles CD, Capone GT, Scher CD, Antoniades HN, Van Wyk JJ, Pledger WJ: Proc Natl Acad Sci USA 76:1279, 1979.
26. Cherington PV, Pardee AB: J Cell Physiol (in press).
27. Gerrard JM, Phillips DR, Rao GHR, Plow EF, Walz DW, Ross R, Harker LA, White JG: Blood 54:241A, 1979.
28. Tepperman J: "Metabolic and Endocrine Physiology." Chicago: Year Book Publ, 1973.
29. Levi-Montalcini R, Booker B: Proc Natl Acad Sci USA 46:384, 1960.

Preparation of Animal Cell Culture Media, Growth Factors, and Substrata,
pages 167–197

12
Purification of Pituitary and Brain Fibroblast Growth Factors and Their Use in Cell Culture

Denis Gospodarowicz

Following the demonstration that both pituitary and brain are an excellent source of mitogenic material for cells derived from the mesoderm [Gospodarowicz, 1974], a number of growth-promoting factors [Gospodarowicz, 1975; Gospodarowicz et al., 1975, 1978a] to which a broad range of target cells are responsive [Gospodarowicz et al., 1978b,c,d] have been identified. One class of growth factors preferentially extracted at neutral or basic pHs has been given the generic name of myoblast growth factors (MGF, also called endothelial cell growth factor) [Gospodarowicz et al., 1975]. It consists of neutral or acidic mitogens. The other class of mitogens, composed primarily of basic peptides, is mostly extracted at acidic pHs and has been termed fibroblast growth factors (FGF) [Gospodarowicz, 1974, 1975; Gospodarowicz et al., 1978a]. In the following review, we will outline the purification procedure of both growth factors and their use in cell culture.

EXPERIMENTAL PROCEDURES

Materials

Freshly excised cow brains or pituitaries are frozen quickly by submersion in liquid nitrogen. Brains and pituitaries are kept in a Revco freezer (−80°C) until extraction (usually within a period of 2 weeks). All reagents used are of analytical grade. Carboxymethyl-Sephadex C-50 and Sephadex G-75 come from Pharmacia. Carboxymethyl cellulose Whatman CM-52 is from What-

Cancer Research Institute and the Departments of Medicine and Ophthalmology, University of California Medical Center, San Francisco, California 94143 man. Dialysis is done with Spectrapor dialysis tubing (molecular weight cutoff 6,000–8,000) from Spectrum Medical Industries. Crystalline bovine serum albumin (BSA) is from Schwarz/Mann. Cytochrome and thymidine are from Sigma. Methyl-^{3}H-thymidine is from New England Nuclear. Calf serum and fetal calf serum are from Hy Clone (Sterile Systems, Inc.). Tissue culture dishes are from Falcon Plastics. Gentamicin is from Schering Co. and Fungizone from Squibb. Dulbecco's modified Eagle's medium (DME, H-16) is from Gibco. Concentrated ampholyte solutions (pH range 9–11, 8–10, and 6–11) are obtained from LKB. Low-molecular-weight standards for SDS-polyacrylamide gel electrophoresis and silver nitrate stain kit are from Bio Rad.

Carboxymethyl Ion Exchange Chromatography and Gel Filtration

Two different ion exchange columns are used to purify FGF [Gospodarowicz, 1975; Gospodarowicz et al., 1978a]. The first column (3 × 20 cm) is prepared with carboxymethyl-Sephadex C-50, which is first submitted to a base and acid wash, followed by equilibration with 0.1 M sodium phosphates, pH 6.0. Although the base and acid wash is not recommended by Pharmacia, we find that washing the carboxymethyl-Sephadex C-50 with 0.5 N NaOH and distilled water followed by a wash with 0.5 N NCl and water greatly increases the flow rate of the column, which can then exceed 500 ml/h. This procedure allows the rapid chromatography of great quantities (40 g) of proteins. The material applied on the column is eluted by increasing the salt concentration stepwise, first to 0.15 M NaCl in 0.1 M sodium phosphates, pH 6.0, and then to 0.5 or 1 M NaCl in 0.1 M sodium phosphates, pH 6.0. The second column (1.5 × 20 cm) is prepared with Whatman carboxymethyl cellulose CM-52 equilibrated in 0.2 M ammonium formate, pH 6.0. The flow rate is adjusted at 20 ml/h, and the material applied on the column is eluted with a salt gradient (0.2–0.4 M ammonium formate).

Gel filtration is performed with a column (5 × 100 cm) of Sephadex G-75 equilibrated with 0.1 M ammonium carbonate (pH 9.2). The column is packed at 40 cm head pressure and operated at 25 cm head pressure with a flow rate of 40 ml/h [Gospodarowicz, 1975; Gospodarowicz et al., 1978a].

Preparative Isoelectric Focusing

Isoelectric focusing is performed with a 110-ml LKB column [Gospodarowicz et al., 1982b]. A 0.2 M NaOH solution in 60% sucrose (wt./vol.) is layered at the cathode, followed by a 40%–5% (wt./vol.) sucrose gradient containing 5.2 ml LKB ampholytes (pH 9–11). When half of the gradient solution has been introduced into the column, pumping is interrupted and 3 ml of gradient solution is collected. Fractions of lyophilized FGF (30–50 mg) are dissolved thereon and insoluble material is removed by centrifugation (50,000g for 30 min). The clear supernatant is then layered on top of the gradient, and pumping is resumed. When a fraction of a previous isoelectric focusing preparation is to be refocused, its sucrose density is determined by refractometry and the fraction then is reinserted into the column at its appropriate sucrose density. A 0.01 M acetic acid solution is layered on top of the gradient, in contact with the cathode. The jacketed column is cooled by a circulating 4°C bath. Focusing is carried out for 48 h or 72 h at 500 V. The content of the column is then eluted over 110 fractions whose pH and A_{280} are determined. Individual tubes are then combined into various fractions, as shown in the figure legends. The protein concentration of each fraction is determined on the basis of an extinction coefficient of $E^{1\%}{}_{1\ cm} = 0.79$. (This value is proved to be correct by dialyzing samples of the various fractions against 0.1 M ammonium carbonate, lyophilizing them, and determining the extinction coefficient of the dry protein dissolved at a final concentration of 1 mg/ml in 0.1 M ammonium carbonate, pH 9.2.) Crystalline BSA (stock solution 5% in DME) is then added to a final concentration of 0.5%. In the case of fractions that are to be refocused, aliquots (50 μl) are taken from each tube, pooled, and then processed as described above. Aliquots of the various fractions to be tested for biological activity are then further diluted to a final protein concentration of 20 μg/ml in DME containing 0.5% BSA.

Bioassays of Growth Factor Activity

Cell proliferation assay. The mitogenic activity of the various fractions can be determined by using as target cells bovine vascular endothelial cells derived from adult aortic arch as already described [Gospodarowicz et al., 1978a,b, 1982b]. Stock cultures, maintained in the presence of DME, H-16 supplemented with 10% calf serum, 50 μg/ml Gentamicin, 0.25 μg/ml Fungizone, and a partially purified FGF fraction (G-75 step, 100 ng/ml), are passaged weekly at a split ratio of 1 to 64. For mitogenic assay, cell monolayers from stock plates are dissociated by exposure (2–3 min, 24°C) to a solution containing 0.9% NaCl, 0.01 M sodium phosphates (pH 7.4), 0.05% trypsin, and 0.02% EDTA (STV solution, Difco). When cells round up, they are resuspended in DME supplemented with 10% calf serum, and an aliquot of the cell suspension is then subjected to counting in a Coulter counter. Cells are seeded at an initial density of 2×10^4 cells per 35-mm dish containing 2 ml of DME H-16 supplemented with 10% calf serum and antibiotics, as described above. Six hours later, a set of triplicate plates is trypsinized and cells are counted to determine the plating efficiency. Aliquots of 10 μl containing the various protein concentrations of the various fractions to be analyzed are then added every other day to the dishes. In the case of fractions purified by isoelectric focusing, the final ampholyte concentrations to which cells are exposed range from 2×10^{-3}% to 10^{-5}%, depending on the original protein content of the various isoelectric focusing fractions. FGF samples are added every other day, and after 5–6 days in culture, triplicate plates are trypsinized and final cell densities are determined by counting of cells in a Coulter counter.

The mitogenic activity of FGF can also be followed by testing its activity on low-density bovine corneal endothelial, vascular smooth muscle, granulosa, adrenal cortex, or lens epithelial cell cultures maintained as previously described [Gospodarowicz et al., 1977a,b,c, 1981b]. Cells are seeded at an initial density of 2×10^4 cells per 35-mm dish. Assays are conducted as described for vascular endothelial cells.

Stimulation of DNA synthesis. The mitogenicity of the purified fractions is also determined by measuring their ability to stimulate the incorporation of methyl-^{3}H-thymidine into Balb/c 3T3 fibroblasts. Assays are conducted as previously described [Gospodarowicz, 1974, 1975]. The use of one 96-well MicroTest II plate permits 32 different sample concentrations to be tested simultaneously at a very low cost in time and materials. A 10-ml syringe fitted with a Hamilton repeating dispenser is used to seed the cells (5×10^3 cells per well in 200 μl of DME medium with 0.2% calf serum). Then 36 h later, various concentrations of the samples to be tested are added in volumes of 2.5, 5, or 10 μl per well. The samples are dissolved in DME medium with 0.5% crystalline bovine serum albumin (to minimize nonspecific absorption of the sample proteins). Twelve hours later, 10 μl of DME medium containing 1 μCi of ^{3}H-thymidine and 2.5 ng of unlabeled thymidine are added to each well with a 0.5-ml syringe fitted with a Hamilton repeating dispenser [Gospodarowicz, 1975].

After 16 h of incubation at 37°C, the medium is aspirated, and the wells are rinsed twice with 200 μl of DME. Fifty microliters of phosphate-buffered saline are then added to each well, and each well is swabbed with a 6-cm-long cotton swab. In our experience, more than 95% of the cells are trapped in the cotton swab. Phase microscopic examination of the wells shows that no cells remain after swabbing. The wooden ends of the swabs are then stuck into another MicroTest II dish in which each well has been filled with Play-Doh (Kenner Products). This second dish is inverted over a stainless steel container containing 10% cold trichloroacetic acid, so that the tip of each swab is immersed in the solution. A magnetic stir bar provides gentle mixing. The swabs are kept in 10% trichloroacetic acid for 20 min. They are then washed in a similar fashion for 10 min in 5% trichloroacetic acid and washed again for 15 min in 95% ethanol. Afterwards, they are dried at 37°C. The cotton tip of each swab is cut and placed in a glass scintillation vial containing 10 ml of a 1% solution of Liquifluor (New England Nuclear) in toluene. The vials are counted in a liquid scintillation counter (Nuclear Chicago Mark I). Since all the ^{3}H-thymidine is incorporated into precipitated DNA that is trapped in the cotton fibers, no labeled DNA is released into the scintillation fluid. The cotton tips can be removed after the vials are counted, and the scintillation fluid can be reused [Gospodarowicz, 1975].

Sodium Dodecyl Sulfate Polyacrylamide Gel Electrophoresis

Protein aliquots from the various fractions and containing 10 μg protein are added to a sample buffer composed of 15% glycerol, 2% SDS, 75 mM Tris-HCl (pH 6.8), 2 mM phenyl methyl sulfonyl fluoride, 2 mM EDTA, 1 mM N-ethylmaleimide, and 1 mM iodoacetic acid. Samples are boiled for 3 min and then applied to an exponential gradient (10–18%) or a 15% polyacrylamide slab gel with a 3% stacking gel [Laemmli, 1970]. Electrophoresis is performed for 4 h at 20 mA. After electrophoresis, the slab gels are fixed and stained with either Coomassie blue or silver nitrate as described by Merril et al. [1981]. When gels are stained with silver nitrate, they are placed for 1 h in 30% methanol, 10% trichloroacetic acid, 3.5% salfosalicyclic; washed for 16 h in several volumes of 30% methanol, 12% trichloroacetic acid to remove the ampholytes; transferred to 40% methanol, 10% acetic acid for 1 h; and washed twice for 30 min with 10% ethanol, 5% acetic acid. Gels are stained with the Bio Rad silver nitrate stain kit. After protein bands appear, development is stopped by immersing the gels in 5% acetic acid, in which they are left overnight, followed by immersion in water for 24 h. Gels are then dried and photographed.

PURIFICATION OF BOVINE BRAIN AND PITUITARY FGF

Extraction

Since acidic and neutral mitogens can best be extracted at neutral or basic pHs, whereas basic mitogens such as FGF are preferentially extracted at acidic pHs (for a review, see Gospodarowicz et al. [1982b]), it is important in order to have minimal contamination by acidic or neutral mitogens to extract brain and pituitary tissue at pHs of 4.5. The pH should be carefully controlled during the whole process of extraction, since solubilized proteins could shift the initial pH of 4.5 to higher values through a buffering effect.

Typically, 1 kg of frozen pituitary is homogenized in 4 liters of 0.15 M NH_4SO_4 in a Waring blender in the cold (4°C). Sixteen milliliters of 6 N HCl is then added to the homogenate under vigorous stirring. The pH should become 4.5. For brain extract, 4 kg of whole brains is homogenized in 8 liters of 0.15 M NH_4SO_4 in the cold (4°C). Since whole brains are kept frozen at −80°C, in order to be able to fragment them easily it is wise to put the frozen brain in the coldroom the night before extraction or to let them stand at room temperature for at least 3-4 h. The brain and pituitary homogenates are stirred for 2 h in the coldroom. Their pH is checked and should be between 4.4 and 4.6. Extracts are then spun at 10,000 rpm (4°C) for 45 min (rotor JA10) in a Beckman J21C centrifuge. Supernatants are then collected and pooled. The pH is then raised to 6.0 by addition of 6 N NaOH. Two hundred grams of $(NH_4)_2SO_4$ per liter is added. After the salt is dissolved under stirring (30 min at 4°C), the suspension is spun at 10,000 rpm for 30 min (rotor JA10) in a Beckman J21C centrifuge. The supernatants are collected, and the small brown pellets are discarded. Then 250 g of NH_4SO_4 is added per liter, and after the salts have dissolved under stirring (30 min at 4°C), the suspension is spun down in a Beckman J21C centrifuge as described above. Supernatants are discarded and the massive pellets, which are reddish and contain most of the FGF activity, are collected into a 250-ml beaker and dissolved into 200 ml of distilled water. The pH is then checked to be sure it is 6.0. If the pH differs from that value, it is adjusted to 6.0 with either 1 N HCl or 1 N NaOH. The solution is then dialyzed overnight against 15 liters of water with two changes of water. After overnight dialysis, the dialysate is checked for pH and centrifuged at 10,000 rpm (rotor JA10) in a Beckman J21 C centrifuge.

Ion Exchange Chromatography on Carboxymethyl-Sephadex C-50

The dialysate is then applied at *room temperature* on a column (3 × 20 cm) of carboxymethyl Sephadex (CMS) C-50 equilibrated with 0.1 M sodium phosphate, pH 6.0 (Fig. 1A,C). Eighty percent of the material (fraction F_1) passes through the column unadsorbed. It does not contain more than 5% of the initial mitogenic activity (Fig. 1B,D). The column is then washed with 0.1 M sodium phosphate, pH 6.0. Further elution with 0.15 M NaCl in 0.1 M sodium phosphate removes a fraction that has a very low mitogenic activity (fraction F_2) (Fig. 1B,D). Most of the FGF activity (Fig. 1B,D) is eluted with 0.5 or 1 M NaCl in 0.1 M sodium phosphate at pH 6.0 (fraction F_3). This eluent is applied onto the column when the optical adsorption of the F_2 fraction reaches a value of 0.4. If one applies it when the optical density of the F_2 fraction has dropped to levels as low as 0.08, little protein will elute in the F_3 fraction. This last fraction (F_3) is then pooled and either concentrated at 4°C to about 10 ml by pressure ultrafiltration (UM-10 membrane) or dialyzed in the cold against 0.1 M ammonium carbonate overnight and then lyophilized.

Exclusion Gel Chromatography on Sephadex G-75

The F_3 CMS preparation, either concentrated by Amicon UM-10 ultrafiltration or lyophilized and dissolved in 10–12 ml of 0.1 M ammonium carbonate, is applied at 4°C to a Sephadex G-75 column (5 × 100 cm) equilibrated with 0.1 M ammonium carbonate, pH 9.2. The column is eluted with the same buffer. A typical elution profile is shown in Figure 2A for brain FGF. The mitogenic activity is contained in fractions 36–50. It covers two ultraviolet-absorbing peaks. A minor peak (FGF-1) elutes between fractions 36 and 41, and a major peak (FGF-2) elutes just before the cytochrome c marker (Mr = 12,700) (fractions 41–49). The material in these two peaks is pooled separately and lyophilized. Figure 2B shows the elution profile and activity of the pituitary FGF F_3 CMS fraction. The mitogenic activity is mostly localized in fractions 64–90 and reaches its maximum between fractions 68 and 80 (Fig. 2B).

The partially purified preparations of brain and pituitary FGF can be purified by either ion exchange chromatography on carboxymethyl cellulose Whatman CM-52 [Gospodarowicz et al., 1978a] or by preparative isoelectric focusing in liquid vein [Gospodarowicz et al., 1982b, 1983].

Ion Exchange Chromatography of Brain and Pituitary FGF on Carboxymethyl Cellulose CM-52

Brain FGF. The two active fractions of brain FGF-1 and -2 are dissolved in 0.2 M ammonium formate (pH 6.0) and further purified by ion exchange chromatography at 4°C on carboxymethyl cellulose CM-52. The partially purified preparations are applied to a column (1.5 × 20 cm) of carboxymethyl cellulose CM-52 equilibrated with 0.2 M ammonium formate, pH 6.0. After the column is washed with 300 ml of buffer, a gradient (0.2–0.4 M ammonium formate buffer) is applied. The gradient is prepared by allowing 150 ml of 0.4 M ammonium formate buffer (pH 6.0) to flow in a 150-ml constant-volume mixing chamber filled with the 0.2 M ammonium formate buffer. A typical elution pattern is shown in Figure 3A,B.

When brain FGF-1 is subjected to chromatography on carboxymethyl cellulose CM-52, most of the mitogenic activity appears between fractions 33 and 40 and corresponds to a protein fraction eluting at 0.35 M ammonium formate, pH 6.0 (Fig. 3A, right panel). When FGF-2 is similarly subjected to chromatography in order to improve the resolution, it becomes necessary to wash the column extensively prior to starting the gradient. A typical elution

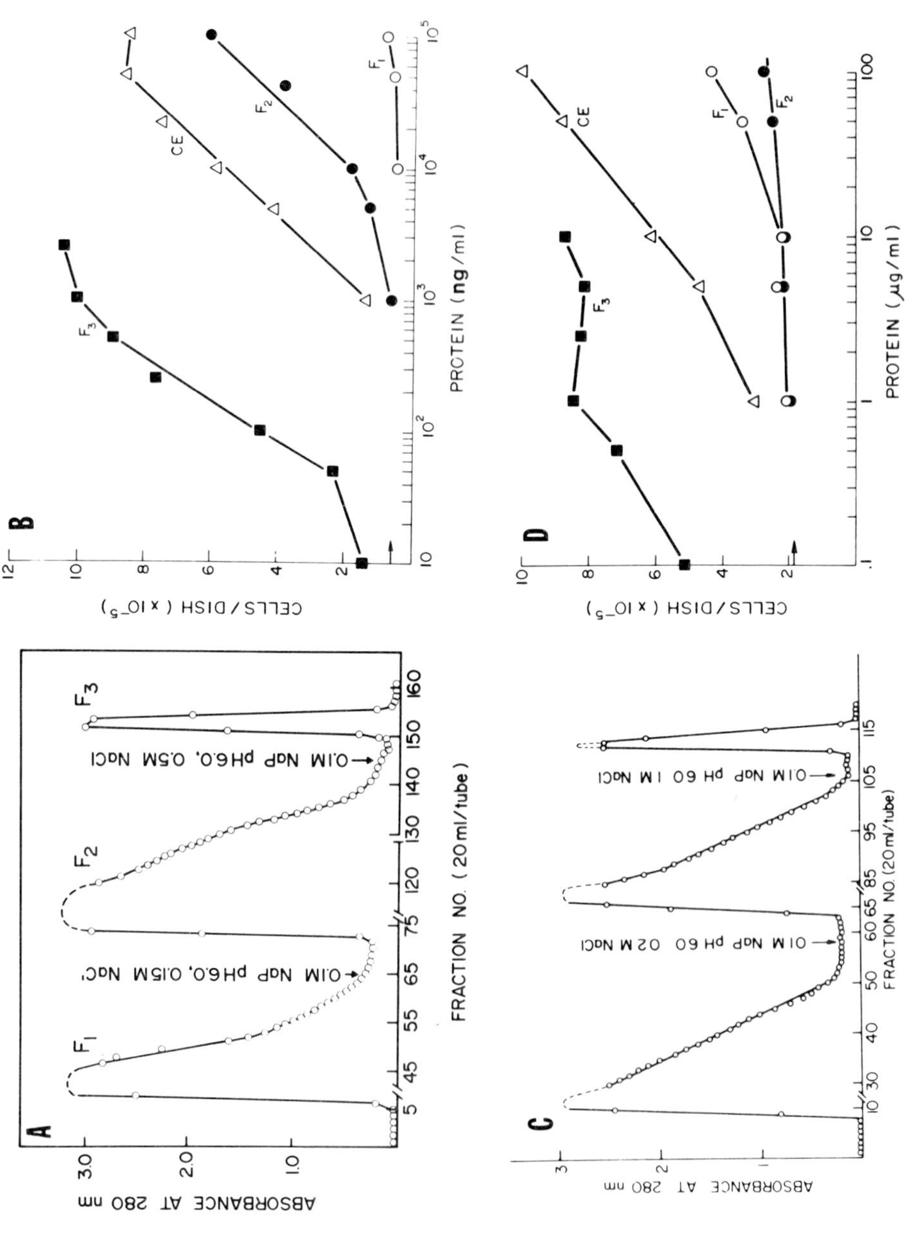
A
ABSORBANCE AT 280 nm
F1
F2
F3
0.1M NaP pH6.0, 0.15M NaCl
0.1M NaP pH6.0, 0.5M NaCl
FRACTION NO. (20 ml/tube)
B
CELLS / DISH (x 10^-5)
PROTEIN (ng / ml)
CE
C
ABSORBANCE AT 280 nm
0.1 M NaP pH 6.0 0.2 M NaCl
0.1 M NaP pH 6.0 1 M NaCl
FRACTION NO. (20 ml/tube)
D
CELLS / DISH (x 10^-5)
PROTEIN (µg / ml)

profile is shown in Figure 3B. The mitogenic activity is concentrated between fractions 25 and 40 and, as for FGF-1, eluted at 0.35 M ammonium formate, pH 6.0 (Fig. 3B, right panel).

The fractions containing the mitogenic activity are then pooled and lyophilized. The homogeneity of fractions FGF-1 and FGF-2 are checked by polyacrylamide gel electrophoresis at pH 4.5 and pH 8.5. At pH 4.5 a single band is observed for FGF-1. FGF-2 shows a major band, which does not run in the same position as FGF-1. In addition to this major band, a minor component present in FGF-1 runs ahead of the major band [Gospodarowicz et al., 1978a]. This minor component does not amount to more than 5% of the total protein content of FGF-2. When the electrophoresis is run at pH 8.5, no bands can be observed with either FGF-1 or FGF-2. The basic nature of the material is thus confirmed. When FGF-1 is analyzed by sodium dodecyl sulfate polyacrylamide slab gel electrophoresis, a single band that runs between the cytochrome c (Mr = 12,700) and the ribonuclease (Mr = 13,700) markers is observed. FGF-2, unlike FGF-1, gives a diffused band that runs at the level of the subunit of chymotrypsinogen (Mr = 11,700). Although FGF-2, when freshly prepared, gives a sharp band on sodium dodecyl sulfate gel, a diffuse pattern develops within a matter of weeks, even when the preparations are lyophilized and stored at −20°C [Gospodarowicz et al., 1978a]. This fuzziness is not due to heterogeneity of the material, since, when it is submitted to an acrylamide electrophoresis at pH 4.5, a single, major, and sharply defined band is observed.

Pituitary FGF. The pooled pituitary FGF F_4 and F_5 fractions partially purified by exclusion gel chromatography are further purified by ion exchange

---

Fig. 1. Elution pattern of the 1.65 to 3.55 M $(NH_4)_2SO_4$ brain or pituitary FGF fraction (40 g of protein) when submitted to ion exchange chromatography on carboxymethyl-Sephadex C-50. After overnight dialysis, as described in the text, the brain FGF fraction is applied to a column (3 × 20 cm) of carboxymethyl-Sephadex C-50 equilibrated in 0.1 M sodium phosphate, pH 6.0. A. Fractions of 20 ml were collected. Fraction F_1 represents unadsorbed proteins. The proteins were eluted stepwise by increasing the NaCl concentration first to 0.15 M (fraction F_2) and then to 0.5 M (fraction F_3). The biological activity was recovered in the last elution step (F_3). A similar elution profile (C) was observed when a 1.65 to 3.55 M $(NH_4)_2SO_4$ pituitary FGF fraction (38 g of protein) was chromatographed on carboxymethyl-Sephadex C-50. B. Distribution of the activities within the different fractions of brain FGF tested at concentrations ranging from 0.1 to 100 μg protein per milliliter in supporting the proliferation of bovine vascular endothelial cells seeded at an initial low density (2×10^4 cells per 35-mm dish; arrow) and exposed to DME supplemented with 10% calf serum. $(NH_4)_2SO_4$ fraction (CE, △) F_1 (○), F_2 (●), F_3 (■). The final density of cultures exposed to 10% calf serum alone was 1.8×10^{-5} cells per 35-mm dish after 6 days in culture. D. Distribution of the activities within the different fractions of pituitary FGF tested at concentrations ranging from 0.1 to 100 μg protein per milliliter in supporting the proliferation of bovine vascular endothelial cells.

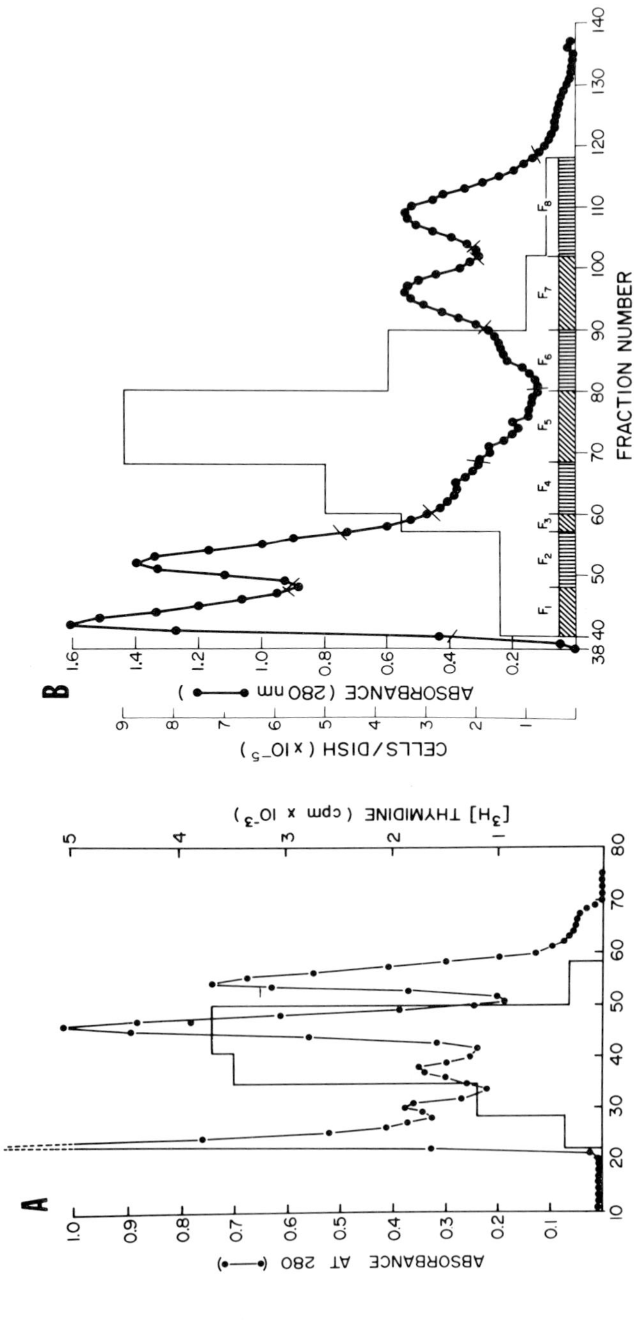
A
ABSORBANCE AT 280 (•—•)
[3H] THYMIDINE (cpm x 10-3)
TUBE NUMBER
B
ABSORBANCE (280 nm •—•)
CELLS/DISH (x 10-5)
FRACTION NUMBER
F1
F2
F3
F4
F5
F6
F7
F8

chromatography on carboxymethyl cellulose CM-52, as described for brain FGF. The partially purified preparations are dissolved in 0.2 M ammonium formate (pH 6.0) and applied to a column (1.5 × 20 cm) of carboxymethyl cellulose CM-52 equilibrated with 0.2 M ammonium formate, pH 6.0. After the column is washed with 300 ml of buffer, a gradient of 0.2-0.4 M ammonium formate (pH 6.0) is applied. A typical elution pattern is shown in Figure 4A.

The unadsorbed fractions do not contain any significant activity when tested for their ability either to stimulate initiation of DNA synthesis in resting cultures of Balb/c 3T3 cells (Fig. 4A) or to stimulate the proliferation of sparse adult bovine aortic endothelial (ABAE) cells (Fig. 4C). Once the column is eluted with an ammonium formate gradient (0.2-0.4 M), three major peaks of proteins are observed eluting at 0.3, 0.325, and 0.35 M ammonium formate, respectively (Fig. 4A). These three protein peaks are very active when tested for their ability to stimulate the initiation of DNA synthesis in resting cultures of Balb/c 3T3 cells (Fig. 4A) as well as the proliferation of vascular endothelial cells (saturation at 5 ng/ml, Fig. 4C), and most if not all of the biological activity recovered from the column is contained within those three protein peaks. When these different fractions are analyzed for their homogeneity by SDS polyacrylamide slab gel electrophoresis, they are observed to yield two primary protein bands that are closely apposed and run between the ribonuclease (Mr 13,700) and cytochrome c (Mr 12,700) molecular weight markers (Fig. 4B).

Isoelectric Focusing of Brain and Pituitary FGF

Brain FGF. The protein distribution and growth-promoting activity of the fraction containing brain FGF-1 activity, obtained by exclusion gel chromatography on Sephadex G-75 (Fig. 2A), can be further analyzed by submitting this fraction to isoelectric focusing over a pH range of 3 to 11 on a sucrose gradient (5-40%). As shown in Figure 5A, when 50 mg of protein was processed in

Fig. 2. Exclusion gel chromatography on Sephadex G-75 of the active bovine brain and pituitary fractions obtained from the carboxymethyl-Sephadex (CMS) C-50 ion exchange chromatography. A. Bovine brain FGF (800 mg) (fraction F_3 CMS) was dissolved in 15 ml of 0.1 M ammonium carbonate (pH 9.2) and applied to a column (5 × 100 cm) of Sephadex G-75 equilibrated in the same buffer. Twenty-two-milliliter fractions were collected. The histogram shows the relative activities of the different fractions on the initiation of DNA synthesis in sparse, resting populations of 3T3 cells. Fractions were tested at 100 ng/ml. B. Bovine pituitary FGF (750 mg) (fraction F_3 CMS) was dissolved in 10 ml of 0.1 M ammonium carbonate (pH 9.2) and submitted to exclusion gel chromatography on a Sephadex G-75 column, as described above. Fifteen-ml fractions were collected. The histogram shows the relative activities of the different fractions in supporting the proliferation of low-density vascular endothelial cells maintained on plastic and exposed to 10% calf serum. Fractions were tested at 50 ng/ml.

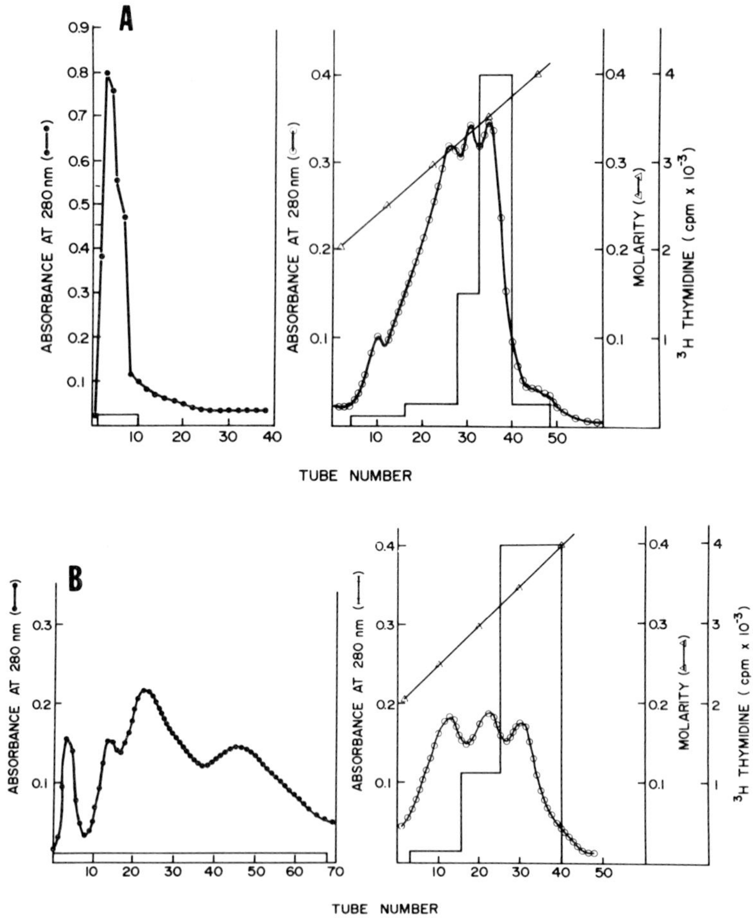

Fig. 3. Ion exchange chromatography of bovine brain FGF on carboxymethyl cellulose CM-52. A. Partially purified FGF-1 (300 mg) (G-75 fraction) was dissolved in 20 ml of 0.2 M ammonium formate (pH 6.0) and applied to a column (1.5 × 20 cm) of carboxymethyl cellulose CM-52 equilibrated in the same buffer. Six-milliliter fractions were collected. The profile of the unadsorbed proteins is shown in the left panel. The column was then eluted (right panel) with a linear salt gradient (0.2–0.4 M ammonium formate, pH 6.0; 300 ml). The flow rate was adjusted to 20 ml/h. The histogram shows the relative activities of the fractions on the initiation of DNA synthesis in sparse, resting 3T3 cell populations. Fractions were tested at 10 ng/ml. B. Elution prolife and mitogenic activity profile of 320 mg of partially purified FGF-2 (G-75 fraction) dissolved in 20 ml of 0.2 M ammonium formate (pH 6) and applied to a column of carboxymethyl cellulose CM-52, as described in A. The left panel shows the unadsorbed protein profile of elution, and the right panel shows the gradient elution and activity profile of the adsorbed protein.

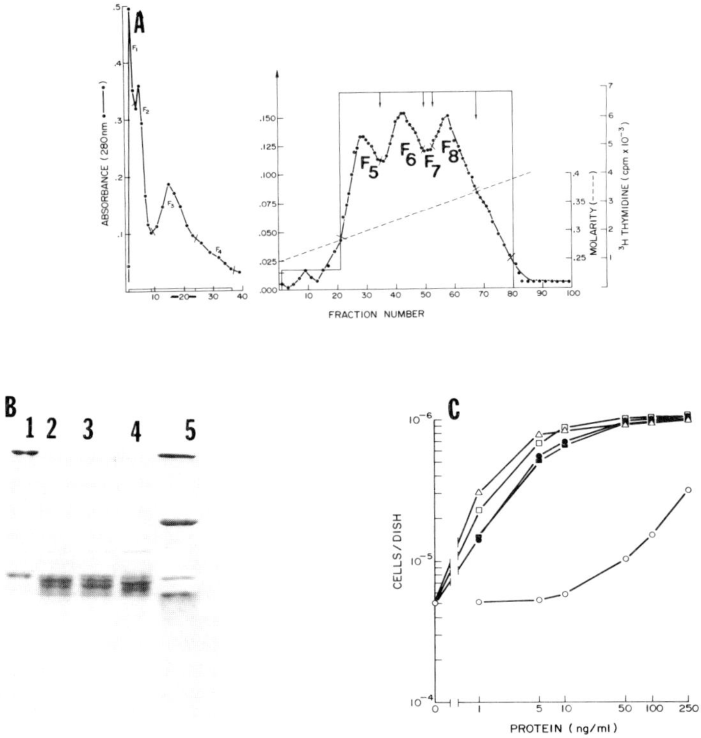

Fig. 4. Ion exchange chromatography on carboxymethyl cellulose CM-52 of a partially purified fraction of pituitary FGF (Sephadex G-75 step). A. Two hundred fifty milligrams of a partially purified fraction of pituitary FGF was dissolved in 20 ml of 0.2 M ammonium formate, pH 6.0, and applied to a column (1.5 × 20 cm) of carboxymethyl cellulose CM-52 equilibrated in the same buffer. Six-milliliter fractions were collected. The profile of the unadsorbed proteins is shown in the left panel. The column was then eluted with a linear salt gradient (right panel) (0.2-0.4 M ammonium formate, pH 6.0; 300 ml). The flow rate was adjusted to 20 ml/h. The histogram shows the relative activities of the fractions on the initiation of DNA synthesis in sparse, resting 3T3 cell populations. Fractions were tested at 10 ng/ml. B. SDS polyacrylamide gel electrophoresis of pituitary FGF (fractions 5, 6, and 8) purified by ion exchange chromatography on carboxymethyl cellulose CM-52. Lane 1 contains the markers bovine serum albumin and ribonuclease; lane 5 contains the marker bovine serum albumin, chymotrypsin, ribonuclease, and cytochrome c. Lanes 2, 3, and 4 contain 10 μg of the pituitary FGF fractions F_5, F_6, and F_8, respectively. C. Comparison of the biological activity of pituitary FGF purified by ion exchange chromatography or carboxymethyl cellulose CM-52. Increasing concentrations of F_1, F_5, F_6, F_7, and F_8 fractions (ranging from 1 to 250 ng/ml) were added to low-density bovine vascular endothelial cell cultures (2 × 10^4 cells per 35-mm dish). After 6 days in culture, the cell density of triplicate cultures was determined. F_1 (○), F_5 (●), F_6 (□), F_7 (■), F_8 (△).

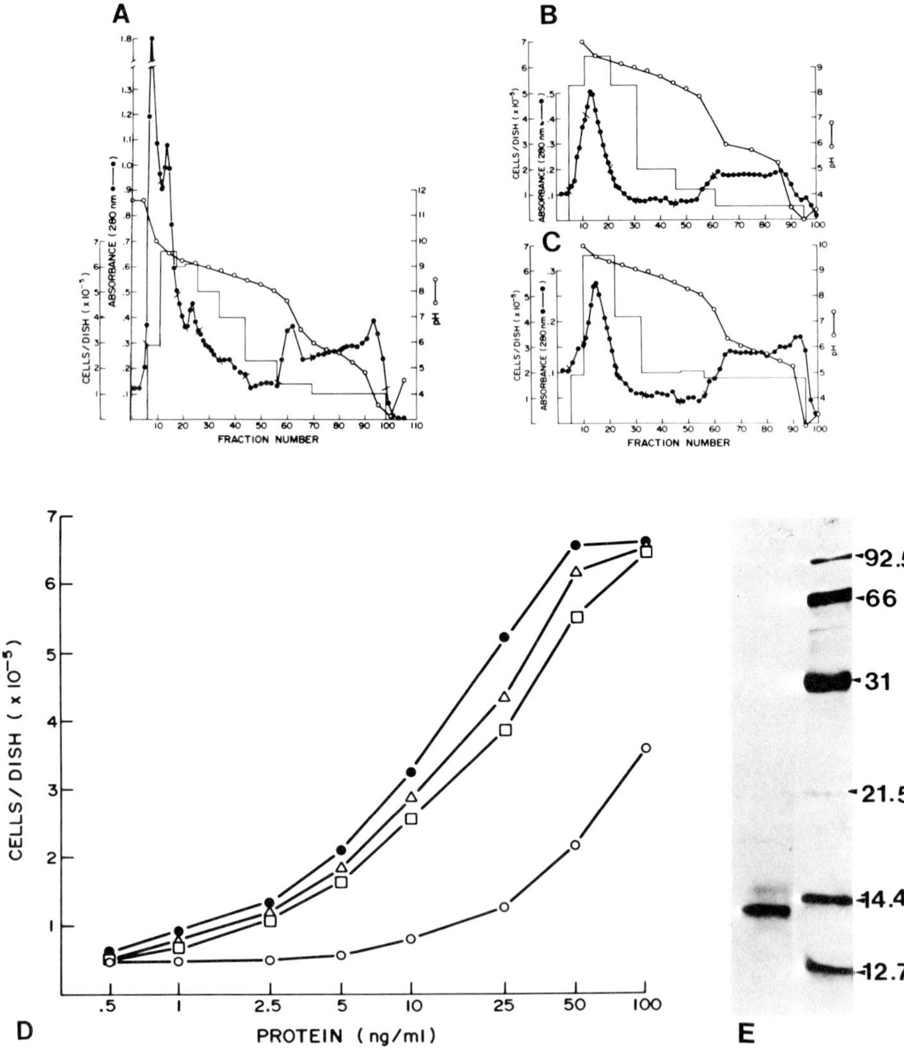
A
B
C
CELLS/DISH (x 10⁻⁵)
ABSORBANCE (280 nm)
pH
FRACTION NUMBER
D
CELLS / DISH (x 10⁻⁵)
PROTEIN (ng/ml)
E
92.5
66
31
21.5
14.4
12.7

Fig. 5. Further purification by isoelectric focusing of bovine brain FGF-1 G75 fractions. A. Fifty milligrams of a partially purified and lyophilized preparation of brain FGF-1 (G75 fraction, Fig. 1A, tubes 46-56) was resuspended in 3 ml of gradient solution and applied at the midposition of a pH density gradient that was formed by 6.2 ml of LKB ampholytes, pH 9-11, and stabilized by a 5-50% sucrose density gradient. The separation was carried out at 4°C for 72 h at 500 V. The gradient was then collected in 1.1-ml fractions. The optical density (A_{280}) of each tube and the pH of every five tubes were determined. Tubes were pooled as indicated in the graph, with the exception of tubes 11-20, covering fractions 2-4. Aliquots of 100 μl from tubes 11-20 were collected, mixed, and tested for biological activity at a final protein concentration of 100 ng/ml. The histogram shows the final cell density after 5 days in culture of vascular endothelial cell cultures seeded at an initial cell density of 2×10^4 cells per 35-mm dish and exposed to 100 ng protein per ml of the various fractions to be tested. The final cell density of vascular endothelial cell cultures exposed to 10% calf serum alone was 6.2×10^4 cells per 35-mm dish. Recovery of biological activity was 78% of the input, and 52% of it was contained between tubes 4 and 25.
B. Tubes 9-18 of the previous isoelectric focusing (A), which contained 22% of the biological activity, were reinserted in sequence on a pH gradient stabilized by a 5-50% sucrose density gradient that was formed by 6.2 ml of LKB ampholytes, pH 9-11. The separation was carried out at 4°C for 72 h at 500 V. The gradient was then collected in 1.1-ml fractions. The optical density of each tube and the pH of every 5 tubes were determined. Tubes were pooled as indicated in the graph, with the exception of tubes 11-17. Aliquots of 100 μl from tubes 11-17 were collected, mixed, and tested for biological activity at a final concentration of 100 ng protein per ml as described in A. Each fraction was tested at a final protein concentration of 100 ng protein per ml. The histogram shows the distribution of the biological activity throughout the elution profile. Recovery of biological activity was 84% of the input, and 75% of it was found between tubes 3 and 21.
C. Tubes 11-17 of the previous isoelectric focusing (B) were reinserted in sequence on a pH gradient stabilized by a 5-50% sucrose density gradient that was formed by 6.2 ml of LKB ampholytes, pH 9-11. The separation was carried out at 4°C for 72 h at 500 V, and the gradient was then collected in 1.1-ml fractions. The optical density of each tube and the pH of every 5 tubes were then determined and the tubes were pooled as indicated in the graph. Each fraction was tested at a final protein concentration of 100 ng protein per ml. The histogram shows the distribution of the biological activity throughout the elution profile. Recovery of biological activity was 83% of the input, and 80% of it was found between tubes 12 and 21.
D. Fractions focusing at a pI of 9.6 after one (tubes 11-20, A), two (tubes 11-17, B), or three (tubes 12-21, C) consecutive isoelectric focusings were analyzed for their ability to stimulate the proliferation of low-density (2×10^4 cells) vascular endothelial cell cultures. Their activity as a function of concentrations (ranging from 0.5 to 100 ng/ml) was compared to the activity of the FGF-1 G-75 fraction from which they originated. The various fractions were diluted at the various appropriate concentrations, and 10-μl aliquots were added every other day to tissue culture dishes, as described in Materials and Methods. Sephadex G-75 FGF-1 fraction (○--○). Fractions focusing at a pI of 9.6 after the first (△--△), second (●--●), and third (□--□) isoelectric focusing. Vascular endothelial cells were counted after 5 days in culture, and the final cell density of cultures exposed to 10% calf serum was 4×10^4 cells per 35-mm dish.
E. SDS polyacrylamide gel electrophoresis of FGF-1 fractions purified by isoelectric focusing. FGF-1 samples containing 2 μg proteins and originated from the third isoelectric focusing shown in Figure 2C (fraction contained between tubes 12 and 21) were subjected to electrophoresis (left lane). The migration of the samples was compared to that of protein standards, which included phosphorylase B (Mr 92,500), bovine serum albumin (Mr 66,200), carbonic anhydrase (Mr 31,000), soybean trypsin inhibitor (Mr 21,500), lysozyme (Mr 14,400), and cytochrome C (Mr 12,700) (right lane). Gels were stained and fixed as described in the text.

this way, the elution profile consisted of two major peaks and one minor peak, which focused in the basic region with pIs of 10.6, 9.6, and 9.2, respectively. Two small peaks focus in the acidic part of the gradient and had pIs of 6.5 and 4.3. These were shown to be artifacts resulting from the high background of adsorption of the ampholytes in the acidic range (for a discussion, see Gospodarowicz et al. [1982b]).

When the distribution of the biological activity throughout the pH gradient is analyzed, the regions with the greatest specific activity are found in the pH range of 9.6–9.2. Little if any activity is found within fractions focusing below pH 8.0. Since all fractions are tested at the same final protein concentration (100 ng/ml), an approximate comparison of their relative specific activities as well as of their total mitogenic contents can readily be made. As shown in Figure 5A, over 90% of the recovered mitogenic activity is concentrated between pH 9 and 10. Of the three protein peaks present in that region, the one with a pI of 9.6 has the highest specific activity. However, there is considerable overlapping between that protein peak and those focusing at pHs of 10.3 and 9.2.

To purify further the protein peak focusing at pH 9.6, it is resubmitted to isoelectric focusing (Fig. 5B). The elution profile of the second isoelectric focusing consisted of a main peak focusing at pH 9.6 and a smaller but broad peak in the acidic range (pH 6.7–4.0). When the mitogenic activity of these fractions is analyzed, it is found only in the main protein peak with a pI of 9.6; none is in the acidic range. To assess still further the association of the mitogenic activity with the twice-focused protein peak with a pI of 9.6, fractions within that pH range are pooled and focused for the third time (Fig. 5C). Biological activity is uniquely concentrated within the peak with a pI of 9.6 (Fig. 5C,D).

When the proliferative responses of low-density vascular endothelial cell cultures exposed to the FGF-1 G-75 fraction and the pI 9.6 protein fraction are compared as a function of protein concentrations (from 0.5 to 100 ng/ml) (Fig. 5D), the isoelectrically focused protein is found to be eight times more potent than the FGF-1 G-75 fraction. Upon refocusing of the fraction with a pI of 9.6, a tenfold increase in potency over the original FGF-1 G-75 fraction was observed. Mitogenicity (as reflected by increased cell proliferation) was readily detectable at 2.5 ng/ml, saturation occurred at 50 ng/ml, and half-maximal stimulation was observed at 10 ng/ml. When this fraction was refocused for the third time, a slight drop (10%) in potency was observed (Fig. 5D).

In order to analyze the homogeneity of the pI 9.6 protein fraction, it is subjected to SDS polyacrylamide slab gel electrophoresis without prior dialysis or lyophilization. After extensive gel washing in order to remove excess ampholytes, the gels are stained with silver nitrate. A major band running below the lyzozyme standard (Mr 14,400), and a minor band, which ran at the level of the lysozyme standard, are observed (Fig. 5E). The minor band did not amount to more than 5% of the total protein content when measured by densitometry. The apparent molecular weight of the major band, considered to be FGF-1, is within the range of 13,200-13,800, as previously reported [Gospodarowicz et al., 1978b]. It could not be determined with more accuracy, since in order to separate proteins of low molecular weight fully, a 10-18% exponential gel gradient is used. Amino acid analysis of that fraction revealed that it has a composition similar to that of FGF-1 previously purified by chromatography on carboxymethyl cellulose CM-52 [Gospodarowicz et al., 1978b].

When the biological activity of the pI 9.6 protein fraction is analyzed with low-density cultures of adrenal cortex, granulosa, vascular smooth muscle, and vascular or corneal endothelial cells, it is observed to be a potent mitogen for all five cell types (Fig. 6). Half-maximal stimulation of cell proliferation is observed at 5 ng/ml for vascular smooth muscle cells, 7 and 10 ng/ml, respectively, for granulosa and vascular or corneal endothelial cells, and 20 ng/ml for adrenal cortex cells (Fig. 6).

When the Sephadex G-75 fraction containing FGF-2 activity is further purified by isoelectric focusing, only the ascending slope of the Sephadex G-75 peak (fractions 40-44, Fig. 2A), which contains most of the biological activity, is used. This results, after isoelectric focusing, in a protein profile in which the protein peak with a pI of 9.54 increased considerably in size relative to that of the protein peak with a pI of 10.3. The acidic part (below pH 7.0) of the gradient, which in previous experiments had been shown to contain little if any activity, is not analyzed.

When the distribution of the mitogenic activity across the elution profile of FGF-2 is analyzed, the fraction of highest specific activity is that with a pI of 9.54 (Fig. 7A). When this fraction is refocused, a single peak with a pI of 9.54 is observed (Fig. 7B). The ability to support the growth of low-density vascular endothelial cell cultures of fractions focusing at pH 9.54 after one or two isoelectric focusings is shown in Fig. 7C and is compared to the activity of the Sephadex G-75 fraction from which they originated. The biological activity of the FGF-2 fraction focusing with a pI of 9.54 is detectable at protein concentrations as low as 1 ng/ml. Half-maximal stimulation is observed at 10 ng/ml, and optimal proliferation of vascular endothelial cells is observed at 25 ng/ml. Similar activity is observed after refocusing, and the final activity of the FGF-2 fraction focusing at pH 9.54 is eight times higher than that of the starting

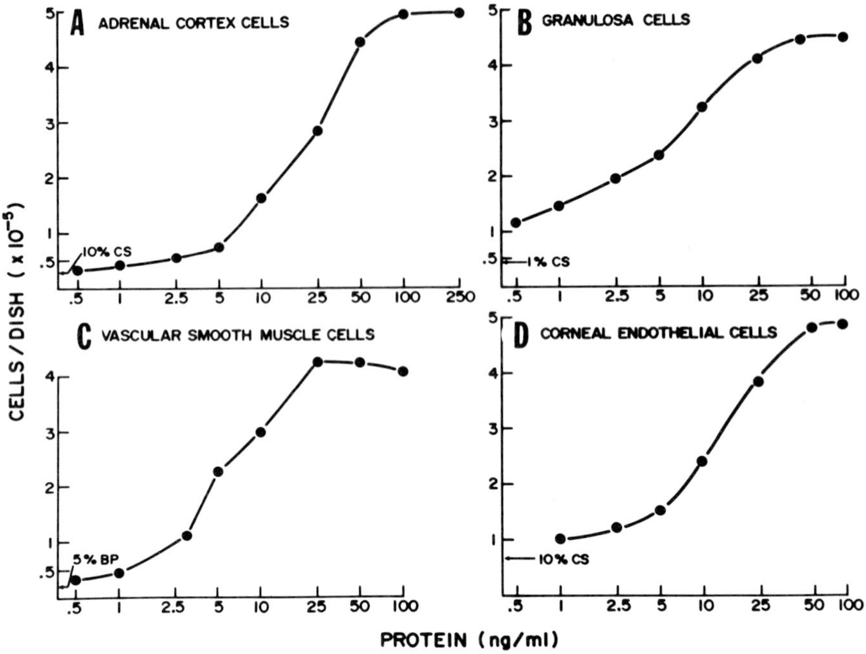

Fig. 6. Comparison of the proliferation of adrenal cortex, granulosa, vascular smooth muscle, and corneal endothelial cells exposed to increasing concentrations of FGF-1 purified by isoelectric focusing. Cells were seeded at an initial cell density of 2×10^4 cells per 35-mm dish. Adrenal cortex cells were maintained in the presence of F-12 medium supplemented with 10% calf serum (A). Granulosa cells (B) and corneal endothelial cells (D) were maintained in DME supplemented with 1% and 10% calf serum, respectively. Vascular smooth muscle cells (C) were maintained in DME supplemented with 10% defibrinated plasma. FGF-1 purified by isoelectric focusing (Fig. 2C, fraction covering tubes 12–21) was added every other day to the dishes at concentrations ranging from 0.5 to 100 or 250 ng/ml. Six days after seeding, triplicate cultures were trypsinized and counted with a Coulter counter. The final cell density of adrenal cortex cells maintained in 10% calf serum (10% CS) and in the absence of FGF-1 was 3.1×10^4 cells per 35-mm dish, that of granulosa cells exposed to 1% calf serum (1% CS) was 4.8×10^4 cells per dish, and that of vascular smooth muscle and corneal endothelial cells exposed to 5% bovine plasma (5% BP) and 10% calf serum (10% CS), respectively, was 2×10^4 and 7×10^4 cells per dish, respectively.

material (Fig. 7C). Brain FGF-2 is also a potent mitogen for human vascular endothelial cells maintained on either fibronectin- or laminin-coated dishes, with half-maximal stimulation at 5 ng/ml and saturation at 20 ng/ml [Gospodarowicz et al., 1983] (Fig. 8).

As already observed with brain FGF-1, FGF-2 is mitogenic for bovine adrenal cortex, granulosa, vascular smooth muscle, corneal endothelial, and bovine and human vascular endothelial cells. When the biological activities of FGF-2 fractions prepared by isoelectric focusing are compared to those of FGF-1 by assaying for initiation of DNA synthesis rather than for proliferation, FGF-2 is twice as potent as FGF-1. Stimulation of ^{3}H-thymidine incorporation into DNA by FGF-2 can be observed at concentrations as low as 0.5 ng/ml, versus 1 ng/ml for FGF-1. Half-maximal and maximal responses are observed at concentrations of 2 and 10 ng/ml respectively, for FGF-2, and concentrations of 4 and 25 ng, respectively, for FGF-1 (Fig. 7D).

When the homogeneity of the FGF-2 fraction focusing with a pI of 9.54 is analyzed by SDS polyacrylamide slab gel electrophoresis (Fig. 9), a prominent band running just above the cytochrome c standard (Mr 12,700) is observed. Above it is a faint band running below the lyzozyme standard (Mr 14,400), which could be FGF-1. When the fraction is refocused, a single band running just above the cytochrome c standard is observed (Fig. 9B). In order to obtain further confirmation of the different electrophoretic mobilities of FGF-1 and -2, samples containing each of these peptides are mixed and analyzed by slab gel electrophoresis (Fig. 9D–G). Two bands are then observed, one running below the lysosome standard and one running just above it. No other major bands can be observed. Amino acid analysis of that fraction indicates that it has a composition similar to that of FGF-2 that had been previously purified by chromatography on carboxymethyl cellulose CM-52 [Gospodarowicz et al., 1978b].

Pituitary FGF. The protein distribution and growth-promoting activity of the pituitary FGF fraction that was partially purified by exclusion gel chromatography on Sephadex G-75 and has the highest activity (Fig. 4, fraction 5, tubes 68–80) can also be further analyzed by submitting it to isoelectric focusing over a pH range of 3 to 11 on a sucrose gradient (5–40%), as described above for brain FGF. As shown in Figure 10A, when 30 mg of protein is processed in this way, a series of sharp peaks all focusing within the basic range is observed. The peak focusing at pH 9.6 has the highest specific activity (Fig. 10A). When this fraction was tested on ABAE cell growth, it had a half-maximal stimulating activity at 0.6 ng/ml and was saturating at 5 ng/ml (Fig. 11B) [Gospodarowicz et al., 1983]. Likewise, when tested for its ability to trigger human umbilical endothelial (HUE) cell growth (Fig. 11A), stimulation was half-maximal at 0.4 ng/ml and saturating at 5 ng/ml [Gospodarowicz et al., 1983]. When the homogeneity of the pituitary fraction was analyzed by slab gel electrophoresis, it was found to be composed of two protein bands closely apposed and running just below the lysozyme marker (Mr 14,400) (Fig. 10B). This profile is not significantly different from that observed when pituitary FGF is purified by ion exchange chromatography of a carboxymethyl cellulose Whatman CM-52 column (Fig. 4A,B).

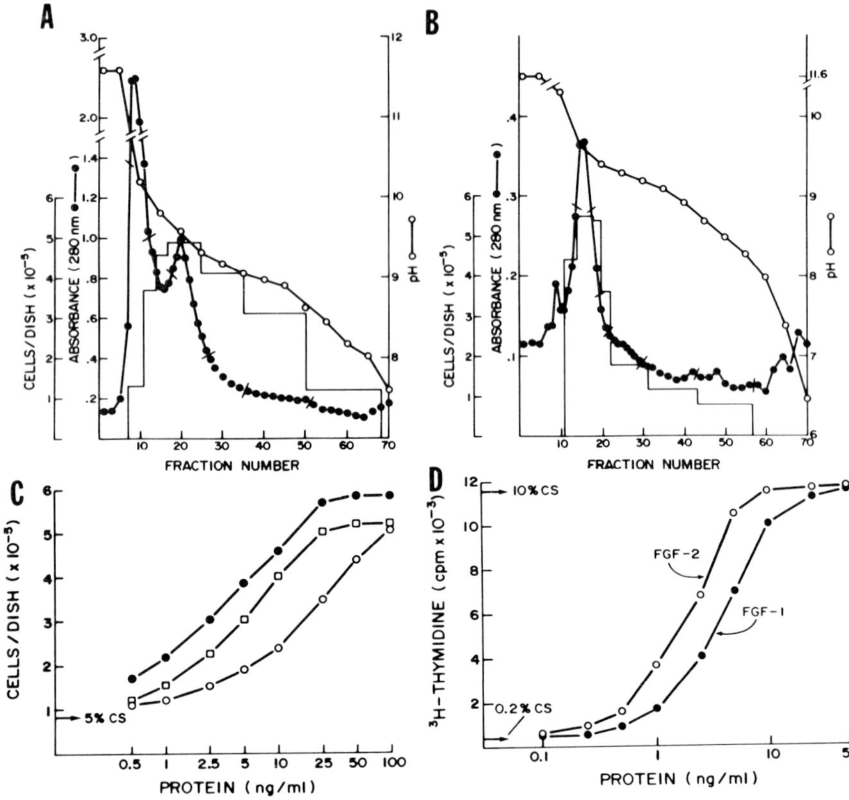

Fig. 7. Further purification by isoelectric focusing of bovine brain FGF-2 partially purified by gel filtration on Sephadex G-75. A. Forty milligrams of a partially purified and lyophilized preparation of brain FGF-2 (G-75 fraction, Fig. 2A, tubes 40-44) was resuspended in 3 ml of gradient solution and applied at the midposition of a pH density gradient that was formed by 6.2 ml of LKB ampholytes, pH 9-11 and stabilized by a 5-50% sucrose density gradient. The separation was carried out at 4°C for 72 h at 500 V. The gradient was then collected in 1.1-ml fractions. The optical density of each tube and the pH of every 5 tubes were determined. Tubes were pooled as indicated in the graph, with the exception of tubes 15-25, covering fraction 2. Aliquots of 100 μl from tubes 15-23 were collected, mixed, and tested for biological activity, as described in Materials and Methods, at a final protein concentration of 100 ng/ml. The histogram shows the final cell density after 5 days in culture of vascular endothelial cell cultures seeded at an initial cell density of 2 × 10^4 cells per 35-mm dish and exposed to 100 ng protein/ml of the various fractions to be tested. The final cell density of vascular endothelial cell cultures exposed to 10% calf serum was 9.6 × 10^4 cells per 35-mm dish. Recovery of biological activity was 87% of the input, and 43.3% of it was contained between tubes 15 and 23.

B. Tubes 15-23 of the previous isoelectric focusing (A) were reinserted in sequence on a pH gradient stabilized by a 5-50% sucrose density gradient that was formed by 6.2 ml of LKB ampholytes, pH 9-11. The separation was carried out at 4°C for 72 h at 500 V. The gradient was then collected in 1.1-ml fractions. The optical density of each tube and the pH of every 5

EXTRACTION OF MYOBLAST GROWTH FACTOR ACTIVITY FROM BRAIN AND PITUITARY

To extract neutral or acidic mitogens from pituitary or brain extract, one has to conduct the extraction at either neutral (pH 7.2) or basic (8.5) pH, as previously described [Gospodarowicz et al., 1975]. Briefly, 1 kg of brain or pituitary tissue is homogenized in 1 liter of 0.15 M ammonium sulfate, and the extract is adjusted with 6 N NaOH to a pH of 7.2 or 8.5. The homogenate is then stirred for 2 h in the cold and then centrifuged (rotor JA10) at 10,000 rpm in a Beckman J21C centrifuge for 30 min. The supernatants are then collected, 450 g of ammonium sulfate is added to them, and the suspension is then stirred for 30 min in the cold. Afterwards, it is centrifuged (rotor JA10) at 10,000 rpm in a Beckman J21C centrifuge. The supernatants are discarded, and the pellets are collected, dissolved in 200 ml of distilled water, and then dialyzed overnight against distilled water, as previously described. The dialyzates are centrifuged again in order to remove insoluble material and are then lyophilized.

The extracts containing mostly neutral and acid mitogens are potent within the range of 10–100 μg protein/ml when tested on either human umbilical cells (Fig. 12A) or adult bovine aortic cells (Fig. 12B). They can easily be further purified by ion exchange chromotagraphy on DEAE Whatman DE-52 and chromatofocusing with PBE 94 gel, a starter buffer of 0.025 M Tris HCl pH 8.3, and an elution buffer of polybuffer 96 (30%) and polybuffer 74 (70%) adjusted to pH 5.0 with 1 N acetic acid.

---

tubes were determined. Tubes were pooled as indicated in the graph and tested for biological activity at a final concentration of 100 ng protein per ml. The histogram represents the distribution of mitogenic activity throughout the elution profile when fractions were tested on low-density vascular endothelial cell cultures, as described in A. Recovery of biological activity was 83% of the input, and 75% of it was contained between tubes 11 and 22.

C. Fractions focusing at a pI of 9.54 after one (tubes 15 to 23, A) or two (tubes 9 to 16, B) consecutive isoelectric focusings were analyzed for their ability to stimulate the proliferation of low density (2×10^4 cells) vascular endothelial cell cultures. Their activity as a function of concentrations (ranging from 0.5 to 100 ng/ml) was compared to that of the Sephadex G75 FGF-2 fraction activity from which they originated (Fig. 1C, tubes 51 to 63). The various fractions were diluted at the various appropriate concentrations, and 10 μl aliquots were added every other day to tissue culture dishes, as described in "Materials and Methods." Sephadex G75 FGF-2 fraction (○--○). Fractions focusing at a pI of 9.6 after the first (□--□) and second (●--●) isoelectric focusings. Vascular endothelial cells were counted after 5 days in culture. The final cell density of cultures exposed to 5% calf serum was 8.5×10^4 cells per 35 mm dish.

D. Comparison of the ability of FGF-1 and -2 purified by isoelectric focusing to stimulate DNA synthesis in resting populations of Balb/c 3T3 cells. Arrows indicate (methyl-^{3}H)thymidine incorporation in cultures exposed to either 0.2% or 10% calf serum (CS). Assays were conducted as previously described [Gospodarowicz, 1975; Gospodarowicz et al. 1978a].

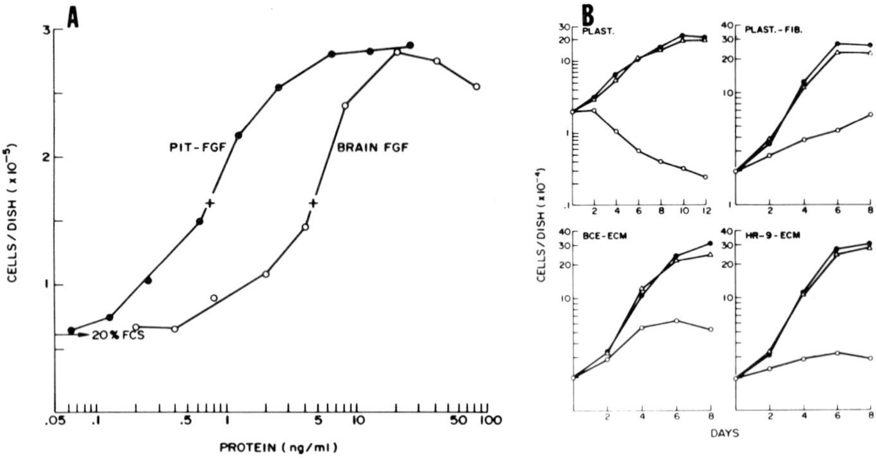

Fig. 8. Proliferation as a function of concentration of brain FGF-2 and pituitary FGF and as a function of time of early-passage human umbilical endothelial (HUE) cells maintained on plastic or fibronectin-, BCE-ECM-, or HR-9-ECM-coated dishes and exposed or not to pituitary or brain FGF. A. Low-density HUE cells (4×10^4 cells per 35-mm dish) in their first passage after primary and previously grown in the presence of pituitary FGF were seeded on fibronectin-coated dishes and exposed to 25 mM HEPES-buffered medium 199 supplemented with 10^{-8} M selenium, 20% fetal calf serum, and increasing concentrations of pituitary FGF (PIT-FGF, ●) or brain FGF (○). Both mitogens were added every other day. After 5 days in culture, triplicate dishes were trypsinized and counted in a Coulter counter. The crosses indicate the concentration at which a half-maximal response was observed for each mitogen. The final cell density of HUE cell cultures exposed only to 20% fetal calf serum was 6×10^4 cells per plate (arrow, 20% FCS). B. Low-density HUE cells (2×10^4 cells per 35-mm dish) in their first passage after primary and previously grown in the presence of pituitary FGF were seeded on plastic (PLAST), or on fibronectin-coated (PLAST-FIB), BCE-ECM-coated, or HR-9-ECM-coated 35-mm dishes and exposed to 25 mM HEPES-buffered medium 199 supplemented with 10^{-8} M selenium, 20% fetal calf serum alone (○) or with brain FGF (50 ng/ml, △) or pituitary FGF (10 ng/ml, ●) added every other day. Triplicate dishes were trypsinized every other day and counted in a Coulter counter.

USE OF FGF AND MYOBLAST GROWTH FACTOR TO GROW CELLS IN CULTURE

The biological effects of either brain or pituitary FGF in vitro can be seen with respect to cell proliferation, differentiation, and senescence. Both pituitary and brain FGF have been shown to be mitogenic for mesoderm-derived cells [Gospodarowicz et al., 1977a,b,c, 1978b,c,d] (Table 1) and have been used primarily to develop new in vitro endothelial cell lines, particularly of vascular and corneal origin (reviewed in Gospodarowicz [1979, 1982, 1983]; see also Gospodarowicz and Giguere [1982]).

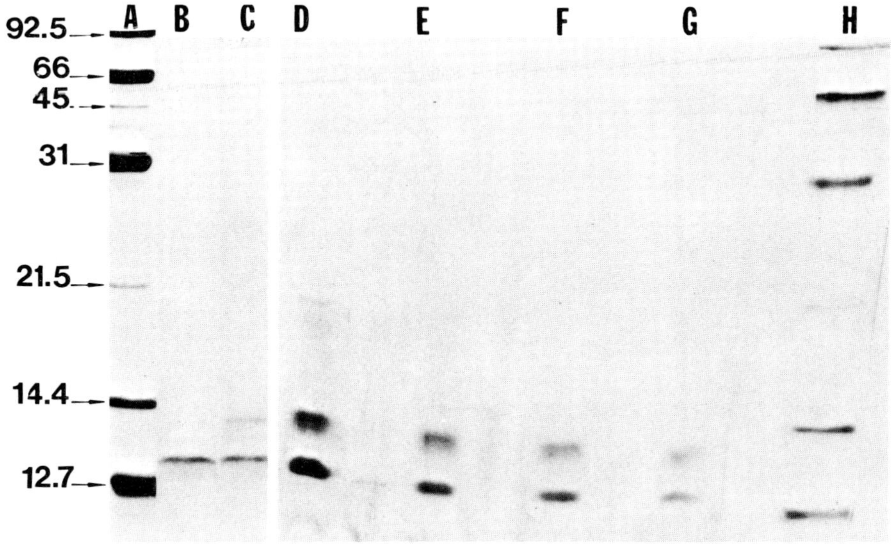

Fig. 9. SDS polyacrylamide gel electrophoresis of FGF-2 fractions purified by isoelectric focusing. A–C. FGF-2 samples originated from the first and second isoelectric focusings shown in Figure 7A and B (fractions contained within tubes 15–23 [7A] and 9–16 [7B], respectively) and containing 2 μg of protein were subjected to electrophoresis, as described in Gospodarowicz et al. [1982b] and Experimental Procedures. Lane B shows the protein distribution after the second isoelectric focusing, and lane C shows the protein distribution after the first isoelectric focusing. The migration of the samples was compared to that of protein standards (lane A), which included phosphorylase B (Mr 92,500), bovine serum albumin (Mr 66,200), ovalbumin (Mr 45,000), carbonic anhydrase (Mr 31,000), soybean trypsin inhibitor (Mr 21,500), lysozyme (Mr 14,400), and cytochrome c (Mr 12,700). D–G. FGF-1 and -2 samples originating from the isoelectric focusing shown in Figures 5B and 7B were mixed at final respective protein concentrations of 6 μg (D), 4 μg (E), 2 μg (F), and 1 μg (G). The migration of both samples was compared to that of the protein standard (lane H). Gels were stained and fixed as described in the text.

FGF can also affect the phenotypic expression of some cell types. This is a particularly interesting characteristic of FGF, since it has made possible the long-term culturing of various cell types that otherwise rapidly lose their normal phenotypes in culture when passaged repeatedly at low cell density. This biological effect of FGF has been best studied with vascular and corneal endothelial cell cultures cloned and maintained in the presence of FGF and then deprived of it for various time periods [Vlodavsky and Gospodarowicz, 1979; Vlodavsky et al., 1979].

Vascular endothelial cells grown and passaged serially in the presence of FGF adopt, upon reaching confluence, the morphology of a polarized cell

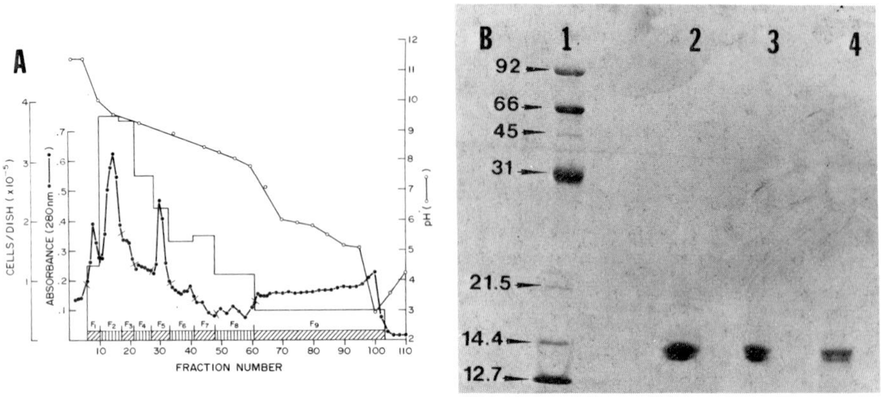

Fig. 10. Further purification by isoelectric focusing of pituitary FGF partially purified by exclusion gel chromatography on Sephadex G-75. A. Thirty milligrams of a partially purified and lyophilized preparation of pituitary FGF (G-75 fraction 5, Fig. 2B) was resuspended in 3 ml of gradient solution and applied at the midposition of pH density gradient that was formed by 6.2 ml of LKB ampholytes, pH 9-11, and stabilized by a 5-50% sucrose density gradient. The separation was carried out at 4°C for 72 h at 500 V. The gradient was then collected in 1.1-ml fractions. The optical density (A_{280}) of each tube and the pH of every 5 tubes were determined. Tubes were pooled as indicated in the graph, and tested for biological activity at a final protein concentration of 100 ng/ml. The histogram shows the final cell density, after 5 days in culture, of vascular endothelial cell cultures seeded at an initial cell density of 2×10^4 cells per 35-mm dish and exposed to 100 ng protein per ml of the various fractions to be tested. The final cell density of vascular endothelial cell cultures exposed to 10% calf serum alone was 5.4×10^4 cells per 35-mm dish.

B. SDS polyacrylamide gel electrophoresis of pituitary FGF purified by isoelectric focusing at a pI of 9.6. FGF samples containing 10, 5, and 2.5 μg proteins and originating from the isoelectric focusing shown in A (fraction contained between tubes 10 and 17, F_2) were subjected to electrophoresis (lanes 2, 3, and 4, respectively). The migration of the samples was compared to that of protein standards, which included phosphorylase B (Mr 92,500), bovine serum albumin (Mr 66, 200), carbonic anhydrase (Mr 31,000), soybean trypsin inhibitor (Mr 21,500), lysozyme (Mr 14,400), and cytochrome c (Mr 12,700) (lane 1). Gels were stained with silver nitrate and fixed as described in Experimental Procedures.

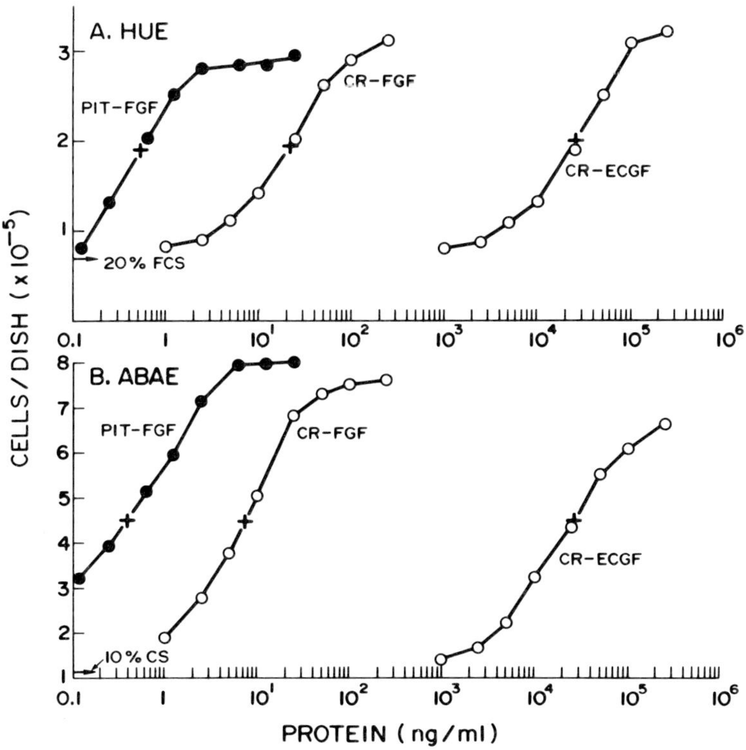

Fig. 11. Effect of increasing concentrations of pituitary FGF, CR-FGF, and CR-ECGF on the proliferation of low density HUE and ABAE cell cultures. A. Low-density HUE cells (4×10^4 cells per 35-mm dish) in their second passage were seeded on fibronectin-coated dishes and exposed to HEPES (25 mM) buffered medium 199 supplemented with 10^{-8} M selenium, 20% fetal calf serum, and increasing concentrations of pituitary FGF (PIT-FGF, ●), CR-FGF (lot 83-225, ○), or CR-ECGF (lot 82-356, ○). PIT-FGF or CR-FGF were added every other day in 10-μl aliquots of medium. Owing to the high concentration tested, CR-ECGF was added in 25-μl aliquots. After 5 days in culture, triplicate dishes were trypsinized and counted in a Coulter counter. The crosses indicate the concentrations at which a half-maximal response was observed for each mitogen. The final cell density of cultures exposed to 20% fetal calf serum was 7.2×10^4 cells (arrow, 20% FCS). B. Low-density ABAE cells (2×10^4 cells per 35-mm dish) were seeded on plastic dishes and exposed to DME supplemented with 10% calf serum and increasing concentrations of pituitary FGF (PIT-FGF, ●), CR-FGF (lot 83-225, ○), or CR-ECGF (lot 82-356, ○) added every other day, as described above. After 6 days in culture, triplicate dishes were trypsinized and cells were counted in a Coulter counter. The crosses indicate the concentrations at which a half-maximal response was observed for each mitogen. The final cell density of cultures exposed to 10% calf serum alone was 1.1×10^5 cells (arrow, 10% CS).

monolayer composed of tightly packed and flattened cells. This cell layer shows, as it does in vivo, an asymmetry of cell surfaces. The apical cell surface is nonthrombogenic, whereas the basal cell surface is involved in the synthesis of a highly thrombogenic basal lamina composed of interstitial collagen type III and basement membrane collagen types IV and V [Vlodvasky and Gospodarowicz, 1979]. Two glycoproteins involved in cell adhesion, laminin and fibronectin, are associated with the ECM produced by vascular endothelial cells [Gospodarowicz et al., 1981a] and are absent from their apical surface.

In contrast, when low-cell-density cultures are passaged repeatedly in the absence of FGF, they lose within three passages their ability to form a confluent monolayer of closely apposed and flattened cells. Instead, the cultures adopt a multilayer configuration consisting of large and overlapping cells that are no longer polarized [Vlodavsky and Gospodarowicz, 1979; Gospodarowicz et al., 1980]. Parallel to these changes in cell morphology, a loss of cell surface properties and polarities is observed. Both the apical and basal cell surfaces become covered by an extracellular matrix composed predominantly of interstitial collagen types I and III together with lesser amounts of basement membrane collagen types IV and V [Vlodavsky and Gospodarowicz, 1979; Gospodarowicz et al., 1981a]. Fibronectin, which in cells expressing their proper phenotype cannot be detected in the apical surface of cells, became present when cells started to express an aberrant morphology [Vlodavsky and Gospodarowicz, 1979; Gospodarowicz et al., 1981a]. These observations suggest that vascular endothelial cells maintained in the absence of FGF exhibit, in addition to a much slower growth rate, morphological and structural alterations that primarily involve changes in the composition and distribution of their ECM [Vlodavsky and Gospodarowicz, 1979]. These changes result in no less than the loss of the expression of the differentiated phenotype, since cells are no longer contact-inhibited and have a thrombogenic apical cell surface. Vascular endothelial cell cultures can regain their characteristic phenotypic expression if sparse or subconfluent cultures previously maintained without FGF are reexposed to it [Vlodavsky et al., 1979]. It may therefore be concluded that FGF stabilizes and can reestablish the phenotypic expression of vascular endothelial cells.

FGF can also significantly delay the ultimate senescence of cultured cells, as well as their ultimate differentiation. Thus, addition of FGF to myoblast cultures delays their fusion and myotube formation (reviewed in Gospodarowicz et al. [1982a]). In the case of granulosa cells, addition of FGF to either a clonal stain of granulosa cells or to low-density cultures can extend their lifespan in culture from 10 generations to 60 generations [Gospodarowicz et al., 1982a]. Adrenal cortex cell lines cloned in the presence of FGF show a

TABLE I. Cell Types Sensitive to the Mitogenic Effect of FGF When Maintained In Vitro on Plastic Dishes and Exposed to Serum-Supplemented Medium

Cell type	Animal from which cells are derived
Swiss 3T3	Mouse
Balb/c 3T3	Mouse
3T3 thermosensitive mutant	Mouse
Foreskin fibroblasts	Human
Glial cells	Human
Kidney fibroblasts	Cow
Amniotic cells (fibroblasts)	Human, cow
Articular and ear chondrocytes	Rabbit
Myoblasts	Cow (fetus), rats
Vascular smooth muscle (aortic arch)	Monkey, human, cow
Vascular endothelial cells (umbilical vein)	Cow, human
Vascular endothelial cells (fetal and adult aortic arch)	Cow
Endocardium	Cow fetus
Lens epithelial cells	Cow, cat
Corneal endothelial cells	Cow, rabbit
Y1 Adrenal cortex cells	Mouse
Adrenal cortex cells	Cow
Granulosa cells	Cow, pig, rabbit
Luteal cells	Cow
Blastemal cells	Frog
Blastemal cells	*Triturus viridescens*

similar dependence during their limited in vitro lifespan, and removal of FGF from the culture medium results not only in a greatly extended doubling time but also in rapid senescence [Gospodarowicz et al., 1982a; Gill et al., 1983]. In the case of vascular and corneal endothelial cells, FGF has been shown to extend the lifespan of the cultures. The effect is best observed with corneal endothelial cells passaged at low cell density, which have a lifespan of 20–30 generations in the absence of FGF, versus 200 generations in its presence [Gospodarowicz et al., 1981c; Giguere et al., 1982].

In summary, the following effects of FGF have been observed with various cell types: It allows cells to survive and grow at the low densities required for cloning; it increases the rate of cellular proliferation; it stimulates cell migration; it stabilizes their phenotypic expression and extends their "in vitro" lifespan. All these effects correlate with the ability of the cells to produce an extracellular matrix.

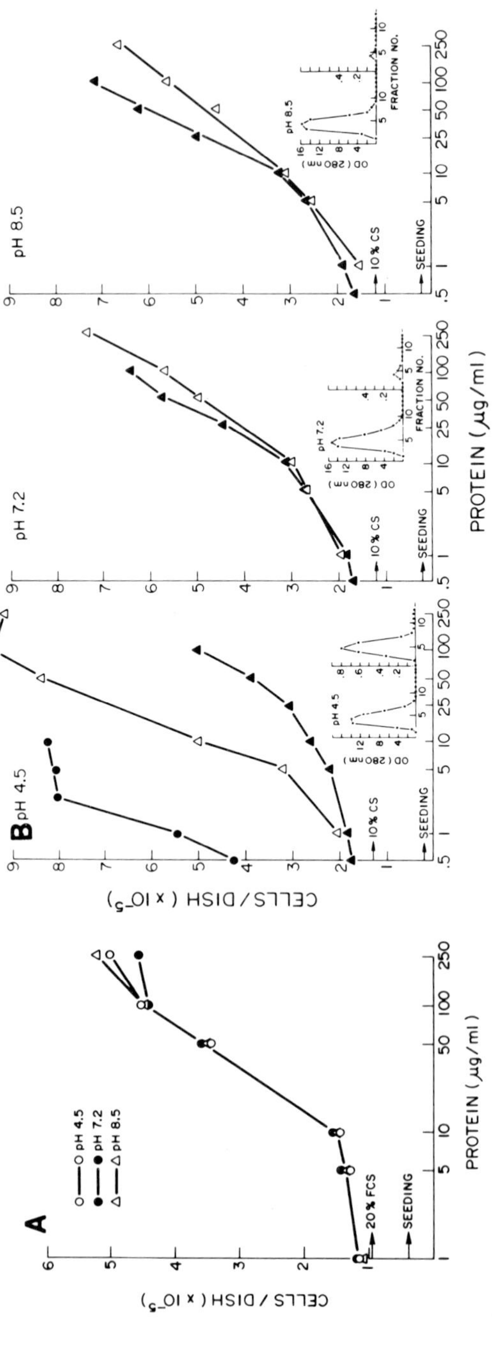
A
B
pH 4.5
pH 7.2
pH 8.5
CELLS / DISH (x 10^-5)
PROTEIN (μg/ml)
20% FCS
10% CS
SEEDING
OD (280 nm)
FRACTION NO.

Fig. 12. Effect of increasing concentrations of MGF prepared at pH 4.5, 7.2, or 8.5 on the proliferation of low-density HUE cells and ABAE cells. MGF were prepared as described in the text and Gospodarowicz et al. [1975].
A. Low-density HUE cells (4 × 10^4 cells per 35-mm dish, arrow, SEEDING), in their second passage were seeded on fibronectin-coated dishes. Cells were exposed to HEPES (25 mM) buffered medium 199 supplemented with selenium (10^{-8} M), 20% fetal calf serum, and increasing concentrations of MGF ranging from 1 to 250 μg/ml. MGF was added every other day, and after 6 days in culture triplicate dishes were trypsinized and counted in a Coulter counter. The cell density of control cultures exposed to 20% fetal calf serum was 9.2 × 10^4 cells after 6 days (arrow, 20% FCS).
B. Effects on the proliferation of low-density HUE cells of increasing concentrations of brain extracts prepared at pH 4.5, 7.2, or 8.5 and chromatographed on a carboxymethyl-Sephadex (CMS) C-50 column. Brain extracts were prepared as described in the text and ref. 7. Twenty milligrams of lyophylized extract was dissolved in 2 ml of 0.1 M Na phosphates buffer, pH 6.0, and applied on a small (3 ml total volume) column of carboxymethyl-Sephadex. The column was washed with 0.1 M Na phosphates buffer, pH 6.0. One-milliliter fractions were collected and their optical density at 280 mm was determined (insert). Protein concentration was determined on the basis of an extinction coefficient of $E^{1\%}{}_{1cm}$ = 0.8. After adequate dilution of the unadsorbed and adsorbed fractions with DME, aliquots (10 μl) containing the various protein concentrations were then added every other day to low-density ABAE cells (2 × 10^4 cells per 35-mm dish, arrow, SEEDING) maintained on plastic dishes and exposed to DME supplemented with 10% calf serum. After 6 days in culture, triplicate dishes were trypsinized and cells counted in a Coulter counter. The cell density of control cultures exposed to 10% calf serum was 1.2 × 10^5 cells after 6 days (arrow, 10% CS).

The biological effect of myoblast growth factor(s), in fact, a crude brain extract prepared at either neutral or basic pH, has heretofore been best studied with regard to the proliferation and differentiation of low-density myoblast cultures, for which it promotes both proliferation and differentiation of even extremely low density cultures. It also has a mitogenic effect on HUE cells, as reported by others [Maciag et al., 1978, 1982], and as recently confirmed by us [Gospodarowicz et al., 1982b; 1983]. Its ability to stimulate the proliferation of other cell types under various culture conditions has yet to be investigated.

REFERENCES

Giguere L, Cheng J, Gospodarowicz D (1982): Factors involved in the control of proliferation of bovine corneal endothelial cells maintained in serum-free medium. J Cell Physiol 110: 72–80.

Gill GN, Crivello JF, Hornsby PJ, Simonian MH (1982): Growth, function and development of the adrenal cortex: Insights from cell culture. In Sato G, Pardee A, Sirbasku DA (eds): "Cold Spring Harbor Conferences on Cell Proliferation: Growth of Cells in Hormonally Defined Media," Vol 9: "Hormones and Cell Culture." Cold Spring Harbor, New York: Cold Spring Harbor Laboratory, pp 461–482.

Gospodarowicz D (1974): Localization of a fibroblast growth factor and its effect alone and with hydrocortisone on 3T3 cell growth. Nature 249:123-127.

Gospodarowicz D (1975): Purification of a fibroblast growth factor from bovine pituitary. J Biol Chem 250:2515-2520.

Gospodarowicz D (1979): Fibroblast and epidermal growth factors. Their uses in vivo and in vitro in studies on cell functions and cell transplantation. Mol Cell Biochem 25:79-110.

Gospodarowicz D (1982): Extracellular matrix and the control of proliferation and differentiation of endothelial cells. In Vogel HJ, Nossel HL (eds): "Pathobiology of the Endothelial Cell." New York: Academic, pp 19-61.

Gospodarowicz D (1983): Control of endothelial cell proliferation and repair. In Cryer A (ed): "Biochemical Interaction at the Endothelium." New York: Elsevier North-Holland, pp 363-403.

Gospodarowicz D, Giguere L (1982): Growth factors, effects on corneal tissue. In McDevitt DS (ed): "Cell Biology of the Eye." New York: Academic, pp 97-135.

Gospodarowicz D, Weseman J, Moran J (1975): Presence in the brain of a mitogenic agent distinct from fibroblast growth factor that promotes the proliferation of myoblasts in low density culture. Nature 256:216-219.

Gospodarowicz D, Ill CR, Birdwell CR (1977a): Effects of fibroblast and epidermal growth factors on ovarian cell proliferation in vitro: I. Characterization of the response of granulosa cells to FGF and EGF. Endocrinology 100:1108-1120.

Gospodarowicz D, Ill CR, Hornsby PJ, Gill GN (1977b): Control of bovine adrenal cortical cell proliferation by fibroblast growth factor. Lack of effect of epidermal growth factor. Endocrinology 100:1080-1089.

Gospodarowicz D, Mescher AL, Birdwell, CR (1977c): Stimulation of corneal endothelial cell proliferation in vitro by fibroblast and epidermal growth factors. Exp Eye Res 25:75-89.

Gospodarowicz D, Bialecki H, Greenburg G (1978a): Purification of the fibroblast growth factor activity from bovine brain. J Biol Chem 253:3736-3743.

Gospodarowicz D, Greenburg G, Bialecki H, Zetter B (1978b): Factors involved in the modulation of cell proliferation in vivo and in vitro: The role of fibroblast and epidermal growth factors in the proliferative response of mammalian cells. In Vitro 14:85-118.

Gospodarowicz D, Mescher AL, Birdwell CR (1978c): Control of cellular proliferation by the fibroblast and epidermal growth factors. In "Gene Expression and Regulation in Cultured Cells," Third Decennial Review Conference, National Cancer Institute Monographs, No. 48, 109-130.

Gospodarowicz D, Moran J, Mescher AL (1978d): Cellular specificities of fibroblast growth factor and epidermal growth factor. In Papaconstantinou J, Rutter WJ (eds): "Molecular Control of Proliferation and Cytodifferentiation." 35th Symposium of the Society for Developmental Biology. New York: Academic Press, pp 33-61.

Gospodarowicz D, Vlodavsky I, Savion N (1980): The extracellular matrix and the control of proliferation of vascular endothelial and vascular smooth muscle cells. J Supramol Struct 13:339-372.

Gospodarowicz D, Greenburg G, Foidart J-M, Savion N (1981a): The production and localization of laminin in cultured vascular and corneal endothelial cells. J Cell Physiol 107:173-183.

Gospodarowicz D, Hirabayashi K, Giguere L, Tauber JP (1981b): Factors controlling the proliferative rate, final cell density, and life span of bovine vascular smooth muscle cells in culture. J Cell Biol 89:568-578.

Gospodarowicz D, Vlodavsky I, Savion N (1981c): The role of fibroblast growth factor and the extracellular matrix in the control of proliferation and differentiation of corneal endothelial cells. Vision Res 21:87-103.

Gospodarowicz D, Cohen DC, Fujii DK (1982a): Regulation of cell growth by the basal lamina and plasma factors: Relevance to embryonic control of cell proliferation and differentiation. In Sato G, Pardee A, Sirbasku DA (eds): "Cold Spring Harbor Conferences on Cell Proliferation: Growth of Cells in Hormonally Defined Media," Vol. 9: "Hormones and Cell Culture." Cold Spring Harbor, New York: Cold Spring Harbor Laboratory, pp 95-124.

Gospodarowicz D, Lui GM, Cheng J (1982b): Purification in high yield of brain fibroblast growth factor by preparative isoelectric focusing at pH 9.6. J Biol Chem 257:12266-12278.

Gospodarowicz D, Cheng J, Lirette M (1983): Bovine brain and pituitary fibroblast growth factors: Comparison of their abilities to support the proliferation of human and bovine vascular endothelial cells. J Cell Biol 97:1677-1685.

Laemmli UK (1970): Cleavage of structural proteins during the assembly of the head of bacteriophage T4. Nature (Lond) 227:680-685.

Maciag T, Cerundolo J, Isley S, Kelley PR, Forand R (1978): An endothelial cell growth factor from bovine hypothalamus. Identification and partial characterization. Proc Natl Acad Sci USA 76:5674-5678.

Maciag T, Hoover GA, Van der Spek J, Stermerman MB, Weinstein R (1982): Growth and differentiation of human umbilical vein endothelial cells in culture. In Sato G, Pardee A, Sirbasku DA (eds): "Cold Spring Harbor Conference on Cell Proliferation: Growth of Cells in Hormonally Defined Media," Vol 9: "Hormones and Cell Culture." Cold Spring Harbor, New York: Cold Spring Harbor Laboratory, pp 525-538.

Merril CR, Goldman D, Sedman SA, Ebert MH (1981): Ultrasensitive stain for proteins in polyacrylamide gels shows regional variation in cerebrospinal fluid proteins. Science 211:1437-1438.

Vlodavsky I, Gospodarowicz D (1979): Structural and functional alterations in the surface of vascular endothelial cells associated with the formation of a confluent cell monolayer and with the withdrawal of fibroblast growth factor. J Supramol Struct 12:73-114.

Vlodavsky I, Johnson LK, Greenburg G, Gospodarowicz D (1979): Vascular endothelial cells maintained in the absence of fibroblast growth factor undergo structural and functional alterations that are incompatible with their in vivo differentiated properties. J Cell Biol 83:468-486.

Methods for Serum-Free Culture of Epithelial and Fibroblastic Cells,
pages 199–207

13
Preparation of Bovine Pituitary Fibroblast Growth Factor

Sandra K. Lemmon and Ralph A. Bradshaw

The bovine fibroblast growth factors (FGF) are mitogenic substances that are known to have a wide variety of effects on cells derived primarily from embryonic mesoderm [Gospodarowicz et al., 1978b]. The two major tissue sources presently appreciated for FGF are brain and pituitary [Gospodarowicz, 1974], and the purification and characterization of the activities from these tissues have been reported [Gospodarowicz, 1975; Gospodarowicz et al., 1978a; Lemmon and Bradshaw, 1983].

The purification method for the pituitary form of bovine FGF that has been developed in our laboratory [Lemmon, 1982; Lemmon and Bradshaw, 1983] utilizes the steps of the earlier published procedures [Gospodarowicz, 1975; Gospodarowicz et al., 1978a], with some modification, with final isolation achieved by gel electrophoresis at acid pH. Pituitary FGF purified by this method has a molecular weight of ~14,500–15,000, has a basic pI, and is labile to acid and heat. These are similar to the properties reported for material isolated in earlier purification procedures [Gospodarowicz, 1975; Gospodarowicz et al., 1978a].

Department of Biological Chemistry, Washington University School of Medicine, St. Louis, Missouri 63110

S.K. Lemmon is now at the Department of Biological Sciences, Carnegie-Mellon University, Pittsburgh, Pennsylvania 15213.

R.A. Bradshaw is now at the Department of Biological Chemistry, California College of Medicine, University of California at Irvine, Irvine, California 92717.

MATERIALS AND METHODS

Source of Pituitaries

Both fresh (Royal Packing Company, East St. Louis, IL) and frozen (Pel-Freez Biologicals, or the gift of Dr. D.A.K. Roncari, University of Toronto) bovine pituitaries have been used. They are generally processed immediately upon receipt or, more conveniently, stored frozen at $\leqslant -20°C$ and used within 3-4 weeks. Differences have been noted in recoveries of FGF from preparation to preparation, but this does not appear to correlate with whether fresh or frozen material is utilized. For most investigators, frozen pituitaries are more readily obtainable. Addition of protease inhibitors in the early steps of the purification (see below) seems to reduce the variation and improve recovery.

Balb/c 3T3 Cell Bioassay

Biological activity is monitored by measuring incorporation of methyl-^{3}H-thymidine into DNA of Balb/c 3T3 fibroblasts as described previously [Lemmon et al., 1982]. One unit of activity is defined as 50% of maximal stimulation by a sample of FGF or 10% calf serum. Test samples from column fractions can usually be added without sterilization; however, samples from the early stages of the preparation may need to be filter-sterilized prior to dilution but after addition of carrier bovine serum albumin (BSA) (Sigma Chemical Company). Individual BSA lots must be screened to ensure that they do not interfere with the assay. Most samples can be diluted sufficiently so that salts and column buffers will not affect the bioassay.

Protein Determination

Protein concentration is determined by the method of Lowry et al. [1951] or can be estimated by absorption at 280 or 220 nm (1 unit at 220 nm is equivalent to ~90 μg/ml based on a BSA standard). Amino acid analyses of acid hydrolysates on a Durrum D-500 amino acid analyzer for the final CM-Sephadex FGF pool have correlated well with estimates by absorption at 220 nm.

ISOLATION PROCEDURE

All steps are performed, and pH and conductivity of solutions have been monitored, at 0-4°C. The columns indicated are appropriate for a typical preparation of approximately 1 kg starting material.

Homogenization and Extraction

Pituitaries are thawed and trimmed, if necessary, and homogenized in a 4-liter Waring blender with three 15-sec bursts in two liters of 0.15 M ammonium sulfate, 0.25 mM phenylmethanesulfonyl fluoride (PMSF), and 1 mM EDTA per kg tissue (wet weight) with 5 mg each of leupeptin and pepstatin A added (Sigma Chemical Company). Pepstatin and PMSF are dissolved in ethanol before addition to the aqueous solution. The pH of the homogenate is adjusted to 4.5 with 2 N HCl, stirred for 2 h, and then spun at 12,000g for approximately 1 h. The supernatant is retained and adjusted to pH 6.5–7.0 with 1 N NaOH.

Ammonium Sulfate Precipitation and Dialysis

The first ammonium sulfate precipitation of the neutralized crude extract is performed by slowly adding 250 g/liter of the finely ground salt, stirring for an additional 15 min, and then centrifuging as above for approximately 45 min. The second precipitation of the resultant supernatant is accomplished by the addition of 290 g/liter ammonium sulfate followed by stirring and centrifugation as in the first treatment. This pellet (P-II), which can be stored at 4°C overnight, is redissolved in distilled H_2O containing 0.25 mM PMSF and 1 mM EDTA and dialyzed extensively against the same solution for 1–1.5 days in dialysis tubing with a 6,000–8,000 MW cutoff (Spectropor 1, Spectrum Medical Industries, Inc.). The final P-II is clarified by centrifugation for 45 min at 12,000g.

Carboxymethyl-Sephadex-C-50 Stepwise Fractionation

The final P-II is adjusted to a pH of 6.0 and a conductivity of 2–2.5 mmho (0–4°C) by addition of 0.2 M sodium phosphate buffer, pH 6.0, and 0.1 M NaH_2PO_4, and the sample is loaded onto a column (4 × 12 cm) of CM-Sephadex-C-50 (Pharmacia Fine Chemicals) in 0.1 M sodium phosphate, pH 6.0. All solutions in this step include 0.25 mM PMSF and 1 mM EDTA. PMSF should be added from a concentrated stock in ethanol sufficiently ahead of time to allow complete dissolution. The column is washed extensively with this buffer and then with 0.1 M sodium phosphate, 0.15 M NaCl, pH 6.0. The active fraction (F-III) is eluted with 0.1 M phosphate buffer, 0.6 M NaCl, pH 6.0; dialyzed extensively against distilled H_2O, as for P-II (omitting PMSF and EDTA from the two final dialysis changes); and lyophilized.

Sephadex G-75 Column Chromatography

A column (2.5 × 90 cm) of Sephadex G-75-superfine (10–40 μm resin) is equilibrated in 0.2 M ammonium bicarbonate, 0.5 mM EDTA, 0.125 mM PMSF, at a flow rate of 10–12 ml/h. The lyophilized F-III fraction is dissolved in 6–8.5 ml of the column buffer and centrifuged (20,000g, 20 min) before loading on the column. Fifteen-minute fractions are collected and the protein profile is monitored by absorption at 280 nm.

An example of a typical gel filtration profile of the active fraction from the CM-Sephadex fractionation is shown in Figure 1. Generally a shoulder of activity (fractions 90-100) elutes prior to the major peak, and smaller amounts of activity can be observed in all of the higher-molecular-weight fractions [Lemmon, 1982]. The major activity peak (pools II-V in this preparation) represents 80-90% of the total activity recovered from gel filtration columns. In order to obtain the most enriched fraction for further purification, it is important to avoid contamination by material found in the large peak absorbing at 280 nm preceding the FGF (fractions 90-105). Therefore, although pool II contained as much activity as pool V in this particular profile, it was not included in the pool of activity prepared for the next step. Note that approximately 20 mg of protein per kg starting material can be obtained at this stage of purification with specific activities of less than 1 ng/ml/unit (Table I).

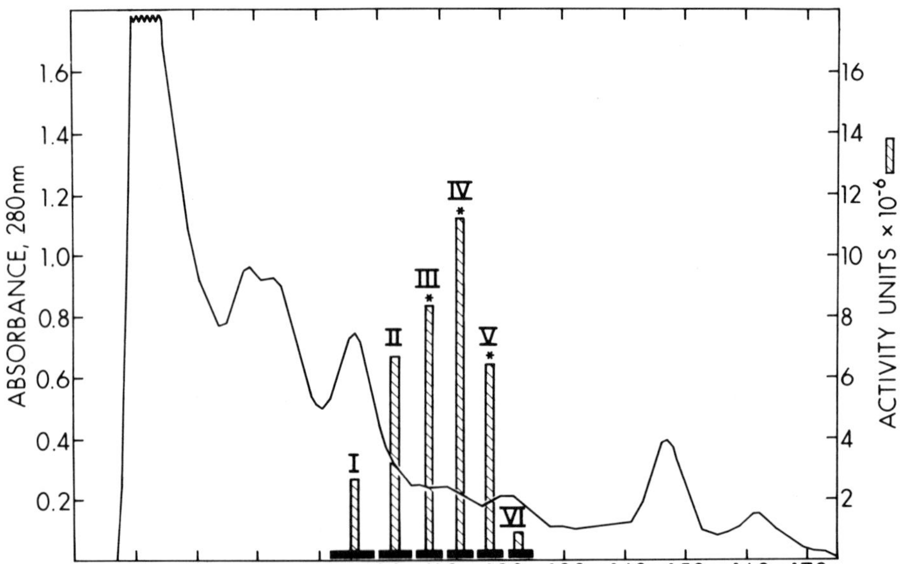

Fig. 1. Elution profile of the gel filtration on a column of Sephadex G-75 superfine of partially purified pituitary FGF. The fraction (~300 mg), corresponding to F-III [Gospodarowicz, 1975], was loaded in 7.6 ml column buffer and eluted at a flow rate of 10 ml/h with 15-min fractions collected. Absorption at 280 nm (——) was read and fractions were pooled. Biological activity was measured and total units (bars) were estimated. Asterisks denote the pools (III-V) that were combined for the next step. From Lemmon and Bradshaw [1983], with permission.

TABLE I. Activity of Fractions From the Sephadex G-75 and Second CM-Sephadex Gradient Purifications of Pituitary FGF

Fraction	Total protein (mg)	ng/ml/unit[a]	Total units × 10^{-6}
G-75[b] (III-V)	19.15	0.737	26
CM-A[c]	0.392	0.35	1.12
CM-B[c]	0.355	0.192	1.82
CM-C[c]	0.395	0.066	5.98
CM-D[c]	0.542	0.125	4.33
CM-E[c]	0.330	0.28	1.18

[a]One unit is defined as 50% of maximal stimulation of incorporation of ^{3}H-thymidine into Balb/c 3T3 cells by FGF.
[b]Sephadex G-75 fraction from chromatography shown in Figure 1.
[c]CM-Sephadex gradient fractions from chromatography shown in Figure 2.

CM-Sephadex-C-50 Gradient Elution

A column (1 × 10 cm) of CM-Sephadex is equilibrated with 0.1 M sodium phosphate, 0.1 M NaCl, pH 6.0, at a flow rate of 6.5-7.5 ml/h. The Sephadex G-75 pool is redissolved in loading buffer, dialyzed against the same for equilibration, and centrifuged to remove particulate matter. The sample is loaded, and the column is washed first with loading buffer and then with 0.15 M NaCl in the sodium phosphate buffer. A gradient from 0.15 M to 0.6 M NaCl (240-280 ml total volume) is developed to elute the FGF activity. Protein is most easily monitored by absorption at 220 nm, since only small amounts of material are eluted in the gradient portion of the column. Fractions are pooled, dialyzed against 0.2 M ammonium bicarbonate, and lyophilized two to three times to remove salts. These are designated CM-FGF.

The conditions of this second CM-Sephadex column are slightly modified from the earlier procedure [Gospodarowicz, 1975] in that a wash with 0.15 M NaCl has been introduced before initiation of the gradient. This serves to remove some of the major protein contaminants that elute early in the direct 0.1-0.6 M NaCl gradient. Figure 2 shows the 0.15-0.6 M NaCl gradient portion of an elution profile.

Acid Gel Electrophoresis

Acid gel disc electrophoresis is performed by the type G modified Reisfeld method of Rodbard and Chrambach [1971]. Ten percent, 7.5- to 8.0-cm, polyacrylamide resolving gels and 3.0% stacking gels are polymerized with light in the presence of riboflavin-5′-phosphate. Samples are dissolved in 10% glycerol and a 1:10 dilution of upper running buffer. Gels are electrophoresed at ap-

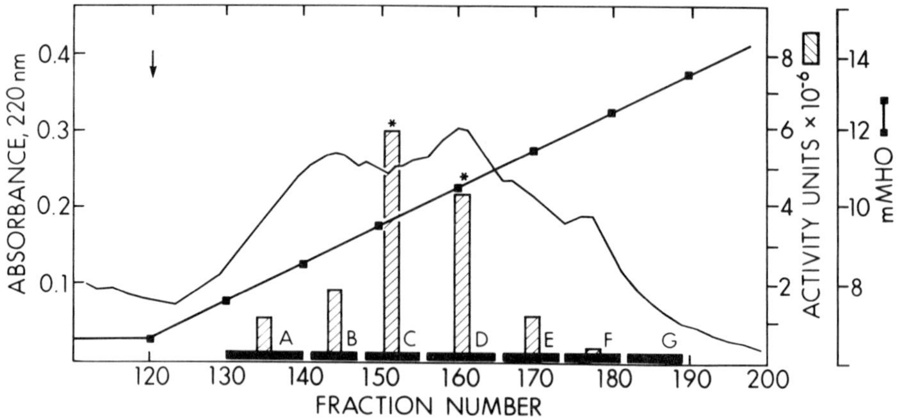

Fig. 2. Elution profile of a portion of the chromatography on a column of CM-Sephadex-C-50 of pituitary FGF. Pools III–V of the Sephadex G-75 column (Fig. 1) (19.15 mg protein) were loaded in 9.7 ml of 0.1 M sodium phosphate, 0.1 M NaCl, pH 6.0, to the CM-Sephadex column (1 × 10 cm), which was washed extensively with this buffer and then with 0.15 M NaCl in the phosphate buffer (not shown). The linear gradient was developed with 120 ml each 0.15 M and 0.6 M NaCl in the phosphate buffer, at a flow rate of 6.5 ml/h with 20-min fractions collected. The arrow indicates the start of the gradient. Fractions were pooled as shown and biological activity (bars) was determined. The pools were dialyzed with 0.2 M ammonium bicarbonate and lyophilized. The starred fractions were used for further purification. Absorption at 220 nm (——) and conductivity at 0°C (■——■) are shown. From Lemmon and Bradshaw [1983], with permission.

proximately 4°C and 10 mA for 2 h (the amount of time it takes for a cytochrome c marker, which runs near the buffer front, to migrate close to the bottom of the gel). Sample lanes to be tested for biological activity are cut into 0.25-cm segments and eluted by gentle shaking overnight at 4°C in 1.0 ml of phosphate-buffered saline (PBS) and 2 mg/ml BSA. Adjacent lanes containing FGF are also sliced and frozen at −20°C, while determination of the segments containing activity is made. Gels are stained with Coomassie blue R.

To obtain samples of homogeneous pituitary FGF free of carrier protein, frozen slices from acid pH gels, which have been selected based on the activity profile of a parallel lane, can be eluted directly into autoclaved distilled water with 1 mM benzamidine and 50 μg/ml pMSF added. Successful elution has been obtained when two thawed slices (0.25 × 0.15 × 2.5 cm) are cut into smaller pieces and shaken gently in a plastic tube in 1.0 ml of the protease inhibitor-containing water at 37°C for 16–20 hours. The first eluate is transferred to another plastic tube and the gel fragments are washed once with 0.2 ml of autoclaved water, which is combined with the first supernatant.

DISCUSSION

The procedure for purification of bovine pituitary FGF is based on the method of Gospodarowicz [Gospodarowicz, 1975; Gospodarowicz et al., 1978a]. The introduction of protease inhibitors, in particular leupeptin and pepstatin, in the initial extraction at pH 4.5 is an important modification that yields better overall recovery of FGF activity and material of higher specific activity. The use of PMSF and EDTA in column chromatography steps should be viewed as a precaution rather than a requirement.

Table I summarizes the recovery and activities of the fractions from the gel filtration and second CM-Sephadex steps. In both operations, selective pooling is required to obtain material of high specific activity. This is particularly true in the second ion exchange column, where the biological activity elutes very broadly from the column and is generally not associated with a distinct peak absorbing at 220 nm; however, careful pooling reveals that some fractions are significantly more active than the others, CM-C having the highest specific activity and CM-D being somewhat less potent. However, even with conservative pooling of fractions, 15–30% of total activity in the initial crude extract can be recovered by these methods with approximately a 10,000-fold to 15,000-fold purification.

It should be noted that recovery of pituitary FGF from columns and dialysis is a major problem. This property of adhering to support media may explain the broad elution pattern of FGF in purification on the CM-Sephadex gradient column, although some inherent heterogeneity in the FGF molecule cannot be excluded. The use of fairly high salts in gel filtration columns and the use of 0.2 M ammonium bicarbonate for dialysis of dilute protein solutions have been found to be very effective in overcoming these problems. The use of plastic or silanized glassware in later stages of purification may enhance recoveries.

The final purification of bovine pituitary FGF requires elution from acid pH electrophoresis gels. Figure 3 shows the activity profile of a sample (CM-C of Fig. 2) that had been electrophoresed, sliced, and eluted. The single mitogenic component observed migrates with a major staining band on acid gels and migrates essentially as a single component when reanalyzed by acid or SDS gel electrophoresis [Lemmon and Bradshaw, 1983]. Quantitative recovery of activity from gels (80–100%) can be obtained by this procedure when material is eluted into BSA.

Pituitary FGF is quite stable to lyophilization and is best stored in this form, but column fractions can be frozen during the time required to monitor biological activity. It is advisable to carry through the purification procedure as rapidly as possible through at least the first CM-Sephadex chromatography

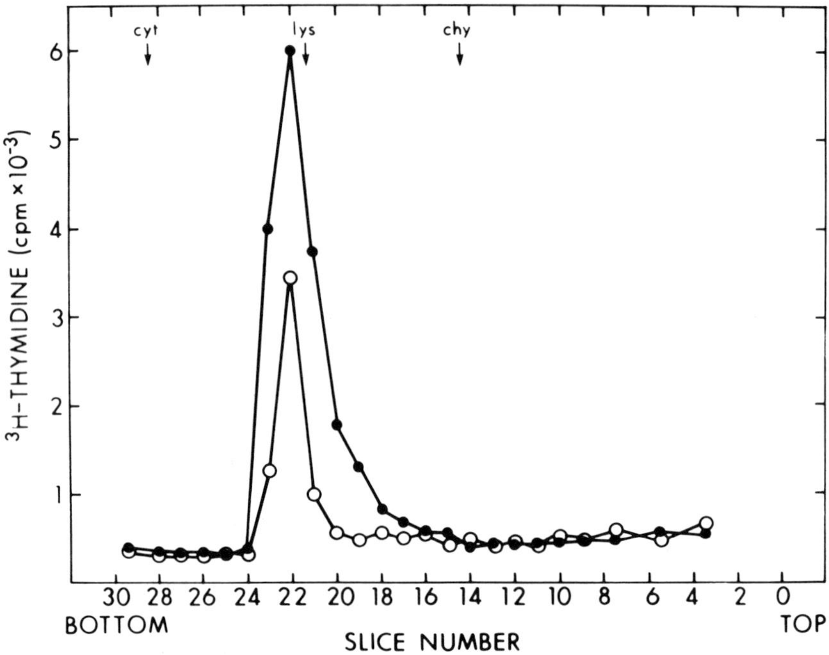

Fig. 3. Activity profile of acid gel electrophoresis of purified FGF. Pituitary FGF (6 μg) (fraction CM-C, Fig. 2) was electrophoresed in a 10% acid pH gel (8 cm). Gels were cut into 0.25-cm slices and eluted in 2.0 mg/ml BSA in PBS for determination of biological activity. Migration of the sample was from right to left and activity is expressed in cpm of ^{3}H-thymidine incorporated into Balb/c 3T3 cells. Sample volume added to assay: 0.2 μl (●—●), 0.04 μl (○—○) of 1.0 ml total per eluted slice. Marker proteins run in a parallel lane are indicated with arrows: cytochrome c (cyt), lysozyme (lys), and chymotrypsinogen (chy).

step. Samples of FGF recovered from the Sephadex G-75 column can be also lyophilized directly and stored at −20°C for long periods of time without noticeable losses of activity. Pituitary FGF is stable to reducing agents, but is inactivated by heat (5 min at 60–70°C), strong acids, and SDS. Mildly acidic conditions such as those used in the acid gel electrophoresis (pH 4.5) do not appear to affect the molecule.

The procedures outlined here for the purification of pituitary FGF provide essentially homogeneous material when an acid gel electrophoresis step is added to a slight modification of the original preparation scheme [Gospodarowicz, 1975; Gospodarowicz et al., 1978a]. Contrary to these reports, we have not been able to obtain homogeneous material when FGF is purified through the final CM-Sephadex column, although comparable specific activities have been obtained [Lemmon, 1982]. Pituitary FGF can also be obtained from commercial sources, but these preparations have often been purified only through the Sephadex G-75 column. For many purposes, FGF of this lesser purity may be suitable, particularly when it is to be used as a media supplement. However, the material purified by acid pH gel electrophoresis will probably be the desired form in studies on the mechanism of action.

Comparison of our preparation of pituitary FGF to the two mitogens present in brain FGF preparations suggests that the basic mitogen obtained from both sources may indeed be identical [Lemmon et al., 1982]; however, none of the components is related to fragments of myelin basic protein [Thomas et al., 1980; Lemmon et al., 1982]. Brain FGF also usually contains another mitogen for 3T3 cells with an acidic pI, and this substance appears to have a different spectrum of biological activity [Thomas et al., 1980; Lemmon et al., 1982]. Investigators may obtain different results when using brain and pituitary FGF preparations as culture supplements.

ACKNOWLEDGMENTS

This study was supported by American Cancer Society research grant BC-273.

REFERENCES

Gospodarowicz D (1974): Localization of a fibroblast growth factor and its effect alone and with hydrocortisone on 3T3 cell growth. Nature (London) 249:123-127.

Gospodarowicz D (1975): Purification of fibroblast growth factor from bovine pituitary. J Biol Chem 250:2515-2520.

Gospodarowicz D, Bialecki H, Greenberg G (1978a): Purification of fibroblast growth factor activity from bovine brain. J Biol Chem 253:3736-3743.

Gospodarowicz D, Greenberg G, Bialecki H, Zetter BR (1978b): Factors involved in the modulation of cell proliferation in vivo and in vitro: The role of fibroblast and epidermal growth factors in the proliferation of mammalian cells. In Vitro 14:85-118.

Lemmon SK (1982): Purification and characterization of bovine fibroblast growth factors. Doctoral Thesis, Washington University.

Lemmon SK, Bradshaw RA (1983): Purification and partial characterization of bovine pituitary fibroblast growth factor. J Cellular Biochem 21:195-208.

Lemmon SK, Riley MC, Thomas KA, Hoover GA, Maciag T, Bradshaw RA (1982): Bovine fibroblast growth factor: Comparison of brain and pituitary preparations. J Cell Biol 95:162-169.

Lowry OH, Rosebrough NJ, Farr AL, Randall RJ (1951): Protein measurements with the Folin phenol reagent. J Biol Chem 192:265-275.

Rodbard D, Chrambach A (1971): Estimation of molecular radius, free mobility and valence using polyacrylamide gel electrophoresis. Anal Biochem 40:95-134.

Thomas KA, Riley MC, Lemmon SK, Baglan NC, Bradshaw RA (1980): Brain fibroblast growth factor: Nonidentity with myelin basic protein fragments. J Biol Chem 255:5517-5520.

Methods for Serum-Free Culture of Epithelial and Fibroblastic Cells,
pages 209–220

14
Preparation of Pituitary Acidic FGF

Angelo A. Gambarini, Mari C.S. Armelin, and Hugo A. Armelin

Fibroblast growth factor (FGF) was first described in bovine pituitary extracts as a peptide distinct from the classical pituitary hormones [Armelin, 1973]. Observations that have accumulated since then led to the belief that pituitary glands possessed a group of unknown growth factors all of which belonged to a family of basic peptides. It was shown that serum could be depleted of its growth-promoting factors by eliminating a minor fraction containing neutral and basic proteins and that the growth-promoting activity of this depleted serum could be restored by the addition of pituitary extracts [Armelin et al., 1974; Nishikawa et al., 1975]. Two different factors have been isolated from bovine pituitaries: ovary growth factor (OGF) [Gospodarowicz et al., 1974] and a fibroblast growth factor [Gospodarowicz, 1975], both of which were basic proteins of pI around 9.5. However, as early as 1973, preliminary isoelectric focusing analyses of pituitary extracts (K. Nishikawa and H. Armelin, unpublished results) suggested that the pituitary gland could be a source of growth factors other than the basic ones. In fact, recently we reported on the isolation of acidic FGF from bovine pituitaries [Gambarini and Armelin, 1979, 1981, 1982; Armelin et al., 1982].

Gospodarowicz et al. [1975] extracted at pH 8.5 neutral to acidic growth factor activity and described it as a myoblast growth factor. More recently Maciag et al. [1979, 1982] have reported the partial purification from hypothalamus of an acidic growth factor for endothelial cells, and Thomas et al. [1980] found an acidic FGF in brain tissue.

The pituitary acidic FGF we described can be found both in fresh, carefully dissected glands and in commercially available frozen tissue [Gambarini and

Instituto de Quimica, Universidade de São Paulo, São Paulo 05508, Brazil

Armelin, 1981, 1982]. FGF is only an operational term to designate growth factors purified by using fibroblasts as the indicator cells in biological activity determinations. The purpose of this article is to describe a relatively simple procedure to purify acidic FGF from pituitaries. In view of the fact that the effects of growth factors in suspended cells have seldom been studied, we also discuss the activity of pituitary growth factors in agarose suspension cultures.

PROCEDURES FOR THE ISOLATION OF PITUITARY ACIDIC FGF

Materials

Bovine pituitaries were either collected fresh at the slaughterhouse and immediately frozen over dry ice or purchased as frozen pituitaries. Large batches of frozen pituitaries (50 kg) were ground (still frozen) in a meat grinder, lyophilized, and stored in plastic bags at −20°C. The chemicals used were: carboxymethyl-Sephadex C-50 (CM-Sephadex, 4.5 meq/g; Pharmacia); diethylaminoethyl-Sephadex A50 (DEAE-Sephadex, 3.5 meq/g; Pharmacia); Bio-Gel P-series, 100–200 mesh (Bio-Rad); LKB Ampholines; fetal calf serum and Dulbecco's modified Eagle's minimal medium (DME) from Gibco; [methyl-^{3}H]thymidine (New England Nuclear). All other reagents were of analytical grade.

Cell Lines

Swiss mouse 3T3 fibroblasts [Todaro and Green, 1963] and rat C_6 glial cells [Benda et al., 1968] were obtained from American Type Culture Collection and Rat-1 fibroblasts [Prasad et al., 1976] came from Dr. R. Hynes's laboratory (M.I.T.). All cells were stored in liquid nitrogen and periodically thawed. The anchorage-dependent C_6 variant (C_6-3D) was isolated from the C_6 line in our laboratory [Armelin et al., 1982, 1983; Armelin and Armelin, 1983]. Stock cultures were grown in screw-cap glass flasks with 95% DME (plus 1.2 g/liter sodium bicarbonate) and 5% fetal calf serum (FCS), flushed with 5% CO_2 and 95% air, and sealed.

Assay for Stimulation of DNA Synthesis Initiation

This assay was used to monitor purification steps during growth factor preparation. 3T3 fibroblasts were plated at 10^4 cells per cm^2 in 24-well Linbro trays in 5% serum-DME. The plating medium was replaced 24 h later by fresh 0.2% serum-DME. After 24 h of serum starvation, cells were growth-stimulated by the addition of pituitary fractions or serum. [^{3}H]thymidine (0.5 μCi/ml; 2×10^{-7} M) was added 12 h after restimulation and incorporation took place for the next 12 h. To measure radioactivity uptake into DNA the medium was removed, cells were washed with cold 10% trichloroacetic acid, dissolved in 0.5 N sodium hydroxide (2 h at 55°C), and applied to pieces of filter paper (0.5 cm^2; Whatman number 17), which were washed with cold 5% trichloroacetic acid-95% ethanol-acetone, dried, and counted in a liquid scintillation spectrometer. Every point was the average of two wells. As an internal standard, 5% serum, which allows maximal DNA synthesis stimulation, was regularly included. We defined 1 unit of activity as 50% of maximal stimulation. Serial dilutions of pituitary protein fractions were made in phosphate-buffered saline containing 0.5 mg/ml ovalbumin; ovalbumin alone had no activity whatsoever. Samples containing ampholytes were assayed directly, since up to 0.02% ampholytes had no effect on the cell response to growth factors.

Assay For Stimulation of Colony Growth in Agarose Suspension Cultures

Cells were plated at different densities (10^2 to 10^3 cells per well) in 24-well Linbro trays containing a layer of 0.6% agarose, 10% fetal calf serum-DME (0.5 ml/well). Plating medium consisted of 0.3% agarose in 10% FCS-DME (1 ml/well). Liquid 10% FCS-DME (1 ml/ per well) was added 24 h after plating and renewed, by gentle aspiration, every 3 days. Plating efficiency was determined at 21 days by counting colonies under a stereo microscope (eight-fold magnification). The activity of growth factors and hormones on growth in suspension was assessed by addition to the liquid overlay medium. Pituitary protein fractions (0.5-5 μg/ml) and hydrocortisone hormone (250 ng/ml) were added every time the liquid overlay was renewed.

Preparation of Acidic FGF

The preparation summarized in Figure 1 is relatively simple and can also be used to prepare basic FGF. All operations were performed at 0-4°C. One kilogram lots of frozen pituitaries were fragmented with a meat grinder, suspended in 2 liters of 0.15 M ammonium sulfate, and further homogenized in a blender. The pH was adjusted to 7.0 with 2 N NaOH and periodically readjusted if necessary. After 1 h stirring, the suspension was centrifuged (10,000 rpm, 20 min), the supernatant was harvested, and the precipitate was reextracted under the same conditions. The supernatants were combined and the final precipitate discarded. Similarly, in the case of lyophilized pituitaries, 200 g of lyophilized powder was extracted twice with 2 liters of 0.15 M ammonium sulfate. The extract obtained (supernatant) was dialyzed against distilled water for 24 h (before use, Visking dialysis tubing was boiled for 15 min, successively, with 0.1% EDTA, distilled water, 0.1 M ammonium car-

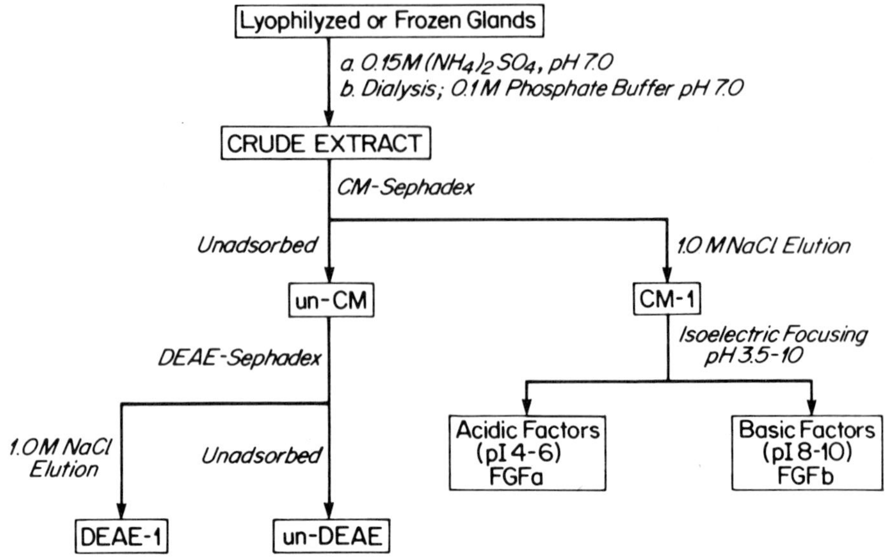

Fig. 1. Scheme for the preparation of acidic and basic growth factors from bovine pituitary glands.

bonate, and distilled water, and was finally extensively washed with water). After dialysis the suspension was made 0.1 M sodium phosphate buffer, pH 7.0. After 30 min stirring, the suspension was centrifuged (10,000 rpm; 20 min), the precipitate was discarded, and the supernatant collected as the *crude extract*.

The crude extract was applied to a CM-Sephadex column equilibrated with 0.1 M phosphate buffer pH 7.0 (PB-pH 7); 1 g protein per gram resin, 20 cm minimal column length, and 10 ml/cm^2 · h flow rate. The column was washed with 2 volumes of PB-pH 7 and the unadsorbed material was pooled in a single protein fraction (un-CM). The material adsorbed to the column was eluted with PB-pH 7 plus 1 M NaCl yielding the *CM-1 fraction*. CM-1 is 10 times more active than the crude extract (Table I). On the basis of the elution conditions used, one would expect that CM-1 was devoid of acidic active species. However, isoelectric focusing demonstrated that this fraction possessed both acidic and basic factors as shown in Figure 2C.

The un-CM fraction was rich in acidic activity (see Fig. 2B) but it had low specific activity owing to the fact that it contained the bulk of the crude extract protein. This fraction was loaded onto a DEAE-Sephadex column (1 g protein per gram resin; 20 cm minimal length) equilibrated with PB-pH 7. Protein retained in the column was eluted with PB-pH 7 plus 1 M NaCl to yield the

TABLE I. Purification of Acidic FGF (FGFa) From Bovine Pituitary Glands

	Protein		Activity		
Fraction	mg/kg tissue	% Recovery	Units/mg	Units/kg tissue	% Recovery
Crude extract	12,000[a]	100	115	1.4×10^6	100
CM-1	300[a]	2.5	1,040	3.1×10^5	22
DEAE-1	900[a]	7.5	714	6.4×10^5	46
un-DEAE	10,000[b]	83	26	2.6×10^5	19
FGFa	15	0.13	10,000	1.5×10^5	11

Procedures are described in detail in the text and summarized in Figure 1; activity was determined by stimulation of DNA synthesis in 3T3 fibroblasts.
[a]Protein estimated by the Peterson method [1977].
[b]Protein estimated by UV absorbance.

DEAE-1 fraction, whose specific activity increased considerably relative to the crude extract (see Table I).

CM-1 fraction is a favorable source for isolation of acidic growth factors because the bulk of its protein is neutral to basic. Isoelectric focusing can efficiently separate both acidic and basic factors (see Fig. 2C). For isoelectric focusing, CM-1 was concentrated by ultrafiltration (Millipore Pellicon Membrane PSAC; molecular weight cutoff 1,000), the buffer was changed to 1% glycine, pH 8.5, and the insoluble material was eliminated by centrifugation. Up to 50 mg protein was focused in LKB columns of 110 ml in a pH range of 3.5-10 (5-50% sucrose gradient; details according to LKB application note 219). Special care was taken throughout the run and fraction collection to avoid contamination with the precipitated protein bands. Typically a complete isoelectric focusing took 20 h. Once finished, the pH of the different fractions was measured immediately without dilution (Ingold surface electrode); pH was then adjusted to 8.5 with NaOH to dissolve precipitated proteins and to allow absorbance determination. Adequate dilutions were made with phosphate-buffered saline plus 0.5 mg/ml ovalbumin to measure biological activity. The active fractions comprising pI ranges 4-6 and 8-10 were pooled to give FGFa and FGFb, respectively.

FGFa is a preparation of acidic growth factors that is very active and is obtained with a reasonable yield (Table I); it can be stored indefinitely at neutral pH and −20°C, the activity being unaffected by repeated freeze-thawing. FGFa contains several acidic factors with different pIs and molecular weights, requiring further steps of isoelectric focusing and molecular sieving for final purification.

To refocus FGFa, a sample containing up to 10 mg protein (as obtained from the first isoelectrofocusing) was made 1.6% and 0.4% ampholytes at

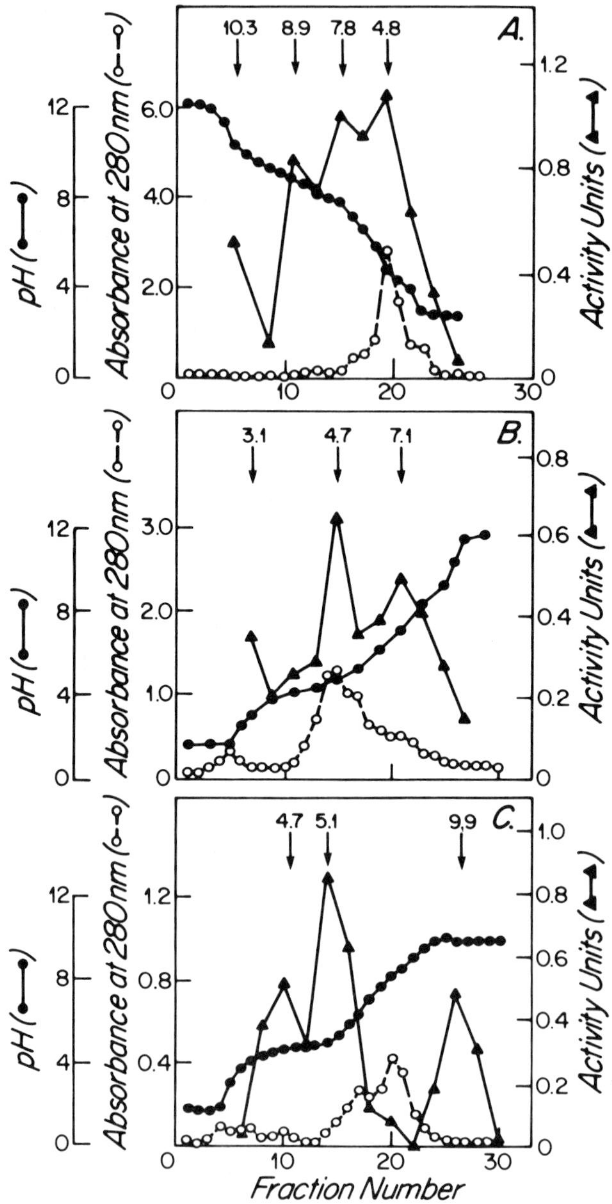

Fig. 2. Isoelectric focusing of crude extract and fractions un-CM and CM-1. Focusing of 8 mg of crude extract (A), 6 mg of un-CM (B), and 4 mg of CM-1 (C) in a 10-ml column in the pH range 3.5–10.0. From Gambarini and Armelin [1982], with permission.

pH ranges 4-6 and 3.5-10, respectively (disregarding the remaining ampholyte from the first focusing). The volume was raised to 50 ml and, if necessary, pH was adjusted to 6-7 with NaOH. Finally this solution was used to make 5% and 50% sucrose solutions, which were loaded onto the 110 ml LKB isoelectric focusing column. As described previously [Gambarini and Armelin, 1982], this procedure separates several acidic active fractions with different pIs. To ensure purification one must pool just a few fractions of the desired pI and then chromatograph it in BioGel P100 column to fractionate by molecular weight. By this procedure we have been able to isolate an acidic factor of pI 4.7, molecular weight 12,000 and specific activity up to 10^6 U/mg; the yield, however, was low, being around 10 μg/kg tissue [Gambarini and Armelin, 1982].

Another possibility is to take FGFa fraction directly to BioGel chromatography. For that purpose, FGFa fraction (as obtained from the first isoelectric focusing; see Fig. 1) was concentrated by ultrafiltration and the buffer was changed to 0.05 M sodium phosphate, pH 7.0 plus 0.5 M NaCl. Up to 10 mg protein in 5-10 ml was applied to a BioGel P100 column equilibrated with the same buffer (2.5 × 80 cm; 2 ml/cm^2 · h flow rate). The activity has been regularly resolved in four fractions of molecular weights 68,000, 43,000, 13,000, and 6,000. Only the 68,000 and the 13,000 molecular weight fractions have been studied further. They are active at nanogram per milliliter levels, but are not homogeneous, and their yield has been in the range of tens of micrograms per kilogram tissue.

PRECIPITATION OF ACIDIC FGF AT pH 4.5

Observations made during isoelectric focusing experiments have shown that acidic pituitary growth factors precipitate near pH 4.5. This fact is exemplified by the following results. A solution of the un-CM fraction (28 mg/ml in 1% glycine, pH 8.5) was adjusted to pH 4.5 with HCl at 0°C, the supernatant was collected by centrifugation, the precipitate was dissolved in 1% glycine, pH 8.5, and 5-mg samples of each were analyzed by isoelectric focusing (Fig. 3). Eighty percent of the acidic activity (pI 4-6) was found in the precipitate (Fig. 3A), though it was contaminated with a considerable proportion of basic activity. The supernatant, on the other hand, was dominated by the basic activity (Fig. 3B). Several cycles of precipitation were tried in an attempt to diminish the contaminant basic activity of the precipitate. A 7.5-g measure of un-CM was successively precipitated at pH 4.5 four times; the final precipitate had 1.3 g, and 46 mg of it were isoelectric focused in the 100-ml LKB column (Fig. 4A). The profile of activity in Figure 4a is dominated by activities of pI

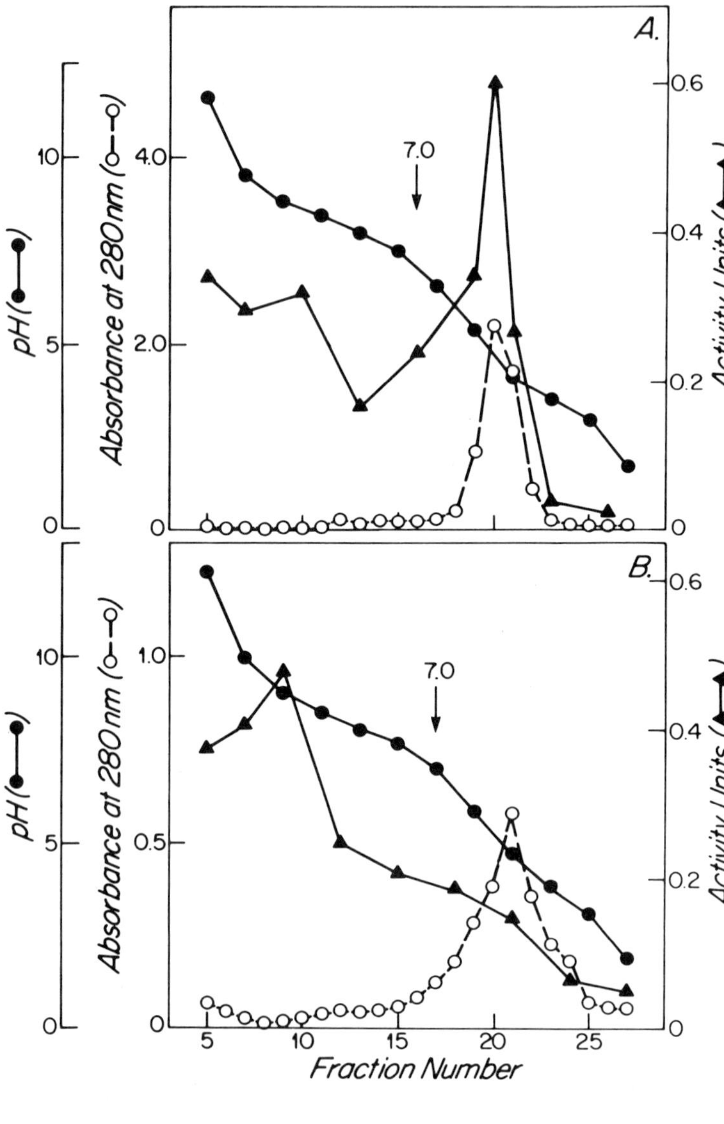

Fig. 3. Isoelectric focusing of fraction un-CM after precipitation at pH 4.5; 5 mg of the pH 4.5 precipitate (A) and supernatant (B) were focused in a 10-ml column. For details, see text.

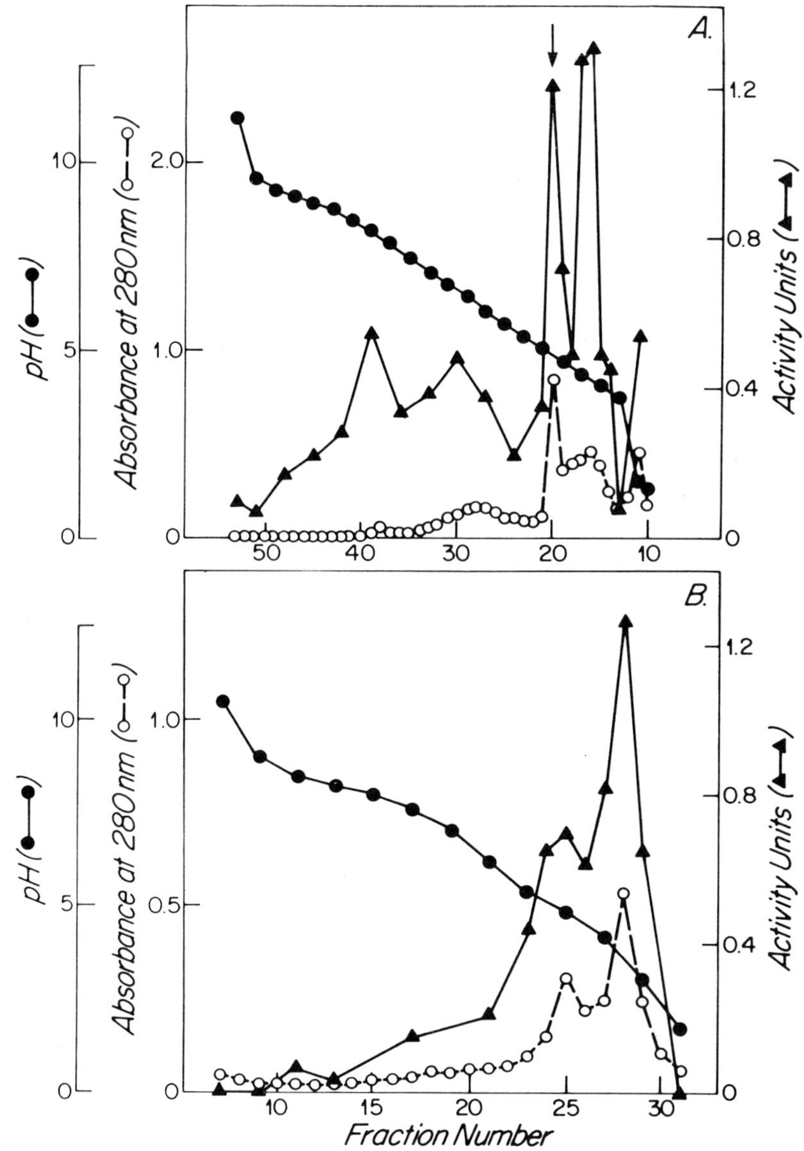

Fig. 4. Isoelectric focusing of un-CM after precipitation at pH 4.5. A. A 46-mg measure of the precipitate obtained from a un-CM fraction by four cycles of precipitation at pH 4.5 was focused in the 110-ml LKB column. B. A 2.3-mg measure of fraction 20 (see arrow in panel A) of electrofocusing presented in A was precipitated at pH 4.5; the supernatant (which had most of the activity) was analyzed by focusing in a 10-ml column after buffer change by ultrafiltration.

4-5; significant activity was found between pH 6 and 8 and no activity above pH 9.0. Fraction 20 (2.3 mg) from isoelectric focusing (pI 4.7, indicated by the arrow in Fig. 4A) was again subjected to precipitation at pH 4.5 and this time 75% of the activity remained in the supernatant. This pH 4.5 soluble activity proved to be only acidic growth factor (Fig. 4B), as one should expect. At this stage of purification the acidic growth factors no longer precipitate at acid pH. Attempts to use the precipitation at pH 4.5 to develop a simple procedure to isolate the acidic growth factors have not resulted in a practical method.

MULTIPLE FORMS OF ACIDIC FGF

The extraction procedure described above yields multiple forms of pituitary acidic growth factors. These factors differ in pI (detected by isoelectric focusing in expanded acidic pH gradients) and in molecular weight (demonstrated by BioGel chromatography). The origin and significance of this heterogeneity are at the moment unknown. These multiple forms of acidic factors could arise from the limited proteolysis of the type used for processing peptide prohormones. We cannot rule out, however, that they could also have arisen from artifactual proteolysis during preparation steps. These two possibilities are presently under investigation.

STIMULATION OF COLONY GROWTH IN AGAROSE SUSPENSION CULTURES

An interesting aspect of the pituitary acidic growth factor's activity is its ability to promote colony growth in agarose suspension. This activity was first detected with agarose suspension cultures of rat C_6 glioma cells [Armelin et al., 1982, 1983]. By virtue of their neoplastic nature C_6 cells grow to very high plating efficiencies in agarose suspension (90%; see Table II). However, hydrocortisone hormone, at physiological concentrations, reversibly sup-

TABLE II. Effects of Hydrocortisone (Hy) and a CM-1 Fraction of Pituitary Growth Factors (PF) on Colony Development in C_6 and 3D Agarose Suspension Cultures

	Plating efficiency			
Cell type	Control	+Hy	+PF	Hy + PF
C_6 (parental)	90 (±5)	0	90 (±5)	10 (±2)
3D (variant)	<0.2	<0.2	6 (±2)	12 (±3)

C_6 cells plated at 10^2 cells per well and 3D at 250 and 10^3 cells per well; 250 ng/ml Hy and 2.5 μg/ml PF. Data are averages (+SEM) of two independent experiments.

presses colony growth. Pituitary growth factor preparations, on the other hand, override hydrocortisone growth suppression and promote colony development (Table II). These experiments were routinely performed with growth factors purified up to the CM-1 step; purified acidic factors are also effective, but purified basic factors have not been tested.

A variant isolated from the C_6 line (clone 3D) was selected, which is dependent on pituitary growth factor for growth in suspension [Armelin et al., 1982]. Table II shows that 3D variant cells develop colonies only in the presence of pituitary growth factors and, interestingly, this effect is enhanced by hydrocortisone.

It is noteworthy that pituitary growth factor preparations that were very active with "normal" Swiss 3T3 fibroblasts anchored to a solid substrate failed to stimulate colony development of these same cells in agarose suspension cultures. Actually these growth factor preparations caused degeneration of 3T3 cells in suspension, whereas no toxic effects could be found in monolayer cultures, even at concentrations as high as 10 μg protein per ml. On the other hand, these same growth factor preparations stimulated colony growth of Rat-1 fibroblasts in agarose suspension. This Rat-1 line by itself showed some ability to grow in suspension ($<0.5\%$ plating efficiency; only microcolonies), whereas the "normal" 3T3 cells were totally unable to grow even into microcolonies of 4-8 cells. Therefore the pituitary growth factors promoted growth in suspension cultures of only those cells that were already somewhat transformed or on the way to a fully transformed phenotype.

ACKNOWLEDGMENTS

This work was supported by Fundacao de Amparo a Pesquisa do Estado de São Paulo (FAPESP), Financiadora de Estudos e Projetos (FINEP), Conselho Nacional de Desenvolvimento Cientifico e Tecnologico (CNPq), and United Nations Developmental Program (UNDP/UNESCO RLA 78/024).

REFERENCES

Armelin HA (1973): Pituitary extracts and steroid hormones in the control of 3T3 cell growth. Proc Natl Acad Sci USA 70:2702-2706.

Armelin MCS, Armelin HA (1983): Glucocorticoid hormone modulation of cell surface and cytoskeleton related to growth control of rat glioma cells. J Cell Biol 97:459-465.

Armelin HA, Nishikawa K, Sato G (1974): Control of mammalian cell growth in culture: Action of protein and steroid hormones as effector substances. Cold Spring Harbor Conf Cell Prolif 1:97-104.

Armelin HA, Gambarini AG, Armelin MCS (1982): Pituitary fibroblast growth factors and growth of rat C_6 glial cell variants. Cold Spring Harbor Conf Cell Prolif 9:987-996.

Armelin MCS, Stocco RC, Armelin HA (1983): Control of rat C_6 glioma cell proliferation: Uncoupling of the inhibitory effects of hydrocortisone hormone in suspension and monolayer cultures. J Cell Biol 97:455-458.

Benda P, Lightbody J, Sato G, Levine L, Sweet W (1968): Differentiated rat glial cell strain in tissue culture. Science 161:370-371.

Gambarini AG, Armelin HA (1979): Growth factors in mammalian cell cultures. Arch Biol Med Exp 12:269.

Gambarini AG, Armelin HA (1981): Pituitary fibroblast growth factors: Partial purification and characterization. Brazil J Med Biol Res 14:19-27.

Gambarini AG, Armelin HA (1982): Purification and partial characterization of an acidic fibroblast growth factor from bovine pituitary. J Biol Chem 257:9692-9697.

Gospodarowicz D (1975): Purification of a fibroblast growth factor from bovine pituitary. J Biol Chem 250:2515-2520.

Gospodarowicz D, Jones KL, Sato G (1974): Purification of a growth factor for ovarian cells from bovine pituitary glands. Proc Natl Acad Sci USA 71:2295-2299.

Gospodarowicz D, Weserman J, Moran J (1975): Presence in brain of a mitogenic agent promoting proliferation of myoblasts in low density culture. Nature 256:216-219.

Maciag T, Cerundolo J, Ilsley S, Kelley PR, Forand R (1979): An endothelial cell growth factor from bovine hypothalamus: Identification and partial characterization. Proc Natl Acad Sci USA 76:5674-5678.

Maciag T, Hoover GA, Weinstein R (1982): High and low molecular weight forms of endothelial cell growth factor. J Biol Chem 257:5333-5336.

Nishikawa K, Armelin HA, Sato G (1975): Control of ovarian cell growth in culture by serum and pituitary factors. Proc Natl Acad Sci USA 72:483-487.

Peterson GL (1977): A simplification of the protein assay method of Lowry et al. which is more generally applicable. Anal Biochem 83:346-356.

Prasad I, Zouzias D, Basilico C (1976): State of the viral DNA in rat cells transformed by polyoma virus. I. Virus rescue and the presence of nonintegrated DNA molecules. J Virol 18:436-444.

Thomas KA, Riley MC, Lemmon SK, Baglan NC, Bradshaw RA (1980): Brain fibroblast growth factor: Nonidentity with myelin basic protein fragments. J Biol Chem 255:5517-5520.

Todaro G, Green H (1963): Quantitative studies on the growth of mouse embryo cells in culture and their development into established lines. J Cell Biol 17:299-313.

Methods for Serum-Free Culture of Epithelial and Fibroblastic Cells,
pages 221–231

15
Growth of SV40 Balb/c-3T3 Cells in Serum-Free Culture Medium

G.A. Rockwell

Tissue culture has come a long way since its inception in the early part of this century. At that time tissue explants were grown in plasma clots and the experimental conditions were not well defined. The first move towards defined medium came in 1911, when M.R. and W.H. Lewis made hanging drop cultures of chick tissues in simple salt solutions to which they added chick bouillon [Lewis and Lewis, 1911]. Since that time great advances have been made by many (including White [1946], Waymouth [1955], Earle [Evans et al., 1956], and Eagle [1957]) in defining and improving the salt solution component of the medium; however, the chick bouillon was merely replaced by "serum bouillon." In recent years the laboratories of Ham and Sato have attempted, in the first instance, to optimally satisfy the nutritional requirements of the cells by careful adjustment of the medium components, and, in the second, to entirely remove, by hormone and growth factor replacement, the need for serum. Both laboratories have met with a significant degree of success [Barnes and Sato, 1980; Walthall and Ham, 1981; Tsao et al., 1982].

This chapter describes the techniques employed in the growth, passage, and colony formation of SV40 virus-transformed Balb/c-3T3 cells in serum-free medium.

In determining the contribution of individual factors to the support not only of viability, but also of the growth of cells in serum-free medium, due regard

C.S.I.R.O. Unit of Cellular and Molecular Biology, North Ryde, N.S.W., 2113, Australia

This work was performed at the Department of Biology, University of California at San Diego, La Jolla, California 92093.

has to be given to a wide range of effects that may not be as critical in serum-containing medium.

Serum proteins probably play a role in detoxification processes in tissue culture and for that reason serum-free medium should be carefully prepared, not stored for extended lengths of time, and changed more frequently. Water used for the preparation of media in these experiments was triple-distilled in glass containers and used immediately after the last distillation. Potassium permanganate was included in the boiler of the second still. Impurities in commercially prepared chemicals and growth factors are always a problem and their possible effects need to be constantly borne in mind. Some of the factors have problems with solubility, storage, and stability. It is unwise to make up and store medium with growth factors present. The factors used in these experiments were stored in polystyrene plastic containers in concentrated stock solutions as indicated below and added to growth medium at the time of experimentation: insulin (2 mg/ml in 0.01 N HCl, 4°C), transferrin (5 mg/ml in phosphate-buffered saline [PBS], −20°C), fatty acid-free (FAF) albumin (50 mg/ml in PBS, −20°C), fatty acids (5 mg/ml in absolute ethanol, −20°C under N_2), and fibronectin (7.5 mg/ml in PBS, −20°C). These factors were not subjected to repeated thawing and freezing.

The relationship between factors is also important as synergism between factors may play a role. Synergism between insulin and transferrin can be seen in SV40 Balb/c-3T3 cells (Fig. 1), and the presence of FAF-albumin is necessary to prevent cytolysis by free fatty acids (Fig. 4).

The cells used throughout this study—namely, normal and SV40-transformed Balb/c-3T3 cells—were obtained from Dr. Robert Holley of the Salk Institute, La Jolla, California, and the work was performed in the laboratory of Professor Gordon Sato, Department of Biology, University of California at San Diego, La Jolla, California [Rockwell et al., 1980].

Chemicals were obtained as follows: DME and Ham's F12 nutrient medium (powdered formulations) were from Gibco; fetal calf serum was from Reheis; crystalline preparations of bovine insulin, soybean trypsin inhibitor, human transferrin (substantially iron-free), bovine serum albumin, as well as linoleic, oleic, and arachidonic acids were from Sigma; linolenic, stearic, and palmitic acids were from Calbiochem; and purified human plasma fibronectin were from Collaborative Research.

Stock cultures were maintained in Dulbecco's modified Eagle's medium containing penicillin (190 IU/ml), streptomycin (0.2 mg/ml), ampicillin (25 μg/ml), sodium bicarbonate (1.2 g/liter), and 15 mM N-2-hydroxyethylpiperazine-N′ ethansulfonic acid (DME) and were supplemented with 10% fetal calf serum (FCS). Cultures were incubated at 37°C in a humidified 5% CO_2

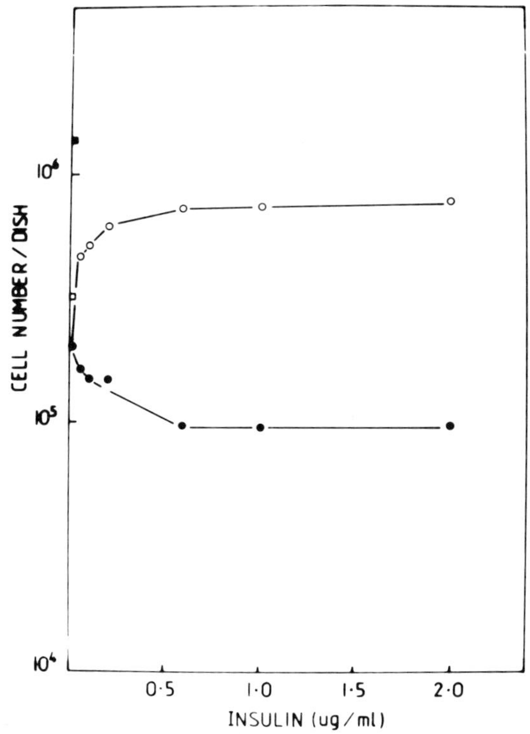

Fig. 1. Synergism between insulin and transferrin in serum-free growth of SV40 Balb/c-3T3 cells. Cells were plated in 10% FCS. This medium was washed out on day 1 and replaced with DME/F12, to which was added insulin (●—●); insulin in the presence of 5 μg/ml transferrin (○—○); 5 μg/ml transferrin (□); 10% FCS (■). Cell count was determined on day 4.

in air atmosphere and were transferred once or twice weekly with 0.1% trypsin-0.03% EDTA in phosphate-buffered saline. Harvested cells were counted in a Coulter counter and all values reported are the means of duplicate dishes.

GROWTH OF CELLS IN SERUM-FREE MEDIUM

Exponentially growing cells were harvested from stock cultures with 0.1% trypsin-0.03% EDTA in PBS, suspended in DME-10% FCS, and plated at the indicated densities in 60-mm Falcon tissue culture dishes containing 5 ml of a 3:1 (vol/vol) mixture of DME and Ham's nutrient mixture F12 containing penicillin, streptomycin, ampicillin, sodium bicarbonate, and HEPES at the above concentrations (DME/F12), and supplemented with 10% FCS to permit proper attachment of the cells. After attachment (the following day) the medium was removed and the cells were washed several times with DME. After washing, the medium was replaced with 3:1 (vol/vol) DME/F12, and various factors were assayed alone and in combination for their growth-promoting activity. A mixture of DME/F12 was superior to either medium used alone for the growth of these cells in serum-free conditions [Rockwell et al., 1980]. Similar mixtures have been shown to be beneficial for other cell types [Barnes and Sato, 1980; Ham and McKeehan, 1979].

Serum is now known to provide not only factors necessary for the growth of cells in tissue culture but also factors involved in the correct attachment and spreading of cells. This permissive attachment, which anchors the cell yet allows the necessary mobility and spreading for cell division, is of fundamental importance in serum-free medium. One can approach the overall problem by allowing cells to attach in the presence of serum and then, after washing the serum out, determining the conditions necessary for the subsequent division of cells. Once those conditions have been determined, serum-free attachment of the cells can be investigated. This approach was the one used in these experiments.

The proliferation of either Balb/c-3T3 or SV40-transformed Balb/c-3T3 cells in the absence of serum was markedly inhibited (Fig. 2). Cells deprived of FCS on day 1 (↑) doubled 2–3 times and then ceased to divide.

This growth inhibition was totally reversed in the case of the transformed cell line when DME/F12 was supplemented with 2 μg/ml insulin, 5 μg/ml transferrin, 1 mg/ml FAF bovine albumin, and 5 μg/ml linoleic acid (Fig. 2).

In supplemented medium, the transformed cells grew exponentially with a generation time of 15–16 h before reaching a saturation density of (3–3.5) $\times$ 10^5 cells per cm^2. These growth parameters were very similar to those obtained in the presence of 10% FCS (Fig. 2). The morphology of cells grown in supplemented medium was identical to that of cells grown in 10% FCS [Rockwell et al., 1980].

Untransformed Balb/c-3T3 cells proliferated poorly in the supplemented medium (Fig. 2), indicating that transformation alters the specific growth requirements of the cells. The growth requirements for Balb/c-3T3 cells have been reported elsewhere [Serrero et al., 1979; McClure et al., 1979].

The effect of each factor can then be determined by individual omission from the mixture (Fig. 3). The deletion of insulin, transferrin, or linoleic acid from the mixture significantly reduced the growth response compared to that produced when all four factors were added simultaneously. The deletion of FAF-albumin caused cytolysis because of the detergent action of the free fatty acid.

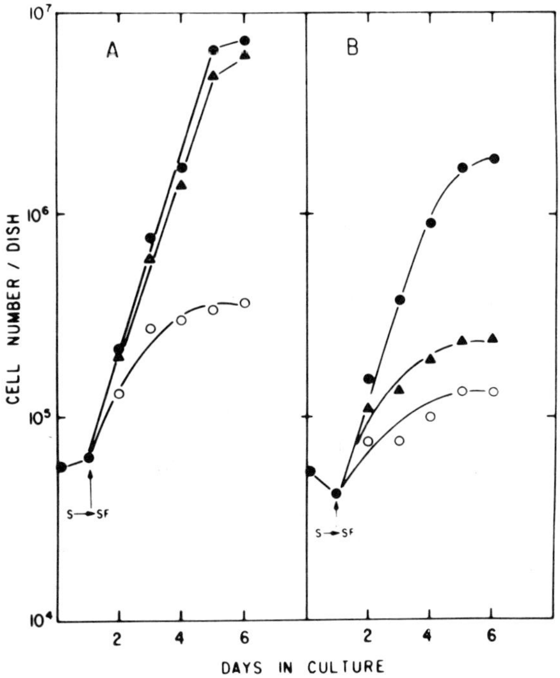

Fig. 2. Growth of SV40-transformed (A) and untransformed (B) Balb/c-3T3 cells in the presence (●—●) and absence (○—○) of 10% FCS and in DME/F12 supplemented with insulin (2 μg/ml), transferrin (5 μg/ml), FAF-albumin (1 mg/ml), and linoleic acid (5 μg/ml) (▲—▲). The cells were plated in DME-10% FCS. This medium was washed out on day 1 (↑) and replaced with the media listed above.

The growth response to each of the supplements varied in a concentration-dependent fashion (Fig. 4). The concentrations of insulin, transferrin, and linoleic acid required for optimal cell growth were 250-300 ng/ml, 200-400 ng/ml, and 4-5 μg/ml, respectively. Protection against the cytotoxic effect of linoleic acid was not achieved until the FAF-albumin concentration reached 200 μg/ml. Maximal protection occurred at 500 μg/ml, corresponding to a molar ratio of fatty acid to FAF-albumin equal to 2.3.

Of the fatty acids tested in combination with FAF-albumin, only the cis-unsaturated fatty acids promoted the growth of SV40-transformed Balb/c-3T3 cells (Table I). Linoleic, linolenic, oleic, and arachidonic acid were equally effective in stimulating cell growth when added at 5 μg/ml in the presence of FAF-albumin. Saturated fatty acids at similar concentrations were either inhibitory (in the case of palmitic) or inactive (in the case of stearic). Commercially

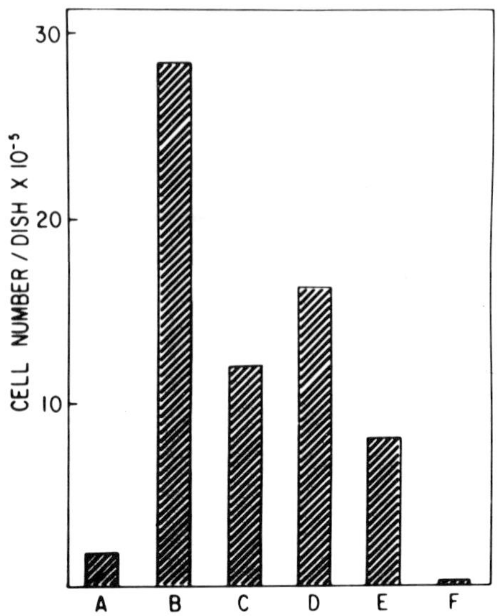

Fig. 3. Response of SV40-transformed Balb/c-3T3 cells to the removal of individual supplements. Cells (4×10^4) were plated in DME-10% FCS, which was removed after one day. After the cells were washed, the medium was replaced with the following: A. DME/F12. B. DME/F12 supplemented with insulin, transferrin, FAF-albumin, and linoleic acid. C. DME/F12 supplemented with transferrin, FAF-albumin, and linoleic acid (without insulin). D. DME/F12 supplemented with insulin, FAF-albumin, and linoleic acid (without transferrin). E. DME/F12 supplemented with insulin, transferrin, and FAF-albumin (without linoleic acid). F. DME/F12 supplemented with insulin, transferrin, and linoleic acid (without FAF-albumin). The concentrations of components used were as in Figure 2. Cells were counted on day 5.

available bovine serum albumin at 2 mg/ml stimulated growth to a similar extent as FAF-albumin plus unsaturated fatty acids (Table I).

Similar lines of evidence suggested that the stimulation of cell growth observed in the presence of fatty acid and FAF-albumin was due to the fatty acid [Rockwell et al., 1980].

PASSAGE AND PLATING OF CELLS IN SERUM-FREE MEDIUM

In order to grow and passage cells in the total absence of serum, several serious problems need to be overcome. First, cells grown in serum-free medium appear to be more susceptible to trypsin-EDTA treatment, and some lesser concentration may be necessary to avoid deleterious effects. In the

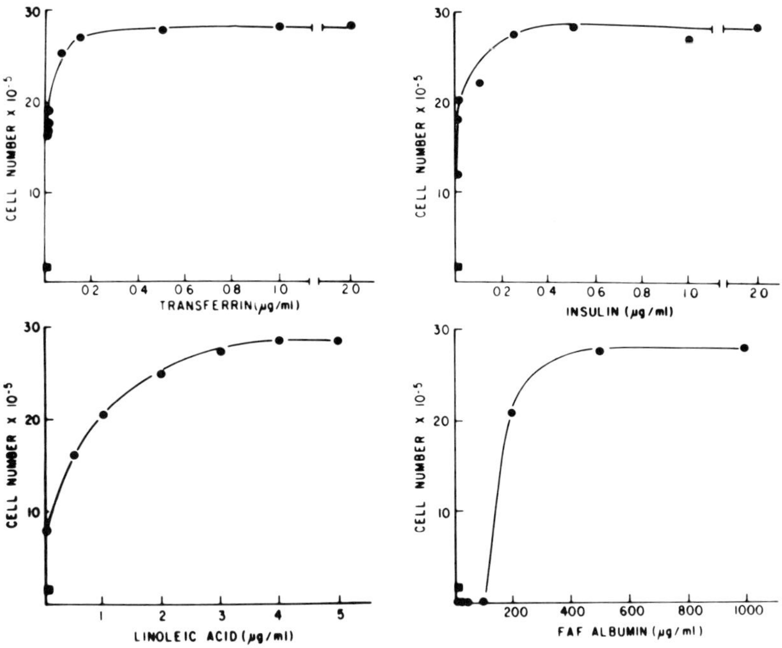

Fig. 4. Dose response of SV40-transformed Balb/c-3T3 cells to individual components in the presence of maximal concentrations of all the other components. In each case, cells (4×10^4) were plated in DME-10% FCS, which was removed on day 1. After the cells were washed, the medium was replaced with DME/F12 in which the concentration of three of the four components was the same as in Figure 2, and the concentration of the fourth component was varied as indicated. DME/F12 with no additions (■). Cell counts were performed on day 5.

absence of serum it becomes necessary to actively inhibit the activity of trypsin. In normal growth medium this function is performed by the protease inhibitors present in serum. In serum-free techniques this can be accomplished by incubating the cells in the presence of soybean trypsin inhibitor. Washed cells can then be inoculated into tissue culture dishes or flasks.

As mentioned previously, serum in tissue culture, as well as supplying essential nutrients and growth factors, has been shown to play an important role in cell attachment (see review article by Yamada and Olden [1979]). Serum fibronectin is necessary for cells to attach and spread on tissue culture dishes [Grinell et al., 1977; Hook et al., 1977]. It is relevant to transformed Balb/c-3T3 cells that cell surface fibronectin is known to decrease in many transformed cells [Yamada and Olden, 1979].

TABLE I. Effect of Fatty Acids on Growth of SV40-Transformed Balb/c-3T3 Cells

Additions	Relative number of cells
None (insulin and transferrin only)	1.0
FAF-albumin	0.9
FAF-albumin + oleic acid (18 cis 9)	3.5
FAF-albumin + linoleic acid (18 cis 9,12)	4.2
FAF-albumin + linolenic acid (18 cis 9,12,15)	3.8
FAF-albumin + arachidonic acid (20, cis 5,8,11,14)	3.5
FAF-albumin + palmitic acid (16)	0.1
FAF-albumin + stearic acid (18)	1.2
Unextracted albumin	3.7

Cells were plated at 4×10^4 cells per 60-mm dish in DME-10% FCS. The medium was removed on day 1. After the cells were washed, the medium was replaced with DME/F12 supplemented with insulin (2 μg/ml) and transferrin (5 μg/ml). Bovine serum albumin (unextracted albumin) and fatty acid-free albumin (FAF-albumin) were added at 2 mg/ml. Fatty acids were added from ethanol stock solutions at a final concentration of 5 μg/ml. Cell counts were performed on day 6. Results are expressed as the relative increase in cell number over that in medium supplemented with insulin and transferrin.

The attachment of SV40-transformed Balb/c-3T3 cells in serum-free medium is therefore an important consideration in the overall problem of serum-free growth of these cells in tissue culture. One further problem arises because ruptured and dying cells probably release into the medium hydrolases that, because of the absence of serum, can have a more deleterious effect on the remaining cells. This can be alleviated by washing those cells that have attached soon after plating, to remove any unattached cells or cellular debris.

In the experiments that follow, trypsinized cells were suspended in a 0.1% solution of soybean trypsin inhibitor in PBS and washed twice with fresh DME. Cells thus treated, when plated into DME/F12 supplemented with insulin, transferrin, FAF-albumin, and linoleic acid, did not adhere to the plastic substratum. The presence of FAF-albumin prevents cell attachment under these conditions. Cells plated into DME/F12 attached but did not proliferate when the medium was supplemented with the four factors subsequent to cell attachment (Fig. 5). The addition of purified human plasma fibronectin at 7.5 μg/ml to medium supplemented with insulin, transferrin, FAF-albumin, and linoleic acid allowed cell attachment and subsequent growth, which was identical to that obtained with serum-precoated dishes and similar to that obtained in medium supplemented with 10% FCS (Fig. 5).

For continuous culture of SV40-transformed Balb/c-3T3 cells in serum-free medium, a modification of the above plating procedure was employed to

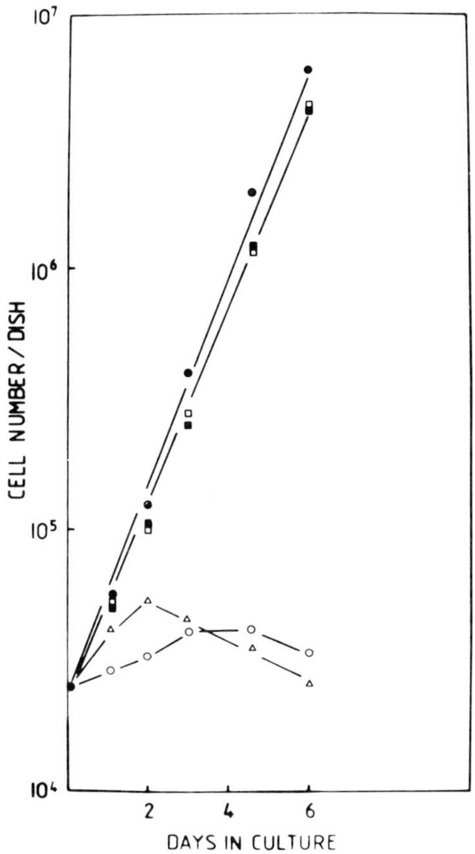

Fig. 5. Effect of human purified plasma fibronectin on the attachment and growth of SV40-transformed Balb/c-3T3 cells. Stock cells were trypsinized and treated with soybean trypsin inhibitor. Cells suspended in fresh DME were plated into DME/F12 (○—○), which was supplemented with 10% FCS (●—●); human fibronectin (7.5 μg/ml), insulin, transferrin, FAF-albumin, and linoleic acid (■—■); insulin, transferrin, FAF-albumin, and linoleic acid in serum-precoated dishes (□—□); and insulin, transferrin, FAF-albumin, and linoleic acid were added subsequent (~1 hr) to cell attachment (△—△). The concentration of supplements was as in Figure 2.

minimize cell damage caused by contaminating cell debris generated during the serum-free transfer procedure. Cells that had been detached with trypsin, treated with trypsin inhibitor, and washed once with fresh DME/F12 were plated into medium containing 7.5 μg/ml of human plasma fibronectin. After cell attachment (2-3 h), the medium was removed and fresh medium, supplemented with insulin (2 μg/ml), transferrin (5 μg/ml), FAF-albumin (1 mg/ml), and linoleic acid (3 μg/ml), was added. By this procedure SV40-transformed Balb/c-3T3 cells have been carried for more than 2 months (~80 generations and 10 passages) without contact with serum. These cells were identical to cells carried in serum in respect to their growth rate in tissue culture and their tumor-promoting capacity when injected into nude mice [Rockwell et al., 1980].

COLONY FORMATION IN SERUM-FREE MEDIUM

Colony formation of the cells can be tested by plating 200 cells per 21-cm^2 plastic dish in medium supplemented with either 10% FCS or insulin, transferrin, FAF-albumin, linoleic acid, and fibronectin. Cells were grown for 14 days and colonies were counted on stained dishes. Cells plated in serum-containing medium formed approximately 70 colonies; those plated in supplemented medium formed 40 discernible colonies. No colony formation was observed when either insulin, transferrin, FAF-albumin, linoleic acid, or fibronectin was individually omitted from DME/F12. Likewise the replacement of DME/F12 with either DME or F12 alone resulted in the virtual absence of colonies.

In summary, SV40-transformed Balb/c-3T3 cells proliferate in medium supplemented with insulin, transferrin, FAF-albumin, and cis-unsaturated fatty acid. Trypsinizing cells, incubating with soybean trypsin inhibitor, and plating in medium containing fibronectin and supplemented with the above factors allows the cells to be grown and continuously subcultured without the use of serum.

The ability to grow these cells in the absence of serum will permit a more realistic and definitive examination of the effect of hormones and growth factors on the metabolism and growth of these cells and will facilitate the search for growth-regulating substances secreted by these cells into the medium.

Some of the diagrams and parts of the text are reproduced with the permission of the publishers from an article published in The Journal of Cellular Physiology 103:323-331, 1980.

REFERENCES

Barnes D, Sato G (1980): Methods for growth of cultured cells in serum free medium. Anal Biochem 102:255.

Eagle H, Oyama VI, Levy M (1957): Amino acid requirements of normal and malignant human cells in tissue culture. Arch Biochem Biophys 67:432.

Evans VJ, Bryant JC, Fioramonti MC, McQuilkin WT, Sandford KK, Earle WR (1956): Studies of nutrient media for tissue cells in vitro. Cancer Res 6:77.

Grinell F, Hays DG, Minter D (1977): Cell adhesion and spreading factor—Partial purification and properties. Exp Cell Res 110:175.

Ham RG, McKeehan WL (1979): Media and growth requirements. In Jacoby WB, Pastan IH (eds): "Methods in Enzymology." New York: Academic, Vol 58, p 44.

Hook M, Rubin K, Oldberg A, Obrink B, Vaheri A (1977): Cold insoluble globulin mediates the adhesion of rat liver cells to plastic petri dishes. Biochem Biophys Res Commun 79:726.

Lewis MR, Lewis WH (1911): The growth of embryonic chick tissues in artificial media, agar and bouillon. Johns Hopkins Hosp Bull 22:126.

McClure D, Rockwell G, Sato G (1979): The growth requirements of normal and SV40 transformed Balb/c-3T3 cells in serum free medium. J Cell Biol 83:96a.

Rockwell GA, McClure DB, Sato GH (1980): The growth requirements of SV40 virus transformed Balb/c-3T3 cells in serum free monolayer culture. J Cell Physiol 103:323.

Serrero GR, McClure DB, Sato GH (1979): Growth of mouse 3T3 fibroblasts in serum free hormone supplemented medium. Cold Spring Harbor Conf Cell Prolif, Vol 6.

Tsao MC, Walthall BJ, Ham RG (1982): Clonal growth of normal human epidermal keratinocytes in a defined medium. J Cell Physiol 110:219.

Walthall BJ, Ham RG (1981): Multiplication of human diploid fibroblasts in a synthetic medium supplemented with EGF, insulin and dexamethasone. Exp Cell Res 134:301.

Waymouth C (1955): Simple nutrient solutions for animal cells. Tex Rep Biol Med 13:522.

White PR (1946): Cultivation of animal tissues in vitro in nutrients of precisely known constituents. J Natl Cancer Inst 17:305.

Yamada KM, Olden K (1979): Fibronectins—Adhesive glycoproteins of cell surface and blood. Nature 275:179.

Methods for Serum-Free Culture of Epithelial and Fibroblastic Cells,
pages 233–248

16
Use of Hormone-Toxin Conjugates and Serum-Free Media for the Isolation and Study of Cell Variants in Hormone Responses

Nobuyoshi Shimizu

In order to facilitate studies on the molecular mechanisms and genetic regulation of polypeptide hormone action, it is necessary to devise a selection procedure by which to isolate cell variants that are defective or extraordinary in hormone response. This has been a difficult task, compared to isolation of drug-resistant mutants, since polypeptide hormones are nontoxic to normal mammalian cells in culture and, instead, are required for their proliferation. However, polypeptide hormones can be used as a selection agent by making them toxic without loss of receptor-recognizing specificity. The approach taken for this was to cross-link a polypeptide hormone, i.e., insulin, to the A fragment of diphtheria toxin (DTa) or the A subunit of toxic ricin (RICa) [Miskimins and Shimizu, 1979; Shimizu et al., 1982b] (Fig. 1). DTa and RICa are extremely potent catalysts for inhibiting protein biosynthesis [Pappenheimer, 1977; Olsnes and Pihl, 1982a,b]. After cross-linking with DTa or RICa, the resulting chimeric insulins have been found to be useful in killing insulin-responsive cells and specifically selecting cell variants that are unable to bind insulin or take up the bound insulin and process it. The general strategy of this hybrid protein technique has been described [Miskimins and Shimizu, 1979, 1981a; Miskimins et al., 1981; Shimizu et al., 1982a]. Two different cross-linking techniques, both through disulfide linkages, have been utilized, one that modifies carboxyl groups [Gilliland et al., 1978] and one that modi-

Department of Molecular and Cellular Biology, The University of Arizona, Tucson, Arizona 85721 and Department of Molecular Biology, Keio University School of Medicine, 35 Shinanomachi, Shinjuku-Ku, Tokyo, Japan

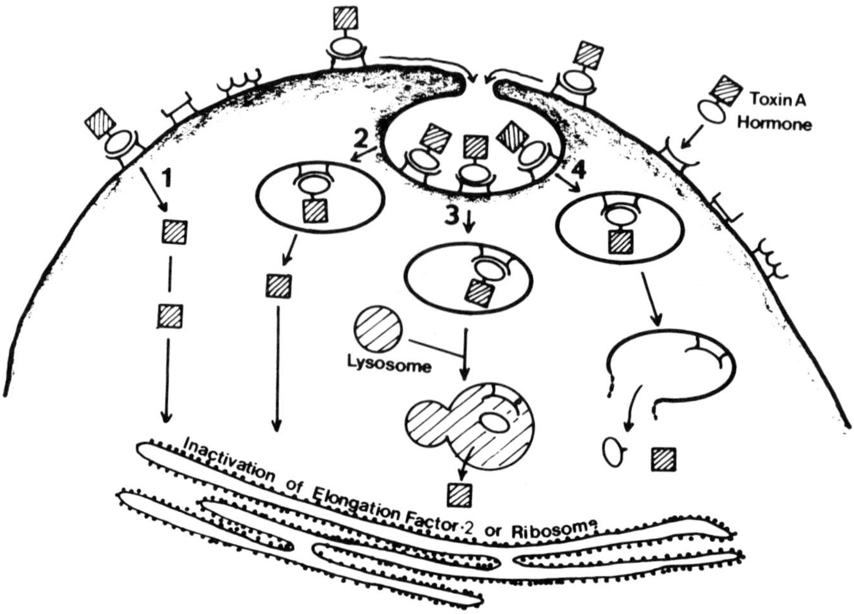

Fig. 1. Schematic representation of the possible routes of entry of hormone-toxin conjugates into cytoplasm; the mutants that lack hormone receptors or those that are defective or amplified in a step of the postreceptor mechanisms will survive. 1) The toxic fragment or subunit A passes directly through the plasma membrane after the conjugate binds the membrane receptor. 2) The toxin A moiety passes through the membrane of the endocytotic vesicle. 3) The toxin A moiety passes through the membrane following fusion with a lysosome, possibly owing to channel formation induced by low pH. 4) The conjugate gains entry into the cytoplasm by rupture of some of the endocytotic vesicles. Pathways 1 and 2 may be likely for the conjugates involving the ricin A subunit, whereas pathways 3 and 4 may be more important in conjugates utilizing the diphtheria toxin A fragment. However, endocytotic mechanisms of each of the hormone-receptor systems would play critical roles both at the initial binding step and the subsequent steps.

fies amino groups of protein [Carlsson et al., 1978]. Similar cross-linking techniques have been utilized with other polypeptide hormones—human placental lactogen [Chang and Neville, 1977; Chang et al., 1977], chorionic gonadotropin [Oeltmann and Heath, 1979], and epidermal growth factor [Shimizu et al., 1980; Cawley et al., 1980; Shimizu et al., 1983]. Most of these conjugates, if not all, were proved to be cytotoxic to various cell types, but only a few cases have so far been utilized as an aid for experimental biology. In this chapter, the methods of constructing the conjugates, including insulin-DTa and insulin-RICa, and some of their properties are described. Procedures for isolating conjugate-resistant cell variants using 3T3-L1 preadipocytes and methods for studying their adipocyte differentiation in serum-free hormone-supplemented medium are also described.

METHODS OF PROCEDURE

Preparation of DTa (Fig. 2A)

All steps for biochemical manipulations were performed at 4°C unless otherwise noted. Purified diphtheria toxin was purchased as a frozen solution from Connaugh Laboratories, Canada. Diphtheria toxin was precipitated by adding ammonium sulfate (ultrapure grade, Schwartzman) to 70% saturation. The precipitate was collected by centrifugation at 10,000g for 15 min, dissolved in a small volume of 10 mM Na phosphate (pH 7.0), and dialyzed against the same buffer. The dialyzed toxin (usually ~35mg/ml) was mixed with two volumes of 150 mM Tris-HCl (pH 8.0) containing 3 mM EDTA. It was then treated with 1 μg/ml trypsin (Sigma) at 25°C for 30 min. The reaction was stopped by adding soybean trypsin inhibitor (Sigma) at 1.5 μg/ml. The nicked toxin was reduced by incubation with 100 mM dithiothreitol (DTT) and 4 M urea at 25°C for 60 min. DTa was then isolated by the method of Kandel et al. [1974] on a Sephadex G-100 column (1.5 cm × 110 cm) equilibrated with 50 mM Tris-HCl (pH 8.0) containing 4 M urea, 1 mM EDTA, and 1% 2-mercaptoethanol. To determine the presence of DTa and its purity, the fractions were analyzed by polyacrylamide (10%) gel electrophoresis (PAGE) in the presence of sodium dodecyl sulfate (SDS) [Laemmli, 1970]. The fractions containing DTa (Mr=21,000) were pooled, concentrated, and dialyzed against 20 mM N-tris-(hydroxymethyl)methyl-2-aminoethanesulfonate (TES, Sigma). For storage the dialysate was dialyzed against the same buffer containing 50% glycerol and kept at −20°C. The absorbance of 1 mg/ml DTa at 280 nm is 1.52. ADP-ribosylation activity of DTa was assayed with partially purified mouse liver elongation factor 2 (EF-2) according to Honjo et al. [1968].

Preparation of RICa (Fig. 2C)

Ricin was extracted from castor beans (*Ricinus communis*, Geo. W. Park Seed Co.) and purified by the method of Lappi et al. [1978]. Briefly, decorticated beans (200 g) were stirred overnight in 500 ml of 5% acetic acid and homogenized in a Waring blender. This was centrifuged at 10,000g for 20 min. The oil was scraped off and ammonium sulfate was added to the supernatant to 80% saturation. The precipitate was collected by centrifugation, dissolved in distilled water, and dialyzed extensively against distilled water. The resulting yellow solution was passed through a DE52 (Whatman)

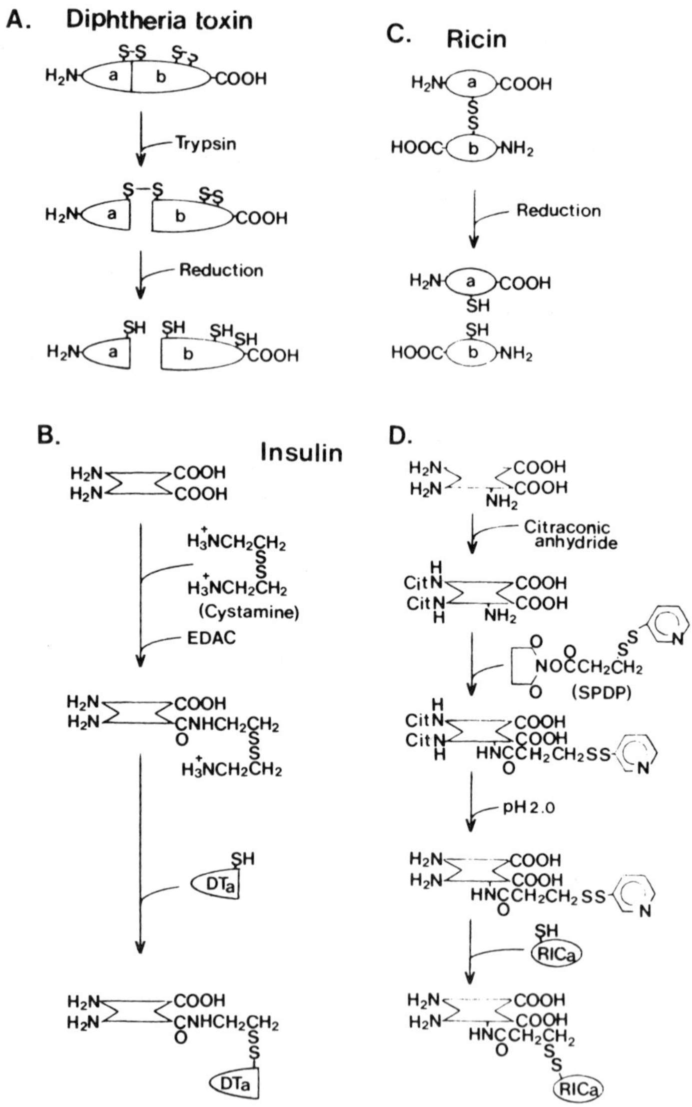

Fig. 2. Diagrammatic representation of the hormone-toxin conjugation. Two different cross-linking techniques, both through disulfide linkages, are utilized: one that modifies carboxyl groups of insulin by cystamine and one that modifies amino group of B29 lysine by SPDP.

column (3.5 cm × 25 cm) equilibrated with 5 mM Na acetate (pH 5.8). The flowthrough fractions were collected and treated with ammonium sulfate at 80% saturation. The precipitate was collected by centrifugation and dissolved in a minimum volume of 10 mM Na phosphate (pH 7.4). An aliquot of this solution was applied to a Sephadex G-100 column (3.5 cm × 60 cm) equilibrated with 10 mM Na phosphate (pH 7.4). The second of the two sharp major peaks was collected as purified toxic ricin.

RICa was separated from ricin B subunits by application of ricin solution to a Sepharose 4B column (2.0 cm × 60 cm) equilibrated with 0.1M Tris-HCl (pH 7.6) containing 0.1 M NaCl. Under these conditions, ricin agglutinin (about 40% of total ricin) passed through the column and the rest (toxic ricin) bound tightly to the galactose moieties of Sepharose through the B-subunit as described by Lin and Li [1980]. The A-subunit (RICa) was eluted with 5% 2-mercaptoethanol in the column buffer. The eluate was dialyzed against 5 mM Na phosphate (pH 6.5). To further purify, the dialysate was applied to a CM-52 column (4.0 × 40 cm) equilibrated with the same buffer. RICa was eluted with a 0–0.1 M NaCl gradient in the same buffer. RICa prepared in this manner revealed a single band of Mr=30,000 on SDS-polyacrylamide gels. It was stored at −20°C after being dialyzed against the same buffer containing 50% glycerol. Catalytic activity of RICa was assayed by an in vitro protein synthesis system according to Sandvig and Olsnes [1979]. The absorbance of 1 mg/ml RICa at 280 nm is 0.77.

Synthesis of Insulin-DTa Conjugate (Fig. 2B)

To activate insulin's carboxyl groups, cystamine was utilized in the presence of a water-soluble carbodiimide reagent according to Gilliland et al. [1978].

Derivatization of insulin with cystamine. To a 2-mg/ml porcine insulin solution (Eli Lilly, in 0.01 N HCl) was added cystamine-diHCl (Sigma) at a final concentration of 1 M, and the pH was adjusted to 4.7. Then, 1-ethyl-3-(dimethylaminopropyl) carbodiimide-HCl (EDAC, Sigma) was added at a concentration of 2 mM and the pH was maintained at 4.7 for 10 min. The reaction was stopped by adjusting the pH to 8.8 by carefully adding 10 N NaOH. The reaction mixture was applied onto a Biogel P4 column (1.6 × 18 cm) equilibrated with 10 mM TES (pH 8.8) to remove unreacted cystamine-diHCl and EDAC. Excess reagents can also be removed by dialysis against the same buffer. The resulting cystaminyl insulin was initially used directly for conjugation. However, since insulin contains six free carboxyl groups, many cystaminyl insulin species were formed and as a consequence many species of conjugate were produced [Miskimins and Shimizu, 1979]. To avoid formation of multiple conjugate species, the cystaminyl insulin products were fractionated by a chromatography on a DE52 column (2.0 cm × 30 cm) equilibrated with 20 mM TES (pH 8.8). When a 0–0.3 M NaCl gradient in the same buffer was applied, multiple peaks of cystaminyl insulin were eluted from the column. Unmodified insulin was eluted last. The extent of cystaminylation was determined by the method of Habeeb [1972] with S,S′-dithiobis-(2-nitrobenzoic acid) (Sigma). The fractions corresponding to a 1:1 molar ratio of cystamine to insulin were pooled.

Disulfide bridge formation and purification. Prior to conjugation, the DTa was completely reduced by incubation with 0.1 M DTT for 4 h and then dialyzed against 20 mM TES (pH 8.0). The cystaminyl insulin was mixed with the reduced DTa at a molar ratio of 8 to 1. The mixture was concentrated, if necessary, to a volume of 2 ml by Amicon membrane (YM 5) filtration and dialyzed against 20 mM TES (pH 8.0) for 2 days. Cross-linking of the two protein species was spontaneously achieved via thiol-disulfide exchange. Purification of the conjugate was accomplished by a Sephadex G-75 column (2.0 cm × 60 cm) equilibrated with 20 mM TES (pH 8.0) containing 0.1 M NaCl. The insulin-DTa conjugate was eluted first, separating from DTa and cystaminyl insulin. The purified conjugate was dialyzed against 50% glycerol in 20 mM TES (pH 8.0) and stored as aliquots at −20°C. The absorbance of 1 mg/ml insulin-DTa at 280 nm is 1.42. The insulin-DTa conjugate prepared by this method proved to be toxic to Swiss/3T3 mouse fibroblasts, 3T3-L1 preadipocytes, and rat hepatoma H35 cells with an ED_{50} of $(7.0–7.7) \times 10^{-9}$ M. A typical dose-dependent killing profile is shown in Figure 3. The toxicity resulted from the inhibition of cellular protein synthesis [Miskimins and Shimizu, 1979].

Synthesis of Insulin-RICa Conjugate (Fig. 2D)

To modify amino groups of insulin, the heterobifunctional reagent N-succinimidyl-3-(2-pyridyldithio)-propionate (SPDP) was utilized according to Carlsson et al. [1978]. Since modification of the two amino termini of insulin results in decreased affinity for the insulin receptor, it is necessary to block these groups, leaving only B29 lysine available for derivatization with SPDP. For this, citraconic anhydride was used, since it can be deblocked after conjugation [Shechter et al., 1978].

Derivatization of insulin with SPDP. Porcine insulin was dissolved in 10 mM Na carbonate at a final concentration of 2 mg/ml, and the pH was adjusted to 6.9 with NaOH. To block the amino termini of insulin, two aliquots (5 μl each) of citraconic anhydride (Sigma) were added to the stirred insulin solution (~ 20 ml) over a period of 1 h, and the pH was maintained at 6.9 for 15 min. The mixture was then applied to a Sephadex G-25 column (1.0

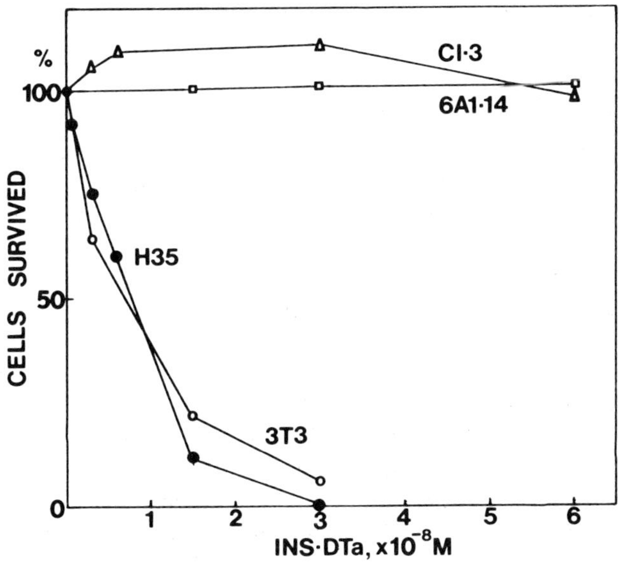

Fig. 3. Dose-dependent profiles of cell killing. ○, Swiss/3T3 mouse fibroblasts; ●, rat hepatoma H35 cells; △, Cl-3 cells isolated from Swiss/3T3 fibroblasts; □, 6A1-14 cells isolated from H35 hepatoma cells.

cm × 15 cm) equilibrated with 10 mM Na phosphate (pH 7.8) containing 0.1 M NaCl. The blocked insulin that was eluted in the void volume was then reacted with a fivefold molar excess of SPDP for 2 h at room temperature. Excess reagent was removed by passing through a Biogel P-6 column (1.6 × 18 cm) equilibrated with 20 mM TES (pH 7.8). These conditions resulted in a derivatized insulin containing 0.7 mole of pyridine-dithiopropionate per mole of insulin, which was determined by the method of Carlsson et al. [1978].

Cross-linking and purification. Completely reduced RICa (1 mg in 20 mM TES, pH 7.8) was mixed with the derivatized insulin at a molar ratio of 1 to 20. It is imperative to use freshly reduced RICa as described for DTa (p.238), since the RICa forms homodimers at a very rapid rate. The mixture was concentrated to a volume of 2 ml if necessary and dialyzed overnight against 10 mM Na phosphate (pH 7.8) containing 0.1 M NaCl. Citraconic blockages at the amino termini were then removed by slowly lowering pH to 2.0 by dialyzing against distilled water adjusted to pH 2.0 with HCl. The insulin-RICa conjugate was separated free of insulin and RICa by a Sephadex G-75 column (2.0 cm × 60 cm) equilibrated with 10 mM Na phosphate (pH 7.8) containing 0.1 M NaCl. The first peak corresponding to the insulin-RICa conjugate was collected, dialyzed against 50% glycerol in 20 mM TES (pH

8.0), and stored as aliquots at −20°C. The absorbance of 1 mg/ml insulin-RICa at 280 nm is 0.81.

Selection of Cell Variants Resistant to Hormone-Toxin Conjugates

Variants were selected from cells with or without mutagenesis. To induce mutations, exponentially growing cells in 75-cm^2 flasks were exposed to 0.15% (vol./vol.) ethylmethanesulfonate (Sigma) for 8 h and washed free of mutagen [Shimizu and Shimizu, 1980]. The surviving cells were grown for several generations in order to allow the expression of any mutations. For variant selection, cells were grown to about 70% confluency in 75-cm^2 flasks. Insulin-DTa or insulin-RICa conjugates were mixed with Dulbecco's modified Eagle's medium (DME) containing 0.1% bovine serum albumin (BSA), and sterilized through a Millipore filter (0.22 μm). The sterilized conjugate was added directly to the 1- to 2-day-conditioned media at a final concentration of 2–3 times the ED_{50}. The medium was changed after 3 days to remove dead cells and debris. To ensure the specific binding of conjugate to insulin receptor, cells were incubated with conjugate in serum-free DME medium at 4°C for 60 min and washed free of unbound conjugate, and the medium was replaced in serum-containing DME medium. Incubation was continued at 37°C for 60 min in a CO_2 incubator to allow internalization of the cell-bound conjugates. Then, cultures were maintained in the serum-containing DME medium. Treatment with conjugate was repeated 2 days after the medium change. The cells were usually treated with the same concentration of conjugate two more times and with 2–3 times higher concentrations one time. Two weeks after the last treatment, surviving colonies were isolated with cloning rings and allowed to expand in DME medium containing 10% fetal calf serum (FCS) for further analysis. For genetic analysis, it is advised that variant isolation should be performed using cells with selectable biochemical markers for cell hybrid formation (see discussion).

Growth of 3T3-L1 Preadipocytes in Serum-Free Medium

3T3-L1 cells (purchased from the American Type Culture Collection) are capable of differentiating into adipocytes when the cultures are confluent [Green and Kehinde, 1976]. The cells were routinely grown in DME medium supplemented with 10% FCS, penicillin (100 units/ml), and streptomycin (100 μg/ml). The cultures were incubated at 37°C in a humidified atmosphere containing 5% CO_2. Serum-free medium used for growth of 3T3-L1 cells is a modification of our previous culture conditions [Shimizu et al., 1982a] by combining the procedures described by Bottenstein et al. [1979], Bettger et al. [1981], and McKeehan et al. [1977]. The ingredients of modified serum-free hormone-supplemented media are listed in Table I. The basal nutrient medium is a 3:1 mixture of DME medium and Ham's F-12 medium, to which are added several amino acids, vitamins, trace elements, bovine crystalline insulin (Sigma), human transferrin (Sigma), BSA (Sigma Fraction V), ethanolamine (Sigma), and epidermal growth factor (EGF, Collaborative Research, culture grade). These supplements are maintained as 100 × stock solutions and added to the medium just prior to plating of cells. Bovine-luteinizing hormone and biochemically undefined gimmel factor [Bottenstein et al., 1979] are eliminated.

Cells detached by trypsinization were first plated at a density of $\sim 1 \times 10^4$ cells per cm^2 in DME medium containing 10% FCS. After 1 day, the cells were washed with serum-free DME and the medium was replaced with serum-free hormone-supplemented medium. Precoating culture dishes with attachment factors such as poly-L-lysine, fibronectin, and collagen (Collaborative Research) are helpful when cells are directly inoculated in serum-free medium. The cultures were fed every other day. Under these conditions, 3T3-L1 cells grew to a saturation density of $\sim 1.0 \times 10^5$ cells per cm^2 with a doubling time of 25 h. The cells have the same morphology as those cultured in DME containing 10% FCS. The cultures can be transferred at least several times.

TABLE I. Serum-Free Hormone-Supplemented Medium for Growth of 3T3-L1 Preadipocytes

Basal medium	DME:F12 (3:1) supplemented with VAT[a]
Insulin	1 μg/ml
Transferrin	5 μg/ml
Ethanolamine	0.1 mM
EGF	20 ng/ml
BSA (Fr. V)	0.1%

[a]V (vitamin solution including 1 μM folinic acid, 0.1 mM pyridoxine HCl, 0.1 mM thiamine HCl, and 1 μM adenine); A (amino acid solution containing 10 μM alanine, 10 μM glutamic acid, 0.1 mM proline, and 1 mM histidine); and T (trace elements including 1 nM $MnSO_4$, 0.1 nM $(NH_4)_6MO_7O_{24} \cdot 4H_2O$, 3 nM $NiCl_2 \cdot 6H_2O$, 10 nM H_2SeO_3, 0.5 nM $SnCl_2 \cdot 2H_2O$, 5 nM NH_4VO_3, 1 μM $ZnSO_4 \cdot 7H_2O$, 10 μM $Na_2SiO_3 \cdot 9H_2O$). The water used in preparation of our medium is obtained as follows: Deionized water is first distilled in a glass still, then processed through a reverse osmosis system, and finally distilled in a glass still.

Selection of Cell Variants With Serum-Free Medium

Cell variants that have lost a particular hormone requirement can be isolated on the basis of their ability to grow as colonies in the serum-free hormone-defined medium in the absence of that hormone. For this, cells were grown in serum-free hormone-supplemented medium from which one hormone (insulin or EGF, for example) had been eliminated. Cells were fed every other day. After 2–3 weeks, cell colonies appeared that no longer required insulin for their growth.

Adipocyte Differentiation in Serum-Free Medium

To induce adipocyte differentiation in serum-free media, 3T3-L1 cells were grown to confluency in DME medium containing 10% FCS. Two days after confluency the cultures were fed with fresh serum-free hormone-supplemented medium containing 1 μM dexamethasone (Sigma) and 0.5 mM methylisobutylxanthine (Sigma). Two days later, the cultures were fed with fresh serum-free hormone-supplemented medium without inducers. The cells were fed every other day. Under these conditions, 3T3-L1 cells differentiated into adipocytes, which were determined by accumulation of lipid droplets stained with Oil Red O solution (Sigma). Both insulin and EGF were required for maximum frequency of differentiation.

APPLICATIONS AND DISCUSSION

Two methods for the isolation of cell variants that are related to mutations in hormone-receptor systems have been presented. The first technique involves the use of hormone-toxin conjugates. The second method relies on the fact that maintenance and growth of cells in serum-free medium requires certain combinations of hormones and growth factors.

For the first method, two different insulin-toxin conjugates have been constructed and utilized to select several series of resistant variants (Table II). When the insulin-DTa conjugate was initially applied to Swiss/3T3 mouse fibroblasts, 2 of 8 isolated variants displayed a deficiency of insulin binding [Miskimins and Shimizu, 1981a,b]. In addition, these variants exhibited alterations in morphology and cell growth properties. These changes were accounted for by the insulin's growth-regulating role and the insulin receptor's possible interaction with other membrane components. One of the variants showed an alteration in the lysosomal system or in the regulation of that system. In this case resistance to the conjugate was correlated with the deficiency of ligand degradation in the lysosomal system [Miskimins et al., 1981b]. Insulin receptor specificity of selection was proven by the finding that

TABLE II. Cell Variants Resistant to Hormone-Toxin Conjugates

Variants	Parental cell line	Conjugate used	Reference
CI-1,2,3,4,6, and 7	Mouse fibroblast Swiss/3T3	Insulin-DTa	Miskimins and Shimizu [1981a]
L1-1a,1b,1d,2b,3b, and 3c	Mouse preadipocyte 3T3-L1	Insulin-DTa	Shimizu et al. [1982a]
IR-1,2,3,33, and 36	Mouse preadipocyte 3T3-L1	Insulin-RICa	See text.
6A1-7,11,14,19, and 24	Rat hepatoma H35-6A1(TG^r)	Insulin-DTa	See text.
ER-1,2,3,4,1-IV-1 ~ 5	Chinese hamster fibroblast CHL-TG2(TG^r)	EGF-RICa	Shimizu et al. [1982b,1983]

the variant that is devoid of insulin-binding activity was found to bind and internalize EGF [Miskimins and Shimizu, 1981a]. In addition, DNA synthesis in this variant was stimulated by EGF but it was not enhanced by insulin. One variant retained only a small number of insulin receptors, and it was found that the remaining receptors are low-affinity type insulin receptors (Shimizu et al., in preparation). Thus, the conjugate allows selection of variants specifically related to insulin's activity.

The same selection using insulin-DTa conjugate was applied to two insulin-responsive cell lines: 3T3-L1 preadipocytes and rat hepatoma H35 cells. The 3T3-L1 cells are able to differentiate into adipocytes under defined induction conditions [Rubin et al., 1978; Lewis et al., 1982]. Several variants selected from the 3T3-L1 cells also showed a substantial loss of insulin-binding capacity. Interestingly, these variants displayed a variety of properties as to ability to differentiate into adipocyte and as to the insulin requirement for growth in serum-free hormone-supplemented medium [Shimizu et al., 1982a].

H35 rat hepatoma cells produce tyrosine aminotransferase (TAT) upon exposure to insulin and/or dexamethasone (DEX). Most of the insulin-DTa-resistant variants of H35 cells revealed substantial loss of insulin-binding capacity. Our preliminary analysis indicates that some variants apparently have lost insulin-inducible TAT activity but retain DEX-inducible TAT activity. Using these variants, it may be possible to study differential induction mechanisms of TAT activity.

The application of insulin-RICa conjugate resulted in several resistant variants from 3T3-L1 cells (Table II). Two variants again showed substantial reduction of insulin binding, whereas others exhibited an increase of insulin-binding capacity. Interestingly, one variant was found to be highly resistant to intact ricin [Shimizu et al., 1982b].

As a second method, serum-free hormone-supplemented medium has been developed for proliferation of 3T3-L1 cells. The availability of completely defined media that support good growth of preadipocytes would facilitate the analysis of the induction mechanism of adipocyte differentiation. Studies on the precise requirements of insulin and EGF for induction and expression of adipocyte differentiation and variant isolation are currently in progress. It is worthy of mention that, with similar serum-free media techniques, variants have been isolated from human epidermoid carcinoma A431 cells and mouse melanoma Cloudman S-91 cells by using the reduction of growth in these cells caused by the presence of EGF and insulin, respectively [Gill and Lazar, 1981; Barnes, 1982; Buss et al., 1982; Pawelek et al., 1982].

The above summary on this topic clearly indicates that the insulin-toxin conjugates not only allow selection of variants specifically related to insulin's activity but also provide variants with a wide array of properties. The reasons for this variety may be twofold: both the properties of hormone-toxin conjugates and the nature of the cell's endocytotic mechanisms. The following discussion may serve as a basis for the future application of the toxic hormone techniques.

Although the exact mechanisms of diphtheria toxin and ricin entry into cells are not yet understood, it is now well established that for hybrid protein formation, DTa and RICa must be linked to the ligand by a disulfide bond in order to exhibit cytotoxicity [Olsnes and Pihl, 1982a,b]. A few more reagents other than those described here are available [Olsnes and Pihl, 1982a,b]. For the choice of toxic fragments, RICa has been more common, since the potency of RICa-conjugates is in general higher than that of DTa conjugates [Olsnes and Pihl, 1982a,b]. It is known, however, that RICa contains a highly hydrophobic region [Funatsu et al., 1978] that may interact with membrane lipids and therefore may facilitate entry of the conjugate. This may in turn cause a reduction of receptor specificity. DTa lacks a hydrophobic region [Delange et al., 1979], and therefore the conjugate of it may rely more on the specific interaction of ligand with its receptor. At present, there is not enough information available to substantiate these hypotheses.

It is now well documented that insulin, related polypeptide hormones, and growth factors are internalized through receptor-mediated endocytosis [Carpenter and Cohen, 1979; Pastan and Willingham, 1981]. The endocytotic vesicles have been shown to fuse with lysosomes; fusion is followed by degradation of the internalized hormones. Recent evidence also suggests that receptor-mediated endocytosis and lysosomal processing may be an essential step in the activation of diphtheria toxin [Sandvig and Olsnes, 1980, 1981; Draper and Simon, 1980]. Therefore, this pathway may be associated with insulin's ability to transport DTa into cells, resulting into cytotoxicity. However, ricin's toxicity appears to be activated through a nonlysosomal pathway and even to be enhanced severalfold by the lysosomal inhibitors methylamine and chloroquine [Cawley et al., 1980; Mekada et al., 1981]. This suggests the importance of nonlysosomal processing of RICa conjugates. Recently, a second pathway for EGF endocytosis has been identified that involves lysosomal enzyme-deficient organelles [Miskimins and Shimizu, 1982a,b]. In this case, EGF-RICa conjugates could best be used to isolate variants of this pathway. There is also evidence that suggests the requirement of the ricin B-subunit for the toxicity of RICa conjugates [Youle et al., 1981]. However, our insulin-RICa conjugate [Shimizu et al., 1982b] and the EGF-RICa conjugates of Cawley et al. [1980] and Shimizu et al. [1982b,1983] did not require copresence of RICb.

Obviously, not only are there many gene products involved in the binding, clustering, internalization, and transport of insulin and other polypeptide hormones, but also the targeting of the toxic moiety of the conjugates to protein synthesis machinery appears complex (Fig. 1). Therefore, a defect in any of these sites might enable the cell to survive in the presence of toxic conjugate. It will be interesting to treat cells with conjugates in the presence of inhibitors that are known to block certain steps of the above pathway.

It is interesting that DTa cross-linked to the amino group of EGF was virtually nontoxic to human epidermoid carcinoma A431 cells [Cawley et al., 1980], whereas the DTa cross-linked to the carboxyl group of EGF was toxic to A431 cells at 10^{-8} M [Shimizu et al., 1980]. Surprisingly, the former conjugate was toxic to rat hepatocytes [Simpson et al., 1982]. These results suggest that different modifications provide different degrees of cell killing and that even the same conjugate may be processed in different ways when applied to different cell types. It may also be possible that the conjugate will allow selection of cells that are defective in a step common to many polypeptide hormones and growth factors. As already seen, defective intracellular processing may be recognized by the conjugates as well as defects in microfilaments involved in aggregation and endocytosis. Since the receptors are often glycoproteins, variants with a defective glycosylation system could be isolated as resistant to the conjugates.

Further application of the hormone-toxin conjugates as selection agents, particularly in combination with serum-free hormone-supplemented medium, cell synchronization methods, and inhibitor treatment, will hopefully allow for the isolation of a series of mutants with defects at each of the steps involved in hormone binding and processing. These variants can then be analyzed by somatic cell hybridization, whether the defect is genetic or epigenetic, in order to identify the various genes involved in hormone action [Shimizu et al., 1980; Behzadian et al., 1982; Kondo and Shimizu, 1983]. In addition, the variants can be analyzed biochemically to identify the lesions at the molecular level and also to provide knowledge as to regulation of the cellular hormone response.

ACKNOWLEDGMENTS

The author thanks W.K. Miskimins, Y. Shimizu, M.A. Behzadian, S. Gamou, and J. Lewis for their contributions to the research project presented in this chapter. The research was supported by Public Health Service grant GM-24375 from the National Institute of General Medical Sciences.

REFERENCES

Barnes DW (1982): Epidermal growth factor inhibits growth of A431 human epidermoid carcinoma in serum-free cell culture. J Cell Biol 93:1-4.

Behzadian MA, Shimizu Y, Kondo I, Shimizu N (1982): Genetics of receptors for bioactive polypeptides: Expression of the human EGF receptor gene and internalization and processing of the receptor-bound EGF in human-mouse cell hybrids. Somat Cell Genet 8:347-362.

Bettger WJ, Boyce ST, Walthall BJ, Ham RG (1981): Rapid clonal growth and serial passage of human diploid fibroblasts in a lipid-enriched synthetic medium supplemented with epidermal growth factor, insulin and dexamethasone. Proc Natl Acad Sci USA 78:5588-5592.

Bottenstein J, Hayashi I, Hutchings H, Masui H, Mather J, McClure DB, Ohasa S, Rizzino A, Sato G, Serrero G, Wolfe R, Wu R (1979): The growth of cells in serum-free hormone-supplemented media. Methods Enzymol 58:94-109.

Buss JE, Kudlow JE, Lazar CS, Gill GN (1982): Altered epidermal growth factor (EGF)-stimulated protein kinase activity in variant A431 cells with altered growth responses to EGF. Proc Natl Acad Sci USA 79:2574-2578.

Carlsson J, Drevin H, Axen R (1978): Protein thiolation and reversible protein-protein conjugation: N-succinimidyl 3-(2-pyridyldithio) propionate, a new heterobifunctional reagent. Biochem J 173:723-737.

Carpenter G, Cohen S (1979): Epidermal growth factor. Ann Rev Biochem 48:193-216.

Cawley DB, Herschman HR, Gilliland DG, Collier RJ (1980): Epidermal growth factor-toxin A chain conjugates: EGF-ricin A is a potent toxin while EGF-diphtheria fragment A is nontoxic. Cell 22:563-570.

Chang T-M, Neville DM Jr (1977): Artificial hybrid protein containing a toxic protein fragment and a cell membrane receptor binding moiety in a disulfide-linked conjugate. Synthesis of diphtheria toxin A-S-S-human placental lactogen with methyl-5-bromovalerimide. J Biol Chem 252:1505-1514.

Chang T-M, Dazord A, Neville DM Jr (1977): Artificial hybrid protein containing a toxic protein fragment and a cell membrane receptor binding moiety in a disulfide-linked conjugate.

Biochemical and biologic properties of diphtheria fragments A-S-S-human placental lactogen. J Biol Chem 252:1515-1522.

DeLange RJ, Williams LS, Druzin RE, Collier RJ (1979): The amino acid sequence of fragment A, an enzymically active fragment of diphtheria toxin. J Biol Chem 254:5838-5842.

Draper RK, Simon MI (1980): The entry of diphtheria toxin into the mammalian cell cytoplasm: Evidence for lysosomal involvement. J Cell Biol 87:849-854.

Funatsu G, Yoshitake S, Funatsu M (1978): Primary structure of ile chain of ricin D. Agric Biol Chem (Tokyo) 42:501-503.

Gill GN, Lazar CS (1981): Increased phosphotyrosine content and inhibition of proliferation in EGF-treated A431 cells. Nature 293:305-307.

Gilliland D, Collier RJ, Moehring JM, Moehring JT (1978): Chimeric toxins: Toxic disulfide-linked conjugate of concanavalin A with fragment A of diphtheria toxin. Proc Natl Acad Sci USA 75:5319-5323.

Green H, Kehinde O (1976): Spontaneous heritable changes leading to increased adipose conversion in 3T3 cells. Cell 1:105-113.

Habeeb AFSA (1972): Reactions of protein sulfhydryl groups with Ellman's reagent. Methods Enzymol 25B:457-464.

Honjo T, Nishizuka Y, Hayaishi O, Kato I (1968): Diphtheria toxin dependent adenosine diphosphate ribosylation of aminoacyl-transferase II and inhibition of protein synthesis. J Biol Chem 243:3533-3554.

Kandel J, Collier RJ, Chung DW (1974): Interaction of fragment A from diphtheria toxin with nicotinamide adenine dinucleotide. J Biol Chem 249:2088-2097.

Kondo I, Shimizu N (1983): Mapping of the human gene for epidermal growth factor receptor (EGFR) on the p13-q22 region of the chromosome 7. Cytogenet Cell Genet 35:9-14.

Laemmli UK (1970): Cleavage of structural proteins during the assembly of the head of bacteriophage T4. Nature 227:680-685.

Lappi DA, Kapmeyer W, Beglan JM, Kaplan NO (1978): The disulfide bond connecting the chains of ricin. Proc Natl Acad Sci USA 75:1096-1100.

Lewis JE, Shimizu Y, Shimizu N (1982): Nicotinamide inhibits adipocyte differentiation of 3T3-L1 cells. FEBS Lett 146:37-41.

Lin TT-S, Li SS-L (1980): Purification and physicochemical properties of ricins and agglutinins from Ricinus communis. Eur J Biochem 105:453-459.

McKeehan WL, McKeehan KA, Hammond SL, Ham RG (1977): Improved medium for clonal growth of human diploid fibroblasts at low concentrations of serum protein. In Vitro 13:399-416.

Mekada E, Uchida T, Okada Y (1981): Methylamine stimulates the action of ricin toxin but inhibits that of diphtheria toxin. J Biol Chem 256:1225-1228.

Miskimins WK, Shimizu N (1979): Synthesis of a cytotoxic insulin cross-linked to diphtheria toxin fragment A capable of recognizing insulin receptors. Biochem Biophys Res Commun 91:143-151.

Miskimins WK, Shimizu N (1981a): Genetics of cell surface receptors for bioactive polypeptides: Variants of Swiss/3T3 fibroblasts resistant to A cytotoxic chimeric insulin. Proc Natl Acad Sci USA 78:445-449.

Miskimins WK, Shimizu N (1981b): Binding and intracellular fate of EGF in Swiss/3T3 and a variant cell line possessing an altered lysosomal system. (Abstract #2.) J Cell Biol 91:216a.

Miskimins WK, Shimizu N (1982a): Dual pathways for epidermal growth factor processing after receptor-mediated endocytosis. J Cell Physiol 112:327-338.

Miskimins WK, Shimizu N (1982b): Involvement of multiple subcellular compartments in intracellular processing of epidermal growth factor. J Cell Biochem 20:41-50.

Miskimins WK, Ferris WR, Shimizu N (1981): Genetics of receptors for bioactive polypeptides: A variant of Swiss/3T3 fibroblasts resistant to a cytotoxic insulin accumulates lysosome-like vesicles. J Supramol Struct Cell Biochem 16:105-113.

Oeltmann TN, Heath EC (1979): A hybrid protein containing the toxic subunit of ricin and the cell-specific subunit of human chorionic gonadotropin. J Biol Chem 254:1028-1032.

Olsnes S, Pihl A (1982a): Chimeric toxins. Pharmacol Ther 15:355-381.

Olsnes S, Pihl A (1982b): Toxic lectins and related proteins. In Van Heyningen S, Cohen P (eds): "Molecular Action of Toxins and Viruses." Amsterdam: Elsevier/North Holland, pp 51-105.

Pappenheimer AM Jr (1977): Diphtheria toxin. Annu Rev Biochem 46:69-94.

Pastan IH, Willingham MC (1981): Receptor-mediated endocytosis of hormones in cultured cells. Annu Rev Physiol 43:239-250.

Pawelek J, Murray M, Fleischmann R (1982): Genetic studies of insulin action in Cloudman melanoma cells. In Sato GH, Pardee AB, Sirbasku DA (eds): "Growth of Cells in Hormonally Defined Media." Cold Spring Harbor, NY: Cold Spring Harbor Laboratory, pp 911-919.

Rubin CS, Hirsch A, Fung C, Rosen OM (1978): Development of hormone receptors and hormonal responsiveness in vitro: Insulin receptors and insulin sensitivity in the preadipocyte and adipocyte forms of 3T3-L1 cells. J Biol Chem 253:7570-7578.

Sandvig K, Olsnes S (1979): Effect of temperature on the uptake, excretion and degradation of abrin and ricin by Hela cells. Exp Cell Res 121:15-25.

Sandvig K, Olsnes S (1980): Diphtheria toxin entry into cells is facilitated by low pH. J Cell Biol 87:828-832.

Sandvig K, Olsnes S (1981): Rapid entry of nicked diphtheria toxin into cells at low pH. Characterization of the entry process and effects of low pH on the toxin molecule. J Biol Chem 256:9068-9076.

Shecter Y, Schlessinger J, Jacobs S, Chang KJ, Cuatrecasas P (1978): Fluorescent labeling of hormone receptors in viable cells: Preparation and properties of highly fluorescent derivatives of epidermal growth factor and insulin. Proc Natl Acad Sci USA 75:2135-2139.

Shimizu Y, Shimizu N (1980): Genetics of cell-surface receptors for bioactive polypeptides: A variant of mouse Balb/c3T3 fibroblasts possessing altered insulin-binding ability. Somat Cell Genet 6:583-601.

Shimizu N, Miskimins WK, Shimizu Y (1980): A cytotoxic epidermal growth factor cross-linked to diphtheria toxin A-fragment. FEBS Lett 118:274-278.

Shimizu N, Miskimins WK, Gamou S, Shimizu Y (1982a): A genetic approach to the mechanisms of insulin action: Use of the insulin-diphtheria toxin fragment A conjugates for isolation of genetic variants. In Pardee A, Sato G, Sirbasku D (eds): "Cold Spring Harbor Conference on Cell Proliferation: Growth of Cells in Hormonally Defined Media." Cold Spring Harbor, NY: Cold Spring Harbor Laboratory, pp 397-402.

Shimizu N, Miskimins WK, Gamou S, Shimizu Y (1982b): Toxin-hormone conjugates and resistant variants. J Cell Biochem, Suppl 6, p 133.

Shimizu N, Shimizu Y, Miskimins WK (1983): Use of EGF-ricin A subunit conjugates for isolation of genetic variants involved in the mechanisms of EGF action. (Manuscript submitted for publication.)

Simpson DL, Cawley DB, Herschman HR (1982): Killing of cultured hepatocytes by conjugates of Asialofetuin and EGF linked to the A chains of Ricin and diphtheria toxin. Cell 29: 469-473.

Youle RJ, Murray GJ, Neville DM Jr (1981): Studies on the galactose binding site of ricin and on the hybrid toxin Man-6P-ricin. Cell 23:551-559.

Methods for Serum-Free Culture of Epithelial and Fibroblastic Cells,
pages 249–264

17
Growth of Human Fibroblast Cultures in Serum-Free Media

Richard G. Ham

Fibroblasts are probably the easiest of all cell types to grow in serum-containing media, and they are regarded by many investigators as "weeds" in the garden of exotic cell types that they are trying to culture. Almost all "conventional" culture media were originally developed for permanent lines of transformed fibroblastic cells [Ham, 1974, 1984a]. In addition, serum (which is often erroneously viewed as a universal cell culture supplement) strongly favors fibroblastic growth over that of many other cell types [Ham, 1984a].

However, despite the almost unavoidable proliferation of fibroblasts in serum-containing media, growth of normal fibroblasts (and permanent lines of "nontransformed" fibroblastic cells) under serum-free conditions has proven to be a much more difficult task. Many serum-free culture systems are actually selective against growth of fibroblasts (reviewed in Ham [1984a]), and those that do support growth of fibroblasts still have a number of incompletely resolved problems. This chapter compares three major systems for "serum-free" growth of human diploid fibroblasts (HDF) that have been described in the literature, and also briefly discusses similar culture systems for fibroblasts from several other species.

The first of the three systems for HDF utilizes medium RITC 80-7, which consists of an extensively modified version of Eagle's minimum essential medium (MEM) plus insulin, triiodothyronine, epidermal growth factor (EGF), and transferrin. For serum-free growth of HDF, medium RITC 80-7 is further

Department of Molecular, Cellular, and Developmental Biology, University of Colorado, Boulder, Colorado 80309 supplemented with bovine serum albumin (BSA), and fibronectin [Yamane et al., 1981a; Kan and Yamane, 1982]. It is the most successful of the three systems, both in terms of growth rate and in terms of number of population doublings that HDF will complete in it. In both cases, the results that are obtained are comparable to those obtained with Eagle's basal medium (BME) supplemented with 10% fetal bovine serum [Yamane et al., 1981a]. Unfortunately, it is also the least defined of the three systems. It contains 5.0 mg/ml fraction V BSA, which gives it a total protein concentration equivalent to 10% serum. In addition, crystalline BSA does not support comparable growth. One of the major functions of the BSA in this system is as a lipid carrier, as will be discussed below, but the possibility that the BSA may also be supplying other undefined growth-promoting substances cannot easily be ruled out.

The second system, developed by Weinstein et al. [1982], employs a 1:1 mixture of Dulbecco's modified Eagle's medium (DME) and medium F12, supplemented with trace elements, hydrocortisone, insulin, EGF, transferrin, and chicken egg white ovalbumin. This system is reported to support a rate of proliferation comparable to that obtained with 5% serum in a similar basal medium, and to support "limited" serial subculture. It also is somewhat less than totally defined, in that it contains 1.0 mg/ml ovalbumin, which gives it a total protein concentration equivalent to that in 2% serum.

The third system, from the author's laboratory, utilizes a basal nutrient medium, MCDB 110, that has been optimized specifically for HDF. It is supplemented with dexamethasone, insulin, EGF, a mixture of lipids in the form of liposomes (designated liposome B), prostaglandins E1 and F2α, dithiothreitol, reduced glutathione, and phosphoenolpyruvate [Bettger et al., 1981]. This system has a total protein concentration of less than 1.0 μg/ml and contains only proteins that are available in highly purified forms (insulin and EGF). It supports rapid clonal growth of HDF (although not as rapid as with 2% dialyzed serum in the same optimized nutrient medium). However, it does not support long-term serial subculture beyond about 12–15 population doublings. In addition, the lipid supplement contains a poorly defined mixture of phospholipids (crude soy lecithin), whose growth-promoting properties, thus far, can only partly be replaced with mixtures of more highly purified phospholipids.

Detailed comparisons of the nutrient media and the supplements used in these three systems are made in Tables I and II, respectively. Table I also includes the compositions of two other optimized nutrient media, MCDB 202 and MCDB 402, which will be discussed later in the chapter.

NUTRIENT REQUIREMENTS

Although the nutrient media for human fibroblasts described in Table I superficially seem to have diverse histories, they are actually rather similar in

TABLE I. Nutrient Composition of Serum-Free Media for Fibroblasts[a]

Component	RITC 80-7[b]	DME: F12[c]	MCDB 110[d]	MCDB 202[e]	MCDB 402[f]
Amino Acids					
L-Alanine	—	5.0E-5	1.0E-4	1.0E-4	—
L-Arginine	6.0E-4	7.0E-4	1.0E-3	3.0E-4	3.0E-4
L-Asparagine	—	5.0E-5	1.0E-4	1.0E-3	1.0E-4
L-Aspartate	1.0E-4	5.0E-5	1.0E-4	1.0E-4	1.0E-5
L-Cysteine	—	1.0E-4	5.0E-5	2.0E-4	—
L-Cystine/2	2.0E-4	2.0E-4	—	—	4.0E-4
Total Cys/2	2.0E-4	3.0E-4	5.0E-5	2.0E-4	4.0E-4
L-Glutamate	1.0E-6	5.0E-5	1.0E-4	1.0E-4	1.0E-5
L-Glutamine	2.0E-3	2.5E-3	2.5E-3	1.0E-3	5.0E-3
Glycine	1.0E-4	2.5E-4	3.0E-4	1.0E-4	1.0E-4
L-Histidine	2.0E-4	1.5E-4	1.0E-4	1.0E-4	2.0E-3
L-Isoleucine	4.0E-4	4.2E-4	3.0E-5	1.0E-4	1.0E-3
L-Leucine	4.0E-4	4.5E-4	1.0E-4	3.0E-4	2.0E-3
L-Lysine	4.0E-4	5.0E-4	2.0E-4	2.0E-4	8.0E-4
L-Methionine	1.0E-4	1.2E-4	3.0E-5	3.0E-5	2.0E-4
L-Phenylalanine	2.0E-4	2.2E-4	3.0E-5	3.0E-5	3.0E-4
L-Proline	3.0E-5	1.5E-4	3.0E-4	5.0E-5	—
L-Serine	1.0E-4	2.5E-4	1.0E-4	3.0E-4	1.0E-4
L-Threonine	4.0E-4	4.5E-4	1.0E-4	3.0E-4	5.0E-4
L-Tryptophan	5.0E-5	4.5E-5	1.0E-5	3.0E-5	1.0E-5
L-Tyrosine	2.0E-4	2.2E-4	3.0E-5	5.0E-5	2.0E-4
L-Valine	4.0E-4	4.5E-4	1.0E-4	3.0E-4	2.0E-3
Vitamins					
d-Biotin	8.0E-8	1.5E-8	3.0E-8	3.0E-8	3.0E-8
Folate	2.3E-6	6.0E-6	—	—	—
Folinate	8.3E-11	—	1.0E-9	1.0E-8	1.0E-6
α-Lipoate	—	5.0E-7	1.0E-8	1.0E-8	1.0E-8
Nicotinamide	8.2E-6	1.7E-5	5.0E-5	5.0E-5	5.0E-5
Pantothenate	4.6E-6	9.0E-6	1.0E-6	1.0E-6	5.0E-5
Pyridoxal	6.0E-6	1.0E-5	—	—	—
Pyridoxine	—	1.5E-7	3.0E-7	3.0E-7	1.0E-4
Riboflavin	2.7E-7	5.8E-7	3.0E-7	3.0E-7	1.0E-6
Thiamine	3.0E-6	6.5E-6	1.0E-6	1.0E-6	1.0E-4
Vitamin B12	1.5E-8	5.0E-7	1.0E-7	1.0E-7	1.0E-8

(continued)

TABLE I. Nutrient Composition of Serum-Free Media for Fibroblasts[a] *(Continued)*

Component	RITC 80-7[b]	DME: F12[c]	MCDB 110[d]	MCDB 202[e]	MCDB 402[f]
Other organic nutrients					
Adenine	—	—	1.0E-5	1.0E-6	1.0E-6
Choline	1.3E-4	6.2E-5	1.0E-4	1.0E-4	1.0E-4
D-Glucose	5.6E-3	7.8E-3	4.0E-3	8.0E-3	5.5E-3
Hypoxanthine	1.8E-6	1.5E-5	—	—	—
i-Inositol	1.1E-5	7.0E-5	1.0E-4	1.0E-4	4.0E-5
Linoleate	—[g]	1.5E-7	—[h]	2.0E-7	3.0E-7
Putrescine	1.2E-7	5.0E-7	1.0E-9	1.0E-9	1.0E-9
Pyruvate	1.0E-3	1.0E-3	1.0E-3	5.0E-4	1.0E-3
Succinate	1.0E-3	—	—	—	—
Tartarate	8.3E-6	—	—	—	—
Thymidine	2.9E-7	1.5E-6	3.0E-7	3.0E-7	1.0E-6
Bulk inorganic ions					
Calcium	1.8E-3	1.0E-3	1.0E-3	2.0E-3	1.6E-3
Magnesium	9.8E-4	7.0E-4	1.0E-3	1.5E-3	8.0E-4
Potassium	5.4E-3	4.2E-3	5.0E-3	3.0E-3	4.0E-3
Sodium	1.4E-1	1.6E-1	1.3E-1	1.4E-1	1.3E-1
Chloride	1.3E-1	1.3E-1	1.2E-1	1.3E-1	1.2E-1
Phosphate	9.7E-4	9.6E-4	3.0E-3	5.0E-4	5.0E-4
Sulfate	—	4.1E-4	1.0E-3	1.5E-3	8.0E-4
Trace elements					
Cadmium	—	5.0E-8	—	—	—
Copper	1.0E-11	5.0E-9	1.0E-9	1.0E-9	5.0E-9
Iron	2.9E-6	1.6E-6	5.0E-6	5.0E-6	1.0E-6
Manganese	1.0E-11	5.0E-10	1.0E-9	5.0E-10	1.0E-9
Molybdenum	7.0E-9	3.5E-9	7.0E-9	7.0E-9	2.1E-8
Nickel	5.0E-11	2.5E-10	5.0E-10	5.0E-12	3.0E-10
Selenium	3.0E-9	1.5E-8	3.0E-8	3.0E-8	1.0E-8
Silicon	—	2.5E-7	5.0E-7	5.0E-7	1.0E-5
Tin	—	2.5E-10	5.0E-10	5.0E-12	—
Vanadium	5.0E-10	2.5E-9	5.0E-9	5.0E-9	5.0E-9
Zinc	—	1.5E-6	5.0E-7	1.0E-7	1.0E-6
Buffers and indicators					
HEPES	1.4E-2	—	3.0E-2	3.0E-2	—
Bicarbonate	1.7E-2	2.9E-2	—[j]	—[j]	1.4E-2
Phenol red	1.6E-5	2.3E-5	3.3E-6	3.3E-6	3.3E-5

[a]This table lists the nutrients in three serum-free media for HDF plus those in media optimized for chicken fibroblasts and Swiss mouse 3T3 cells. Total amounts of specific nutrients are listed irrespective of the molecular forms in which they are added. Supplements used with these nutrients for serum-free growth of HDF are listed in Table II. Except for components of medium F12 and the MCDB media, substances with potential nutrient roles that were classified by the original authors as "supplements" have been listed in Table II. These include ascorbate, vitamin E, glutathione, and all lipids except linoleic acid. Most of the values are from Table 2 of Ham and McKeehan [1979]. In some cases, averaging of previously rounded values may have introduced small errors in the second significant figure. All values are moles per liter, expressed in computer-style exponential notation.

[b]Yamane et al. [1981a]; Kan and Yamane [1982]. The authors use the name "RITC 80-7" to describe all components of their serum-free medium (including those in Table 2 of this chapter) except bovine serum albumin and fibronectin.

[c]Weinstein et al. [1982]. The nutrients listed here plus the supplements listed in Table II constitute "defined medium F." The heading "DME:F12" has been employed for brevity, but the nutrients listed here include trace elements [Hutchings and Sato, 1978] that are not components of DME or F12.

[d]Bettger et al. [1981]. Several of the "supplements" used with medium MCDB 110 for serum-free growth of HDF (Table II) also have potential nutrient roles.

[e]McKeehan and Ham [1977]. MCDB 202 was optimized for chicken embryo fibroblasts, but it also works quite well with human fibroblasts [Ham and McKeehan, 1978; Graves and Ham, 1982].

[f]Shipley and Ham [1981]. MCDB 402 was optimized for Swiss mouse 3T3 cells.

[g]Linoleate is not provided as a separate component in RITC 80-7. However, the bovine serum albumin that is used with it for serum-free growth of HDF provides a substantial amount of bound linoleate and other lipids [Kan and Yamane, 1982].

[h]Linoleate is not included in the formulation of MCDB 110. However, the lipid supplements employed with it (Table II) contain linoleic acid bound in phospholipids and sphingomyelin.

[i]Zinc is not supplied as a deliberately added component of RITC 80-7. However, analysis by atomic absorption has shown the presence of 0.2 mg/liter (3.1E-6M) zinc in the complete medium.

[j]No bicarbonate is added to MCDB 110 or 202 during their formulation. However, bicarbonate is formed rapidly when these HEPES-buffered media are placed in an incubator chamber containing 2% carbon dioxide.

many respects. MCDB 110 and its predecessor, MCDB 104, were developed from medium F12 [Ham, 1965] by a process of qualitative and quantitative optimization for progressively better growth with progressively smaller amounts of dialyzed serum. Except for the addition of trace elements, nearly all of the changes that were introduced were quantitative [McKeehan et al., 1976, 1977; Ham et al., 1977; Bettger et al., 1981]. Medium RITC 80-7 retains the original concentrations of most of the components of MEM (which are generally somewhat higher than in MCDB 110, but still reasonably close to optimal values for HDF). In addition, it has been enriched with many of the components of MCDB 110 that are not in MEM. The F12:DME mixture employed by Weinstein et al. [1982] is enriched with a trace element supplement based on that in MCDB 110 [Hutchings and Sato, 1978]. It also contains all of the specialized components of F12 at reduced concentrations. DME was originally formulated to contain four times the levels of vitamins and amino acids in BME, which has generally lower nutrient concentrations than those in MEM. Since most of the concentrations in F12 are lower yet, an equal mixture of DME and F12 has many nutrient concentrations that are similar to those in MEM.

Thus, despite the diverse histories of these media, they are quite similar qualitatively, and somewhat similar quantitatively. MCDB 110 has the special advantage of being optimized specifically for growth of HDF with minimal amounts of supplementation (see Ham [1984b] for an extended discussion of the advantages of using optimized media). However, both RITC 80-7 and DME:F12 plus trace elements have quantitative compositions that are close enough to optimal for HDF so that they will support substantial growth when provided with sufficiently large amounts of defined supplements.

Medium MCDB 202 was originally optimized for clonal growth of chicken embryo fibroblasts with minimal amounts of dialyzed serum [McKeehan and Ham, 1977]. It is qualitatively similar to MCDB 110, but has a somewhat different quantitative composition. However, the differences are small enough so that MCDB 202 also supports good growth of HDF, both with dialyzed serum and with defined supplements [Ham and McKeehan, 1978; Graves and Ham, 1982]. However, MCDB 110 does not support satisfactory growth of chicken embryo fibroblasts. The other medium in Table I, MCDB 402, was formulated for Swiss mouse 3T3 cells. It does not support satisfactory multiplication of HDF, and is discussed in the section on cells from other species near the end of this chapter.

Several quantitative requirements for growth of HDF that are not fully evident from Table I need to be noted. One is that HDF are severely inhibited by excessive levels of cysteine (or cystine). In a typical growth-response titration for clonal growth in medium F12 or MCDB 104 with small amounts of dialyzed serum, the optimum concentration is around 5×10^{-5} M, with a definite inhibitory effect at concentrations above about 1.0×10^{-4} M [Ham et al., 1977]. Many well-known culture media contain levels of cysteine (or cystine) that are far above these levels (see Table 2 of Ham and McKeehan [1979] for a detailed comparison of the compositions of widely used media). All of the media in Table I, except MCDB 110, could presumably be made better for HDF by reducing their cysteine/cystine levels. However, in view of the possibility that complex nutrient balance relationships [Ham, 1974] could alter the quantitative requirement for cysteine under different background conditions, the efficacy of any such changes should be verified experimentally before they are implemented.

A high extracellular calcium concentration is very important for growth of HDF with minimal amounts of dialyzed serum or with defined supplements [McKeehan et al., 1977]. Analysis of the effects of calcium, serum, and specific growth factors on growth rate has shown that it is necessary to increase the concentration of calcium to maintain optimal growth as the concentrations of serum or growth factors are decreased [McKeehan and Ham, 1978; McKeehan and McKeehan, 1979].

The trace element selenium is also of major importance for serum-free growth of HDF [McKeehan et al., 1976]. This requirement is masked in most serum-containing systems, but becomes very important when the concentration of dialyzed serum drops to 3% or less.

In systems where transferrin is not employed, it is very important to be certain that the iron is not removed from the medium as an insoluble ferric hydroxide precipitate during sterile filtration [Walthall and Ham, 1981; Bettger et al., 1981]. One of the best approaches is to add a freshly prepared sterile solution of ferrous sulfate aseptically after sterile filtration of the medium has been completed.

Most of the other trace elements in Table I tend to be ubiquitous contaminants, such that it is difficult to prepare a test system that is clean enough to demonstrate a definitive requirement for them. For example, Yamane et al. [1981a] note that there is no deliberate addition of zinc to RITC 80-7, but that atomic absorption analysis indicates the presence of 0.2 mg/ml (about 3×10^{-6} M) zinc in the final medium. However, despite their frequent presence as contaminants, marginal responses of cultured HDF to several trace elements have already been demonstrated [McKeehan et al., 1977], and it is likely that additional requirements will become evident as the purity of the chemicals used in medium preparation is gradually improved over the years.

Therefore, in order to avoid possible unanticipated failures of serum-free culture systems in the future, it is highly desirable to include in all serum-free media a relatively complete supplement of the trace elements that are known to be required by intact animals [Underwood, 1977] and that are likely to be required by cultured cells when culture media become sufficiently well defined. One example of such an unanticipated failure that has already occurred is the requirement of Chinese hamster ovary cells for selenium. The original formulation of medium F12 [Ham, 1965] was totally dependent on the presence of selenium as a contaminant in sufficient amounts to support cellular proliferation. When medium F12 is prepared with modern chemicals, selenious acid must be added deliberately in order to obtain satisfactory serum-free growth of CHO cells [Hamilton and Ham, 1977].

Another special requirement for growth of HDF with minimal supplementation is an unusually high concentration of nicotinamide [McKeehan et al., 1977]. MEM and DME both have reasonably high concentrations of nicotinamide, but many other media, including F12, have much lower concentrations [Ham and McKeehan, 1979].

Although MCDB 110 is referred to in the foregoing discussion as an "optimized" medium, it has not been optimized specifically for growth of HDF with defined supplements. The last round of optimization was done with a suboptimal amount of dialyzed serum as the only supplement [McKeehan et al., 1977]. The reduced serum requirement in the resulting optimized medium made it relatively easy to replace the remaining requirements with defined supplements. However, complex holistic interactions among medium components [Ham, 1981] can significantly alter optimum ranges for specific nutrients. Thus, replacing the dialyzed serum with more precisely defined supplements could have affected quantitative requirements for specific nutrients. Ideally, another complete round of optimization should now be done under serum-free conditions.

SERUM-REPLACING SUPPLEMENTS

The supplements that have been used to replace serum in the three "serum-free" culture systems are compared in Table II. It is immediately evident that there is much more diversity among these supplements than among the nutrient mixtures in Table I. The only components shared by all three systems as originally described are insulin and EGF, and even for them, there are considerable quantitative differences. However, a more recent report from Dr. Yamane's laboratory has also added glucocorticoids to the list of components held in common. The addition of 1.0 μg/ml of hydrocortisone to medium RITC 80-7 is distinctly beneficial to growth of human fetal lung fibroblasts, particularly in the absence of BSA [Hoshi et al., 1982].

One of the most difficult and least understood aspects of serum-free growth of fibroblasts (and certain other mesodermally derived cells such as chondrocytes) is the problem of supplying a satisfactory source of preformed lipids. Three very different approaches to this problem are evident in Table II. Kan and Yamane [1982] have utilized large amounts of incompletely purified BSA (Cohn fraction V) to supply the mixture of lipids that is naturally present in such preparations. Delipidation destroys the growth-promoting activity of the BSA, which can be restored to 90% of the original level with a mixture of oleic and linoleic acids, cholesterol, and lecithin.

Weinstein et al. [1982] included chicken egg white ovalbumin in their serum-free medium, presumably as a lipid source, although their paper does not specifically identify its function.

Bettger et al. [1981] avoided the need for a specific lipid carrier by the use of sonicated phospholipid vesicles (liposomes) into which other appropriate lipids had been incorporated. The resulting supplement, designated "liposome B," was quite effective for short-term clonal growth and limited serial subculture. However, the defined medium supplemented with liposome B failed to support longer-term growth. The reasons for this failure have not been specifically identified, but a careful comparison with the system of Yamane et al. [1981a], which does support long-term growth, suggests that the critical difference probably lies either in the way that lipids are supplied to the cells, or else in some undefined function of the fraction V BSA that is employed by Yamane et al. to supply lipids.

The apparent requirements of fibroblasts for lipids are surprisingly complex. Kan and Yamane [1982] found that linoleic and oleic acids, cholesterol, and a crude lecithin preparation were all needed to restore growth in their serum-free medium with delipidated BSA. Bettger et al. [1981] found that cholesterol, sphingomyelin, and mixed phospholipids all contributed to growth in their serum-free medium, and that a mixture of several highly purified phospholipids only partially replaced the crude lecithin. Fibroblasts from chickens [Graves and Ham, 1982] and sheep [Broad and Ham, 1983] and chondrocytes from rabbits and humans [Jennings and Ham, 1983a,b] also exhibit lipid requirements for serum-free growth that are at least superficially similar to those of HDF.

There is a critical need for a detailed study of the lipid requirements for serum-free growth of fibroblasts and related cell types. Development of a truly defined medium for HDF will require a complete understanding of the fatty acids, phospholipid head groups, steroids, and possibly other lipids that are essential for their proliferation, as well as a detailed understanding of their lipid biosynthesis capabilities and the extent to which biosynthesis of one type

TABLE II. Supplements for Serum-Free Growth of Human Fibroblasts[a]

Nutrients (Table I)		RITC 80-7[b] Yamane et al. [1981]; Kan and Yamane [1982]	DME:F12[c] Weinstein et al. [1982]	MCDB 110[d] Bettger et al. [1981]
Nonpeptide hormones				
Dexamethasone	(M)	—	—	5.0E-7
Hydrocortisone	(M)	—[e]	5.0E-5	—
Triiodothyronine	(M)	3.1E-10	—	—
Peptide hormones and growth factors				
EGF	(ng/ml)	10.0	100.0	30.0
Insulin	(μg/ml)	1.0	0.10	0.95
Carrier proteins				
Bovine serum albumin	(μg/ml)	5,000.0	—	—
Chicken ovalbumin	(μg/ml)	—	1,000.0	—
Transferrin	(μg/ml)	10.0[f]	10.0[g]	—
Lipids				
Cholesterol	(M)	—[h]	—	7.6E-6[i]
Lecithin (soy)	(μg/ml)	—[h]	—	6.0[i]
Sphingomyelin	(μg/ml)	—	—	1.0[i]
Antioxidants				
Ascorbate	(M)	—	5.6E-5	—
Vitamin E	(M)	—	—	5.6E-7[i]
Dithiothreitol	(M)	—	—	6.5E-6
Glutathione (SH)	(M)	—	—	6.5E-7
Miscellaneous				
Fibronectin	(μg/ml)	10.0	—[j]	—
Phosphoenolpyruvate	(M)	—	—	1.0E-5
Prostaglandin E1	(M)	—	—	2.5E-8
Prostaglandin F2α	(M)	—	—	2.0E-6
Thrombin	(μg/ml)	—	1.0	—

[a]This table lists substances regarded by the original authors as supplements, including several substances with potential nutrient effects. It also lists hormones and growth factors that are considered to be part of medium RITC 80-7. Molar units, expressed in computer-style exponential notation, are used for all well-defined low-molecular-weight components. Weight units are used for all proteins, and for lecithin and sphingomyelin, whose fatty acid compositions are not precisely known.

[b]Medium RITC 80-7, as defined in the original references, included several components that have been listed in this table, rather than Table I, for easier comparison.

[c]Trace elements that are not components of DME or F12 are also included in the nutrient medium.

[d]MCDB 202 can be used almost interchangeably with MCDB 110 for serum-free growth of HDF.

[e]Hydrocortisone was not included in the original formulation of RITC 80-7, but a more recent paper recommends its addition at 2.8 E-6 M (1.0 μg/ml) [Hoshi et al., 1982].

of lipid may be inhibited by the presence of another in the extracellular environment. Ideally, such a study should be undertaken collaboratively by a laboratory group skilled in lipid biochemistry and a laboratory group skilled in serum-free cell culture, since each of these disciplines is so specialized that few laboratory groups are sufficiently skilled in both.

Another aspect of cellular lipid requirements that needs more emphasis is the development of water-soluble synthetic lipid carriers that are capable of supplying adequate amounts of lipids to cells growing in serum-free media. At present, lipids are usually supplied bound to serum albumin, lipoproteins, or other proteins that are extremely difficult to prepare in a form sufficiently pure to be certain that they are not affecting cellular growth in other ways. The liposomal supplement of Bettger et al. [1981] avoids the problem of carrier proteins, but it does not provide good control over the amount of lipids taken up by the cells, since the liposomes can fuse directly with cellular membranes and "force feed" slowly multiplying cells with excessive amounts of lipids. Yamane et al. [1981b] achieved limited success in providing oleic and linoleic acids to cultured fibroblasts with α-cyclodextrin, but the resulting growth was not fully equivalent to that obtained with BSA, perhaps because of the broader spectrum of lipids that seem to be involved in the growth-promoting activity of BSA [Kan and Yamane, 1982].

Oxidation of lipids is another major problem. Bettger et al. [1981] incorporated vitamin E into liposome B to minimize oxidative damage to unsaturated fatty acids. Kan and Yamane [1982, 1983] found that serum-free growth was substantially better in an atmosphere of 7% oxygen than in a fully aerobic environment. Declining growth under fully aerobic conditions was accompanied by the accumulation malondialdehyde, a product of lipid peroxidation, in the cultures [Kan and Yamane, 1983]. It should also be noted that the results reported by Bettger et al. [1981] were achieved in a laboratory located 1,650 m above sea level, with an ambient oxygen tension comparable to approximately 17% oxygen at sea level. Even with serum present, normal

---

[f]Bovine transferrin.

[g]Human transferrin.

[h]Experiments in which the bovine serum albumin is delipidated and reconstituted with defined lipids indicate that fatty acids (linoleic and oleic), cholesterol, and lecithin are all beneficial to growth of HDF in RITC 80-7 with delipidated BSA plus fibronectin [Kan and Yamane, 1982].

[i]Cholesterol, sphingomyelin, soy lecithin, vitamin E, and vitamin E acetate are added as sonicated phospholipid vesicles (liposomes) at a total lipid concentration of 10 μg/ml.

[j]Culture dishes that have been precoated with fibronectin are used.

atmospheric oxygen tension can be severely inhibitory to HDF in media that do not provide adequate protection against oxidative damage [Taylor et al., 1974].

Ferric iron is extremely insoluble at neutral pH and does not pass through sterilizing filters. Transferrin is, therefore, essential for serum-free growth of HDF under conditions where stringent precautions are not taken to be certain that the iron in the medium formulation has remained soluble and is available to the cells. However, by careful addition of freshly prepared ferrous iron just before the medium is sterilized or, better yet, by addition of a sterile stock solution of ferrous sulfate just before cells are inoculated into the medium, good growth of HDF in a serum-free medium can be achieved without transferrin [Walthall and Ham, 1981; Bettger et al., 1981].

It is important to note that the ferrous sulfate in dry powder media normally does not survive the ball milling process in a form that will pass through a sterilizing filter and support cellular proliferation [Walthall and Ham, 1981; Bettger et al., 1981]. Thus, if a serum-free medium is prepared from a dry powder, a freshly prepared ferrous sulfate solution should be added to it, even if the formula of the medium indicates that it already contains an adequate amount of ferrous iron.

Yamane et al. [1981a] included fibronectin in their serum-free medium, primarily to counteract inhibitory effects on cellular attachment that were observed when BSA was present from the beginning of the culture period. More recently, Hoshi et al. [1982] have reported that the inhibitory effects of BSA are greatly reduced when hydrocortisone is present. Weinstein et al. [1982] employed fibronectin-coated culture dishes for serum-free growth. Bettger et al. [1981] did not employ fibronectin, but they did use polylysine-coated culture dishes to improve attachment and growth [McKeehan and Ham, 1976; Walthall and Ham, 1981].

The need for the various other supplements listed in Table II appears to be marginal at best. Prostaglandins E1 and F2α may be slightly beneficial, but substantial serum-free growth was obtained without them [Bettger et al., 1981]. Phosphoenolpyruvate, dithiothreitol, and reduced glutathione were marginally beneficial during early testing in suboptimal media. They have been retained, but are of questionable importance in the complete serum-free medium [Bettger et al., 1981]. Deletion of triiodothyronine from medium RITC 80-7 appears to have little effect on growth of HDF [Kan and Yamane, 1982]. Ascorbate was a carryover from an earlier medium for a different cell type, and was retained by Weinstein et al. [1982] in their serum-free medium for HDF, although their data suggest that it has little effect.

Platelet-derived growth factor (PDGF) is conspicuously absent from the list of supplements in Table II. Weinstein et al. [1982] looked specifically for a requirement for PDGF in their system and concluded that it was not needed. Phillips and Cristofalo [1981] examined this question in a serum-free culture system consisting of MCDB 104 supplemented with EGF, hydrocortisone, insulin, and transferrin, and concluded that PDGF was needed for growth of HDF only when the cellular density was high. A preliminary study in the author's laboratory [Boyce and Shipley, 1982] reached a similar conclusion. Recently, Clemmons [1983] has reported that HDF from fetal donors, but not from older donors, produce competence factors whose effects on Balb/c 3T3 cells are similar to those of PDGF.

An important requirement for serum-free growth of HDF that is not reflected in Tables I or II is the need for very gentle trypsinization and the complete removal of residual trypsin. Serum contains potent trypsin inhibitors that are not included in most serum-free media. The problem becomes even more critical when medium formulations with very low protein levels are used [Walthall and Ham, 1981]. Weinstein et al. [1982] used soybean trypsin inhibitor to minimize the effects of residual trypsin. Kan and Yamane [1982] and Bettger et al. [1981] both employed variations of the low-temperature trypsinization technique of McKeehan [1977], which has the added advantage of preventing endocytosis of trypsin during the trypsinization process.

FIBROBLASTS FROM OTHER SPECIES

As described above, basal nutrient medium MCDB 202, which was originally optimized for chicken embryo fibroblasts, can be used interchangeably with MCDB 110 for serum-free growth of human fibroblasts. When supplemented with insulin, EGF, and liposome B, MCDB 202 supports serum-free clonal growth of chicken embryo fibroblasts [Graves and Ham, 1982]. It is also the most "universal" of the MCDB media and has been found to be a good starting point for development of serum-free media for fibroblasts from other species. When supplemented with insulin, dexamethasone, fibroblast growth factor, and liposome B, MCDB 202 supports clonal growth of sheep fibroblasts, including both growth and adipose differentiation of preadipocyte fibroblasts [Broad and Ham, 1983]. Rat mammary stromal fibroblasts can also be grown in MCDB 202 with defined supplements (B. Van der Haegen and R. Ham, unpublished observations).

Medium MCDB 402 (Table I) was originally optimized for Swiss mouse 3T3 cells [Shipley and Ham, 1981]. When supplemented with fibroblast growth factor, insulin, dexamethasone, and fibronectin, MCDB 402 supports limited serum-free proliferation of Swiss 3T3 cells [Shipley and Ham, 1983a]. Although an additional undefined factor is still needed for long-term growth,

MCDB 402 supplemented with insulin and dexamethasone has proven to be a very useful background medium for studying the mitogenic action of FGF on sparse cultures of 3T3 cells in the complete absence of serum, plasma, or other undefined additives, and without the usual complications due to density-dependent inhibition [Shipley and Ham, 1983b]. Exposure of FGF-starved 3T3 cells to FGF for as little as 30 min results in initiation of DNA synthesis by essentially all of the cells, beginning approximately 9 h later and proceeding with kinetics that closely fit the predictions of the transition probability model of Smith and Martin [1973]. There also appears to be no requirement for any type of "progression" factor activity in this system [Shipley and Ham, 1983b].

Species specificity is clearly evident in the requirements for serum-free growth of fibroblasts. This is sometimes manifested as a requirement for a different basal nutrient medium (e.g., chicken fibroblasts do not grow well in MCDB 110), and sometimes as a requirement for different supplements. As an example of the latter, serum-free cultures of normal human, chicken, and rat fibroblasts are responsive to EGF, whereas normal sheep fibroblasts and sparse cultures of mouse 3T3 cells are responsive to FGF. On the basis of a very limited sampling of species, it appears to be of critical importance to custom-tailor serum-free culture systems for fibroblasts specifically for the species of origin, and probably, in some cases, also for the tissues from which the fibroblast cultures were developed.

SUMMARY AND FUTURE PROSPECTS

There have been a number of reports of growth of fibroblasts from humans (and other species) in serum-free media. However, in all of the cases examined in this chapter, the serum-free media that were employed were not as highly defined as would be desirable. In addition, with the exception of the system developed by Yamane et al. [1981a], the population-doubling potential in the serum-free media was far less than in serum-containing media. The most critical needs at this time are for a better understanding of cellular lipid requirements and for a better way to present lipids to the cells. In addition, the possibility that bovine serum albumin may be providing a longevity factor needs to be examined. There is also a need to explore the effects of combining the optimized basal nutrient medium used by Bettger et al. [1981] with the supplements used by Yamane et al. [1981a].

On the basis of the progress that has already been made, the prospects for developing a serum-free medium for human fibroblasts that is both highly defined and highly effective seem to be very good. However, the amount of work that remains to be done is far from trivial. Because of the multiplicity of disciplines that will be involved, it can probably best be accomplished through the collaborative efforts of a team of experts in serum-free cell culture, lipid biochemistry, and purification techniques for the various types of medium components.

REFERENCES

Bettger WJ, Boyce ST, Walthall BJ, Ham RG (1981): Rapid clonal growth and serial passage of human diploid fibroblasts in a lipid-enriched synthetic medium supplemented with EGF, insulin, and dexamethasone. Proc Natl Acad Sci USA 78:5588-5592.

Boyce ST, Shipley GD (1982): Comparison of mitogenic activities of PDGF, FGF, and EGF on human diploid fibroblasts and Swiss 3T3 cells in serum-free culture. In Vitro 18:305 (Abstract No. 128).

Broad TE, Ham RG (1983): Growth and adipose differentiation of sheep preadipocyte fibroblasts in serum free medium. Eur J Biochem 135:33-39.

Clemmons DR (1983): Age dependent production of a competence factor by human fibroblasts. J Cell Physiol 114:61-67.

Graves DC, Ham RG (1982): Serum-free clonal growth of chicken embryo fibroblasts. In Vitro 18:305 (Abstract No. 127).

Ham RG (1965): Clonal growth of mammalian cells in a chemically defined, synthetic medium. Proc Natl Acad Sci USA 53:288-293.

Ham RG (1974): Nutritional requirements of primary cultures. A neglected problem of modern biology. In Vitro 10:119-129.

Ham RG (1981): Survival and growth requirements of nontransformed cells. Hdbk Exp Pharmacol 57:13-88.

Ham RG (1984a): Selective media. In Pretlow TG II, Pretlow T (eds): "Cell Separation: Methods and Selected Applications." New York: Academic, Vol 3, pp 209-236.

Ham RG (1984b): Formulation of basal nutrient media. In Barnes D, Sirbasku D, Sato G (eds): "Cell Culture Methods for Molecular and Cell Biology." Vol 1, New York: Alan R. Liss, pp 3-21.

Ham RG, McKeehan WL (1978): Development of improved media and culture conditions for clonal growth of normal diploid cells. In Vitro 14:11-22.

Ham RG, McKeehan WL (1979): Media and growth requirements. Methods Enzymol 58:44-93.

Ham RG, Hammond SL, Miller LL (1977): Critical adjustment of cysteine and glutamine concentrations for improved clonal growth of WI-38 cells. In Vitro 13:1-10.

Hamilton WG, Ham RG (1977): Clonal growth of Chinese hamster cell lines in protein-free media. In Vitro 13:537-547.

Hoshi H, Kan M, Minamoto Y, Yamane I (1982): Hydrocortisone potentiates cell proliferation and promotes cell spreading on tissue culture substrate of human diploid fibroblasts in a serum-free hormone-supplemented medium. Biomed Res 3:546-552.

Hutchings SE, Sato GH (1978): Growth and maintenance of HeLa cells in serum-free medium supplemented with hormones. Proc Natl Acad Sci USA 75:901-904.

Jennings SD, Ham RG (1983a): Clonal growth of primary cultures of rabbit ear chondrocytes in a lipid-supplemented defined medium. Exp Cell Res 145:415-423.

Jennings SD, Ham RG (1983b): Clonal growth of primary cultures of human hyaline chondrocytes in a defined medium. Cell Biol Int Rep 7:149-159.

Kan M, Yamane I (1982): In vitro proliferation and lifespan of human diploid fibroblasts in serum-free BSA-containing medium. J Cell Physiol 111:155-162.

Kan M, Yamane I (1983): Oxygen-dependent growth declining and effect of vitamin E for human diploid fibroblasts in serum-free, BSA-containing culture. Tohoku J Exp Med 139:389-398.

McKeehan WL (1977): The effect of temperature during trypsin treatment on viability and multiplication potential of single normal human and chicken fibroblasts. Cell Biol Int Rep 1:335-343.

McKeehan WL, Ham RG (1976): Stimulation of clonal growth of normal fibroblasts with substrata coated with basic polymers. J Cell Biol 71:727-734.

McKeehan WL, Ham RG (1977): Methods for reducing the serum requirement for growth in vitro of nontransformed diploid fibroblasts. Dev Biol Stand 37:97-108.

McKeehan WL, Ham RG (1978): Calcium and magnesium ions and the regulation of multiplication in normal and transformed cells. Nature 275:756-758.

McKeehan WL, McKeehan KA (1979): Epidermal growth factor modulates extracellular Ca^{2+} requirement for multiplication of normal human skin fibroblasts. Exp Cell Res 123:397-400.

McKeehan WL, Hamilton WG, Ham RG (1976): Selenium is an essential trace nutrient for growth of WI-38 diploid human fibroblasts. Proc Natl Acad Sci USA 73:2023-2027.

McKeehan WL, McKeehan KA, Hammond SL, Ham RG (1977): Improved medium for clonal growth of human diploid fibroblasts at low concentrations of serum protein. In Vitro 13:399-416.

Phillips PD, Cristofalo VJ (1981): Growth regulation of WI-38 cells in a serum-free medium. Exp Cell Res 134:297-302.

Shipley GD, Ham RG (1981): Improved medium and culture conditions for clonal growth with minimal serum protein and for enhanced serum-free survival of Swiss 3T3 cells. In Vitro 17:656-670.

Shipley GD, Ham RG (1983a): Multiplication of Swiss 3T3 cells in a serum-free medium. Exp Cell Res 146:249-260.

Shipley GD, Ham RG (1983b): Control of entry of Swiss 3T3 cells into S phase by fibroblast growth factor under serum-free conditions. Exp Cell Res 146:261-270.

Smith JA, Martin L (1973): Do cells cycle? Proc Natl Acad Sci USA 70:1263-1267.

Taylor WG, Richter A, Evans VJ, Sanford KK (1974): Influence of oxygen and pH on plating efficiency and colony development of WI-38 and Vero cells. Exp Cell Res 86:152-156.

Underwood EJ (1977): "Trace Elements in Human and Animal Nutrition." 4th Ed. New York: Academic.

Walthall BJ, Ham RG (1981): Multiplication of human diploid fibroblasts in a synthetic medium supplemented with EGF, insulin, and dexamethasone. Exp Cell Res 134:303-311.

Weinstein R, Hoover GA, Majure J, van der Spek J, Stemerman MB, Maciag T (1982): Growth of human foreskin fibroblasts in a serum-free, defined medium without platelet-derived growth factor. J Cell Physiol 110:23-28.

Yamane I, Kan M, Hoshi H, Minamoto Y (1981a): Primary culture of human diploid cells and its long-term transfer in a serum-free medium. Exp Cell Res 134:470-474.

Yamane I, Kan M, Minamoto Y, Amatsuji Y (1981b): α-Cyclodextrin, a novel substitute for bovine albumin in serum-free culture of mammalian cells. Proc Jpn Acad 57B:385-389.

Methods for Serum-Free Culture of Epithelial and Fibroblastic Cells,
pages 265–276

18
Serum-Free Cell Culture for Growth of NIH 3T3 and 10T1/2 Mouse Embryo Fibroblast Cell Lines, SV40 Virus Propagation, and Selection of SV40-Transformed Cells

Lin-Chang Chiang, Janet Silnutzer, James M. Pipas,
and David W. Barnes

Historically, fibroblastic cells in culture, particularly those of mouse embryo origin, often have been used in the development of model systems for the study of oncogenic transformation and growth control in vitro [Todaro and Green, 1963, 1964; Aaronson and Todaro, 1968; Holley and Kiernan, 1968; Cooper et al., 1980; Kennedy et al., 1980; Pledger et al., 1982; Reznikoff et al., 1973; Shih et al., 1979; Smith et al., 1971]. Many studies of this type use as an oncogenic agent the DNA tumor virus SV40 [Tooze, 1981]. This virus can be grown in vitro in permissive host cells (several cell lines commonly used are derived from monkey kidney) and it is capable, with low frequency, of stably transforming mouse fibroblasts in culture. Transformation, as judged by a number of assays of malignancy, occurs through expression of the multifunctional SV40 large T antigen [Tooze, 1981; Clark et al., 1983; Peden et al., 1980; Pipas et al., 1983; Smith et al., 1971; Soprano et al., 1983; Todaro and Green, 1964]. We have developed serum-free culture conditions for the continuous growth of several cell types that are useful in studies of this kind [Pipas et al., 1984]. These include 10T1/2 and NIH 3T3, two lines established from mouse embryos and widely used in studies of chemically induced and radiation- and virus-induced malignancy, in identification of cellular oncogenes, and in studies of hormonal and nutritional control of cell

Department of Biological Sciences, University of Pittsburgh, Pittsburgh, Pennsylvania 15260 proliferation [Shih et al., 1979; Reznikoff et al., 1973; Pledger et al., 1982; Kennedy et al., 1980; Cooper et al., 1980]. We also have examined the growth of BSC40, an established line of monkey kidney cells that was derived from the BSC1 line often used for propagation of SV40 virus. Serum-free cell culture techniques, when applied to these cells, provide a number of advantages [Barnes, 1984a,b; Barnes and Sato, 1980a,b; Barnes et al., 1981]. In particular, we have been involved in the use of modifications of serum-free culture conditions for these cell types in order to create growth-restrictive conditions that are selective for transformed or variant cell types within the population [Pipas et al., 1984]. In this chapter are described culture methods for the serum-free growth of the (C3H)10T1/2, NIH-3T3, and BSC40 cell lines and for selection of SV40 virus-transformants from the mouse embryo lines.

MATERIALS

Bovine insulin, human transferrin, 4-(2-hydroxy-ethyl)-1-piperazine-ethanesulfonic acid (HEPES), highly purified soybean trypsin inhibitor, and antibiotics can be obtained from Sigma Chemical Company, St. Louis. Powdered formulations of Ham's F12 and Dulbecco-modified Eagle's medium (DME) can be obtained from Grand Island Biological Company, Grand Island, NY. Fetal calf serum was obtained from Reheis, Irvine, CA. Many other serum sources are acceptable. Sodium selenite can be obtained from Fisher Chemical Company, Pittsburgh, or Difco. Epidermal growth factor (EGF) and human plasma fibronectin (cold-insoluble globulin: CIg) can be obtained from Meloy Laboratories, Springfield, VA; Bethesda Research Laboratories, Gaithersburg, MD; or Collaborative Research, Inc., Waltham, MA. Human high-density lipoprotein (HDL) can be obtained from Meloy Laboratories or isolated from human plasma as described elsewhere [Gospodarowicz, 1984; Gospodarowicz et al., 1982]. Water for culture medium is freshly prepared, either by triple distillation in glass containers [Barnes and Sato, 1980a] or by passage through a Milli-Q water purification system (Millipore).

PREPARATION OF CULTURE MEDIUM AND SUPPLEMENTS

The basal nutrient formulation to which are added supplements that stimulate mouse fibroblast or monkey kidney cell growth is a one-to-one mixture of DME and Ham's F12 [Mather and Sato, 1979] supplemented with 1.2 gm/liter sodium bicarbonate, 15 mM HEPES, pH 7.4, 200 IU/ml penicillin, 200 μg/ml streptomycin, and 25 μg/ml ampicillin (F12:DME). The HEPES and the antibiotics are maintained as 100-fold-concentrated stock solutions in highly purified water. For the preparation of 2 liters of F12:DME, powdered medium in 1-liter preweighed packets (one F12 packet and one DME packet) is added to approximately 500 ml highly purified water and stirred until dissolved. To this is added 20 ml of concentrated HEPES stock, followed by concentrated stock solutions of the antibiotics. The solution is then brought nearly to 2,000 ml with additional highly purified water, sodium bicarbonate is added, and the final adjustment to 2,000 ml is made.

The medium is immediately membrane-filtered for sterilization by passage through plastic disposable 0.2 μ filter units and stored frozen in 200-ml aliquots in sterile, disposable tissue culture flasks of 75-cm^2 surface area (T75). In general, storage of basal nutrient medium for long periods of time, even in the frozen form, is not recommended. In practice, our laboratory makes medium in 2-liter batches as needed, and the medium is usually used within 2 weeks. While this approach is time-consuming, it minimizes the chances of inadvertant use of F12:DME in which some decomposition of components has occurred during storage. This problem is of little concern when the basal nutrient medium is supplemented with serum, but it may have serious consequences for serum-free cell culture [Ham, 1984]. Storage of medium in glass containers that are reused is not recommended. Use of disposable plastic pipets also is preferred.

Growth-stimulating supplements are prepared and maintained as sterile concentrated stock solutions. Insulin is stored at 4°C at 1 mg/ml in 0.05 M HCl. Transferrin is stored frozen in 1-ml aliquots of 1–5 mg/ml in phosphate-buffered saline (PBS). Both of these solutions can be membrane-filtered for sterilization after the commercially available powders are dissolved. EGF and fibronectin can be obtained commercially as frozen sterile solutions or as sterile lyophilized powders that can be reconstituted with sterile solutions. Alternatively, these components can be isolated from primary sources as discussed elsewhere [Yamada and Akiyama, 1984; Rubin and Bradshaw, 1984; Savage and Harper, 1984; Carpenter and Cohen, 1979; Hynes and Yamada, 1982]. Concentrated stocks of these factors can be maintained at 0.5–1.0 mg/ml (fibronectin) and 50–100 μg/ml (EGF). Care must be taken to prevent frequent freezing and thawing of fibronectin solutions in order to avoid aggregation, and it is best to store fibronectin in small aliquots of frozen stocks. General approaches to the use of fibronectin and similar factors in serum-free cell culture are described elsewhere [Barnes, 1984c]. Sodium selenite is prepared as a concentrated (100 ×) stock in highly purified water and is membrane-filtered for sterilization and stored refrigerated. HDL is obtained commercially as a sterile concentrated solution that is stored refrigerated or prepared and handled as described elsewhere [Gospodarowicz, 1984]. HDL preparations may become toxic for cells upon extended storage after preparation. Plastic (polypropylene) disposable tubes are used in virtually all situations for storage of concentrated stock solutions.

TRYPSINIZATION AND PLATING OF CELLS IN SERUM-FREE MEDIUM

Passage of mouse embryo or monkey kidney cell lines is accomplished first by incubation of stock cultures with 0.1% crude trypsin in PBS with 0.1 mM ethylenediaminetetraacetate (EDTA) at 37°C until the cells are detached. The cell suspension is then diluted into F12:DME containing 0.1% highly purified soybean trypsin inhibitor, and the cells are centrifuged from suspension at low speed and resuspended in fresh F12:DME. Some preparations of soybean trypsin inhibitor that are not sufficiently purified contain cytotoxic components. For serum-free plating in 35-mm-diameter dishes, cells are diluted in F12:DME to give the final cell density desired and added in 1-ml aliquots to dishes containing 1 ml of F12:DME previously equilibrated for 30 min in a 37°C incubator in a 5% CO_2-95% air atmosphere. Dishes that are to receive substratum modifications, such as exposure to fibronectin, also should have such factors added to the medium prior to preequilibration and cell seeding. Other growth-stimulating supplements are then added to the cells immediately after plating. Supplements are added as small volumes (e.g., 20 μl per plate) of sterile concentrated stocks.

Care should be taken to prevent exposure of the cells to extremes of pH and temperature during the passaging process. Trypsinization, centrifugation, and resuspension should be accomplished as quickly and efficiently as possible so that cells spend as little time exposed to these conditions as can be reasonably achieved. Liberties that can be taken in the passage of these and other cell types when serum-containing media are used may result in greatly reduced cell survival upon passage when the cells are cultured under serum-free conditions. It may be helpful to change the medium the day after the cells are passaged under serum-free conditions in order to reduce the level in the medium of hydrolase activity released from cells damaged in the passaging procedure. Effects on cell viability of hydrolytic enzyme activities may not be of consequence in serum-containing media because these activities may be inhibited by serum factors, but the presence of these enzymes may represent a problem when cells are passaged under serum-free conditions [Rockwell, 1984].

GROWTH OF BSC40 MONKEY KIDNEY CELLS AND SV40 IN SERUM-FREE CULTURE

BSC40 cells can be grown in long-term, multipassage culture in F12:DME supplemented with insulin (10 μg/ml), transferrin (10 μg/ml), fibronectin (2 μg/ml), EGF (25 ng/ml), and sodium selenite (10^{-8} M). Most important among these supplements for the growth of BSC cells at low plating densities are fibronectin, selenium, and insulin (Table I). At high cell densities, on the order of 5×10^5 per 35-mm-diameter culture dish, no supplements are required, and cells can be maintained and will proliferate in unsupplemented F12:DME, although they do so with a generation time that is greater than that of cells plated at similar densities in F12:DME supplemented with 10% fetal calf serum. Under serum-free conditions, SV40 virus will infect BSC40 cells and replicate. Virus titers under these circumstances are equivalent to those obtained from cells at the same density infected at the same multiplicity in serum-containing medium, although onset of cytopathic effects and production of infectious virus occurs at a somewhat later time after infection under serum-free conditions than in serum-containing medium. Multiple passage of virus on BSC cells at high density in unsupplemented F12:DME allows the isolation of infectious virus stocks that are free from contamination by serum or exogenous supplements. These serum-free stocks are useful for many types of experiments in which the effects of permissive or transforming infections of SV40 on cells in serum-free medium are examined.

GROWTH OF C3H 10T1/2 MOUSE EMBRYO FIBROBLASTS IN SERUM-FREE CULTURE

The 10T1/2 mouse cell line can be grown in long-term, multipassage culture in F12:DME supplemented with insulin (10 μg/ml), transferrin (30 μg/

TABLE I. Serum-Free Cell Culture of BSC40 Monkey Kidney Cells

Medium supplements	Number of cells per 35-mm-diameter plate
Insulin, transferrin, fibronectin, selenium, EGF	7.6×10^4
Transferrin, fibronectin, selenium, EGF	3.9×10^4
Insulin, fibronectin, selenium, EGF	6.7×10^4
Insulin, transferrin, selenium, EGF	3.0×10^4
Insulin, transferrin, fibronectin, EGF	3.4×10^4
Insulin, transferrin, fibronectin, selenium	6.4×10^4

Cells were seeded in F12:DME with the indicated supplements at a cell density of 2,500 per plate. Number of cells was determined 8 days after plating. Concentrations of the supplements were: insulin, 10 μg/ml; transferrin, 10 μg/ml; fibronectin, 2 μg/ml; selenium (sodium selenite), 10^{-8} M; EGF, 50 ng/ml.

ml), EGF (10 ng/ml), and fibronectin (10 μg/ml) (Figs. 1, 2). Cells in this medium grow with a generation time that is longer than that of cells in serum-supplemented medium, but eventually they reach the same saturation density as cultures maintained in serum-supplemented medium. Deletion of either insulin or EGF from the serum-free medium results initially in greatly reduced cell growth and eventually in cell death.

SV40 infection and subsequent culture of 10T1/2 cells in F12:DME supplemented with insulin, transferrin, EGF, and fibronectin in various combinations designed to create media suboptimal to varying degrees for cell growth results in improved growth of cells of the cultures compared to mock-infected cultures under identical serum-free conditions (Fig. 3). Serum-free SV40 virus stocks were prepared by multiple passage of the virus through serum-free cultures of high-density BSC40 cells grown in unsupplemented F12:DME. Infection of 10T1/2 cells in unsupplemented F12:DME does not result in

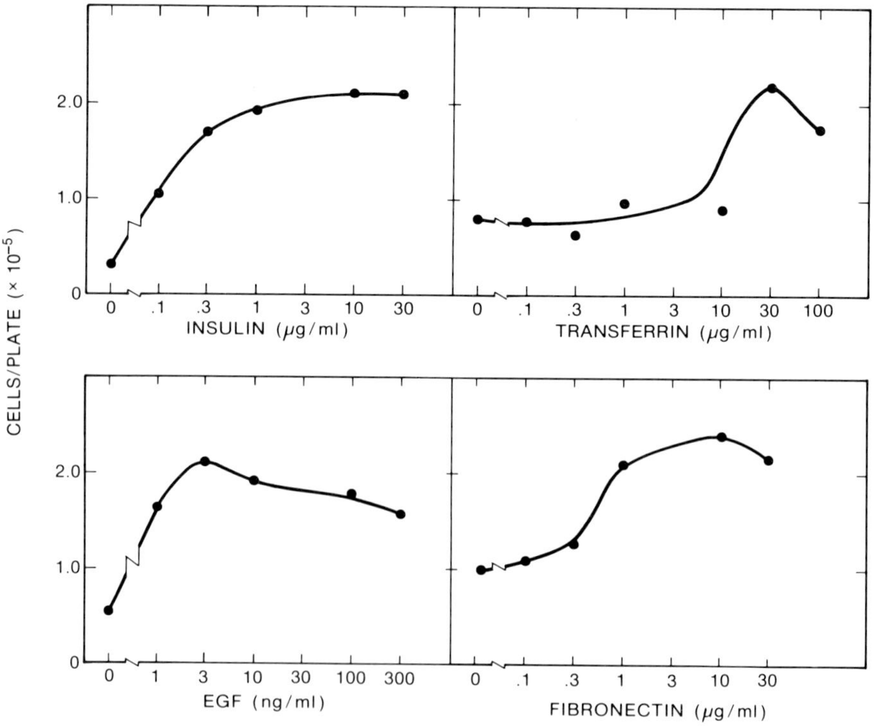

Fig. 1. Dose-response curves of 10T1/2 cells to basal nutrient medium supplements. Cells were plated (2 × 10^4 cells per 35-mm dish) in media containing constant amounts of three supplements, whereas the concentration of the fourth was varied. Number of cells was determined 6 days after plating. From Pipas et al. [1984], with permission.

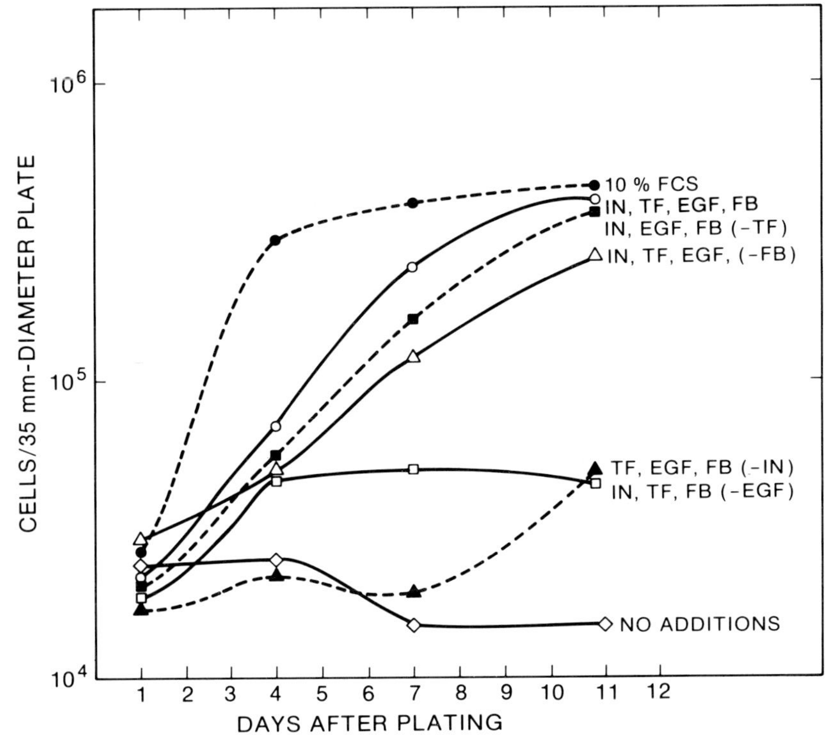

Fig. 2. Growth curves of 10T1/2 cells in basal nutrient media containing various supplements. Number of cells was determined periodically following plating in medium supplemented with 10% fetal calf serum (●), the four supplements of Figure 1 (○), or three of the four supplements: without fibronectin (−FB, △), without transferrin (−TF, ■), without insulin (−IN, △), without EGF (−EGF, □), or no supplements, ◇. From Pipas et al. [1984], with permission.

improved cell growth, compared to mock-infected cultures. Long-term growth and passage of SV40-infected 10T1/2 cultures under growth-restrictive conditions created by the individual omission of serum-free medium supplements (e.g., medium containing insulin, transferrin, and fibronectin, but not EGF, as supplements to F12:DME) results in selection of cells from the cultures that express nuclear SV40 T antigen as determined by immunofluorescence techniques [Pope and Rowe, 1964; Pipas et al., 1984].

GROWTH OF NIH 3T3 MOUSE EMBRYO FIBROBLASTS IN SERUM-FREE CULTURE

The NIH 3T3 mouse cell line can be grown in long-term multipassage culture in F12:DME supplemented with insulin (10 μg/ml), transferrin (25 μg/

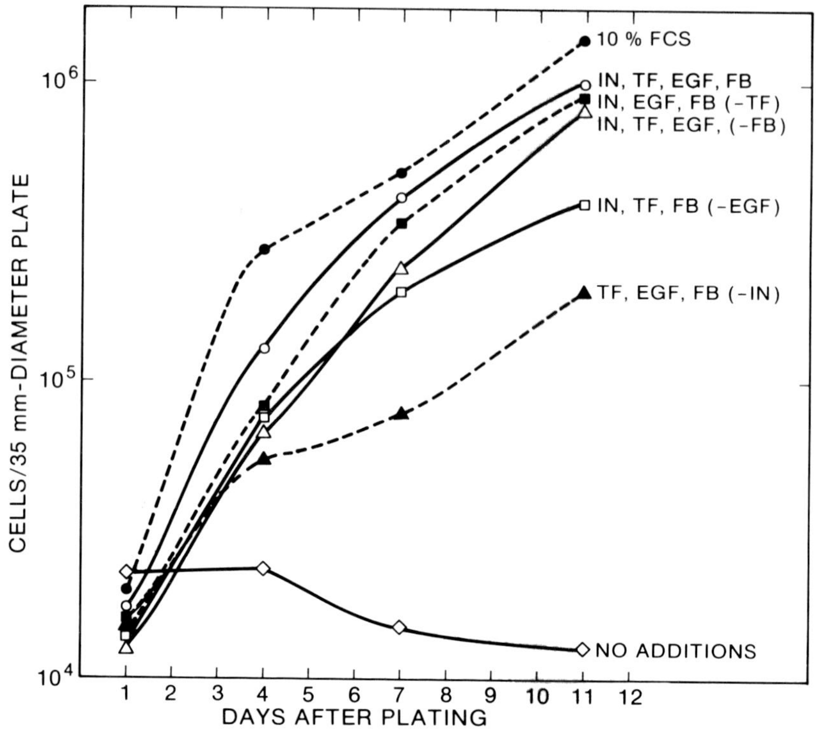

Fig. 3. Growth curves of SV40-infected 10T1/2 cells in media containing various supplements. Experimental protocol and figure symbols are the same as for Figure 3 except that cells were infected with SV40 serum-free virus stock (moi = 100) at initiation of the experiment. From Pipas et al. [1984], with permission.

ml), EGF (10 ng/ml), fibronectin (5 μg/ml), and HDL (25 μg/ml) (Table II). Cells in this serum-free medium grow as well as or better than cells in medium containing 10% fetal calf serum. Individual deletions of HDL, EGF, or insulin result in major reductions in cell growth within the cultures, and modifications of the complete serum-free medium for NIH 3T3 in which one of these components has been omitted can be used as selective, growth-restrictive conditions for the selection of oncogenically transformed cells. Figure 4 shows an example of such a selection, in which NIH 3T3 cells were cultured in F12:DME containing as supplements insulin, transferrin, EGF, and fibronectin (HDL omitted). In this medium large colonies appear in plates cultured for several weeks after infection with serum-free stocks of SV40. Cells of these colonies express SV40 T antigen and are anchorage-independent, as deter-

TABLE II. Serum-Free Cell Culture of NIH 3T3 Mouse Embryo Fibroblasts

Medium supplements	Number of cells per 35-mm-diameter plate
Insulin, transferrin, fibronectin, EGF, HDL	3.6×10^5
Transferrin, fibronectin, EGF, HDL	1.0×10^5
Insulin, fibronectin, EGF, HDL	1.5×10^5
Insulin, transferrin, EGF, HDL	2.0×10^5
Insulin, transferrin, fibronectin, HDL	0.84×10^5
Insulin, transferrin, fibronectin, EGF	0.9×10^5

Cells were seeded in F12:DME with the indicated supplements at a cell density of 10^4 per plate. Number of cells was determined 6 days after plating. Concentrations of the supplements were: insulin, 10 μg/ml; transferrin, 10 μg/ml; fibronectin, 10 μg/ml; HDL, 50 μg/ml; EGF, 10 ng/ml.

mined by a soft-agar growth assay [Fredman and Shin, 1974; MacPherson and Montagnier, 1964].

CONCLUSIONS

In this chapter we have described methods for serum-free growth of two of the mouse embryo fibroblast cell lines commonly used in studies of chemically induced and virus- or radiation-induced carcinogenesis and methods for the serum-free infection and propagation of SV40, a commonly used DNA tumor virus, on the BSC40 monkey kidney cell line. It has been recognized for some time that SV40-transformed cells selected by various means in serum-containing medium exhibit different growth requirements for specific factors from those of the transformed cells from which they were derived [Cherington et al., 1979; Pardee et al., this volume; Rockwell, this volume; Kaplan et al., 1982; McClure, 1983; McClure et al., 1982]. Among the advantages of serum-free culture for these systems is the ability to devise modifications of the serum-free media that provide growth-restrictive, selective conditions for the identification and isolation of cells with altered growth requirements after treatments such as SV40 virus infection [Barnes and Sato, 1980a,b; Shimizu, this volume]. We have given examples of procedures through which SV40-transformed cells can be isolated, either by continuous culture of the total population or by clonal selection under growth-restrictive conditions. The model systems described should provide new approaches to in vitro studies of cancer and cancer-causing agents by providing novel methods for selection of cells altered by these agents.

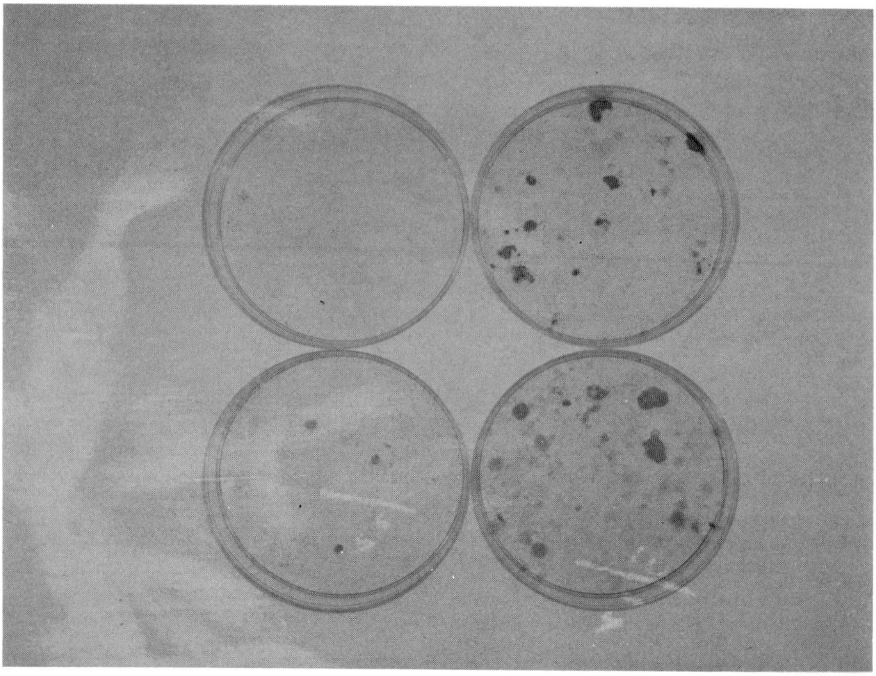

Fig. 4. Isolation of colonies of SV40 virus-transformed NIH 3T3 cells by selection in growth-restrictive, serum-free medium. NIH 3T3 were infected with SV40 virus or mock-infected and plated in 60-mm dishes in F12:DME supplemented with insulin (10 μg/ml), transferrin (10 μg/ml), EGF (10 μg/ml), and fibronectin (10 μg/ml). This medium was suboptimal for NIH 3T3 growth because HDL was not included among the supplements added. Medium was changed every 7 days. Plates were stained with crystal violet to identify colonies after 2 weeks (top plates) and 3 weeks (bottom plates) of culture. Mock-infected cultures (left plates) gave rise to a few small colonies after 3 weeks of incubation. SV40-infected cultures gave rise to many large colonies less than 2 weeks after infection. Colonies arising from SV40-infected cultures and examined further were found to express nuclear SV40 T antigen, as determined by immunofluorescence techniques, and were capable of anchorage-independent growth in a soft agar assay in serum-containing medium.

ACKNOWLEDGMENTS

This work was supported by grants BC-368 from the American Cancer Society and CA 35214-01 from the National Institutes of Health to D.W.B. and PCM 80-21649 from the National Science Foundation and MV-182 from the American Cancer Society to J.M.P. Dr. L.-C. Chiang is a Visiting Research Scientist from PPG Industries, Inc.

REFERENCES

Aaronson SA, Todaro GJ (1968): Basis for the acquisition of malignant potential by mouse cells cultivated in vitro. Science 162:1024-1026.

Barnes DW (1984a): Hormonally-defined serum-free media for epithelial cells in culture. In Taub M (ed): "Tissue Culture in the Study of Epithelial Transport." New York: Plenum (in press).

Barnes D (1984b): Nutritional and hormonal requirements of mammalian cells in culture. World Rev Nut Diet (in press).

Barnes D (1984c): Attachment factors in cell culture. In Mather J (ed): "The Use of Serum-Free and Hormone-Supplemented Media." New York: Plenum, pp 195-237.

Barnes D, Sato G (1980a): Methods for growth of cultured cells in serum-free medium. Anal Biochem 102:255-270.

Barnes DW, Sato GH (1980b): Serum-free cell culture: A unifying approach. Cell 22:649-655.

Barnes DW, van der Bosch J, Masui H, Miyzaki K, Sato G (1981): The culture of human tumor cells. Methods Enzymol 79:368-391.

Carpenter G, Cohen S (1979): Epidermal growth factor. Annu Rev Biochem 48:193-216.

Cherington PV, Smith BL, Pardee AB (1979): Loss of epidermal growth factor requirement and malignant transformation. Proc Natl Acad Sci USA 76:3937-3942.

Clark R, Peden K, Pipas JM, Nathans P, Tjian R (1983): Biochemical activities of T-antigen proteins encoded by simian virus 40 A gene deletion mutants. Mol Cell Biol 3:220-230.

Cooper GM, Okenquist S, Silverman L (1980): Transforming activity of DNA of chemically transformed and normal cells. Nature 284:418-421.

Freedman VH, Shin SI (1974): Cellular tumorigenicity in nude mice: Correlation with cell growth in semi-solid medium. Cell 3:355-359.

Gospodarowicz D (1984): Preparations and uses of lipoproteins to culture normal diploid and tumor cells under serum-free conditions. In Barnes DW, Sirbasku DA, Sato GH (eds): "Cell Culture Methods for Molecular and Cell Biology." New York: Alan R. Liss, Vol 1, pp 69-86.

Gospodarowicz D, Lui GM, Gonzales K (1982): High-density lipoprotein and the proliferation of human tumor cells maintained on extracellular matrix-coated dishes and exposed to defined medium. Cancer Res 42:3704-3713.

Ham R (1984): Formulation of basal nutrient media. In Barnes DW, Sirbasku DA, Sato GH (eds): "Cell Culture Methods for Molecular and Cell Biology." New York: Alan R. Liss, Vol 1, pp 3-21.

Holley RW, Kiernan JA (1968): Contact inhibition of cell division in 3T3 cells. Proc Natl Acad Sci USA 60:300-304.

Hynes RO, Yamada KM (1982): Fibronectins: Multifunctional modular glycoproteins J Cell Biol 95:369-378.

Kaplan PL, Anderson M, Ozanne B (1982): Transforming growth factor(s) production enables cells to grow in the absence of serum: An autocrine system. Proc Natl Acad Sci USA 79:485-490.

Kennedy AR, Fox M, Murphy G, Little JP (1980): Relationship between X-ray exposure and malignant transformation in C3H 10T1/2 cells. Proc Natl Acad Sci USA 77:7262-7266.

MacPherson I, Montagnier L (1964): Agar suspension culture for the selective assay of cells transformed by polyoma virus. Virology 23:291-294.

Mather JP, Sato GH (1979): The use of hormone-supplemented serum-free media in primary cultures. Exp Cell Res 124:215-221.

McClure DB (1983): Anchorage-independent colony formation of SV40 transformed Balb/c-3T3 cells in serum-free medium: Role of cell- and serum-derived factors. Cell 32:999-1002.

McClure DB, Hightower MJ, Topp WC (1982): Effect of SV40 transformation on the growth-factor requirements of the rat embryo cell line REF52 in serum-free medium. Cold Spring Harbor Conf Cell Prolif 9:345-356.

Peden KWC, Pipas JM, Pearson-White S, Nathans D (1980): Isolation of mutants of an animal virus in bacteria. Science 290:1392-1396.

Pipas JM, Peden KWC, Nathans D (1983): Mutational analysis of simian virus 40 T antigen: Isolation and characterization of mutants with deletions in the T-antigen gene. Mol Cell Biol 3:203-213.

Pipas JM, Chiang L-C, Barnes DW (1984): Influence of genetic, cellular and hormonal factors on simian virus 40-induced transformation. Cold Spring Harbor Conf Cell Prolif (in press).

Pledger WJ, Leof EB, Chou BB, Olashaw N, O'Keefe EJ, Van Wyk JJ, Wharton WR (1982): Initiation of cell-cycle traverse by serum-derived growth factors. Cold Spring Harbor Conf Cell Prolif 9:259-273.

Pope JH, Rowe WP (1964): Detection of specific antigen in SV40-transformed cells by immunofluorescence. J Exp Med 120:121-125.

Reznikoff CA, Brankow DW, Heidelberger C (1973): Establishment and characterization of a cloned line of C3H mouse embryo cells sensitive to postconfluence inhibition of division. Cancer Res 33:3231-3238.

Rubin JS, Bradshaw RA (1984): Preparation of guinea pig prostate epidermal growth factor. In Barnes DW, Sirbasku DA, Sato GH (eds): "Cell Culture Methods for Molecular Cell Biology." New York: Alan R. Liss, Vol 1, pp 139-145.

Savage CR, Harper RA (1984): Purification of human epidermal growth factor from urine. In Barnes DW, Sirbasku DA, Sato GH (eds): "Cell Culture Methods for Molecular and Cell Biology." New York: Alan R. Liss, Vol 1, pp 147-158.

Shih C, Shilo B-Z, Goldfarb MP, Dannenberg A, Weisberg RA (1979): Passage of phenotypes of chemically transformed cells via transfection of DNA and chromatin. Proc Natl Acad Sci USA 76:5714-5718.

Smith HS, Scher CD, Todaro GJ (1971): Induction of cell division in medium lacking serum growth factor by SV40. Virology 44:359-370.

Soprano KJ, Galanti N, Jonak GJ, McKercher S, Pipas JM, Peden KWC, Baserga R (1983): Mutational analysis of simian virus 40 T antigen: Stimulation of cellular DNA synthesis and activation of rRNA genes by mutants with deletions in the T-antigen gene. Mol Cell Biol 3:214-219.

Todaro GJ, Green H (1963): Quantitative studies of the growth of mouse embryo cells in culture and their development into established lines. J Cell Biol 17:299-331.

Todaro GJ, Green H (1964): An assay for cellular transformation by SV40. Virology 23:117-119.

Tooze J (1981): "DNA Tumor Viruses: Molecular Biology of Tumor Viruses." 2nd Ed, revised. Cold Spring Harbor, NY: Cold Spring Harbor Laboratory.

Yamada K, Akiyama S (1984): Preparation of cellular fibronectin. In Barnes DW, Sirbasku DA, Sato GH (eds): "Cell Culture Methods for Molecular and Cell Biology." New York: Alan R. Liss, Vol 1, pp 215-230.

Index

ABAE cells and FGFs, 185, 191, 194-195
Adelberg, E.A., 157
Adenocarcinoma, lung, 65-66, 69-70
Adrenal cortex, FGF mitogenic, 184-185, 192, 193
A549 cell line (lung cancer), 59, 65; *see also* Lung cancer
A431 human epidermoid carcinoma cell culture, 95-103
- DME:F12, 96-103
- EGF, 95-97
  - inhibition, 99-103
  - reversal, 102
- ethanolamine, 96, 100
- FCS, 96, 98, 99, 101, 103
- fibronectin, 59, 97, 98, 100
- glycyl-histidyl-lysine, 100
- growth characteristics, 99-101
- hormone-toxin conjugates, 244
- insulin, 95, 97, 100
- medium, 96-97
- parathyroid hormone, 100
- pH and temperature extremes, 98
- serum spreading factor, 100
- somatostatin, 100
- transferrin, 95, 97, 100
- trypsinization and plating, 97-99

AFP, 143
Agarose suspension culture, pituitary acidic FGF, 218-219
Ammonium sulfate precipitation, pituitary FGF, 201
Anchorage-dependent rat glioma cell line C_6-3D, 210, 218-219
Antibiotics
- kidney cell culture, 10
- mouse embryo fibroblasts, 266-267
- SV40 Balb/c-3T3 cell culture, 222, 223

APUD, small cell lung cancer, 58, 59
Arginine-free medium, stock solutions listed, 45
Arrest test, regulators cf. nutrients, 158-160

BACE cells, 145-146
Balb/c 3T3 cells, 159, 163, 200; *see also* Fibroblast culture in SFM; SV40 Balb/c-3T3 growth in SFM
BCE-ECM, 188
BHK cells, 161
Bladder carcinoma, rat squamous cell, 93
Bombesin, small cell lung cancer culture in HITES, 63, 66
Brain. *See* Pituitary and brain FGFs (bovine)
Bronchogenic epidermoid carcinoma, growth and differentiation, 83-94

cell growth in primary culture, 87-89
cell preparation, 84-85
DMEM and Ham's F12, 84-93
EGF, 90
glucagon, 84, 87, 89-91
insulin, 84, 87, 89-91, 93
keratinization (squamous or keratin pearls), 83, 89-92
antikeratinization by retinoids, 92
serial passage, 92-94
cf. other squamous cell cancers, 83, 93
primary and sub-cultures, 85-86
quantitative assay of growth and keratinization, 86-87
retinoic acid or T_3, 84, 87-93
retinoids, 91, 94
selenium, 84, 87, 89-91, 93
serum-free cf. serum-supplemented medium, 84, 85, 92
T_4, 91
T222 tumor transplant in nude mice, 85
see also Lung cancer
BSA
fatty-acid free
embryonal carcinoma cell culture, 110, 111
SV40 Balb/c-3T3 cell culture, 222, 224-226, 228-230
fibroblast culture in SFM, 250
hormone-toxin conjugates, 241
pituitary FGF, 200, 204, 205
BSC40 monkey kidney cells, 266, 268-270, 273
Buffer solutions, hepatocytes, primary culture, 47, 48

Calcium, 255
Capillary endothelial cell migration, modulation, 145-147
Cell cycle, distinguishing regulators from nutrients, 158
Cell growth. *See* Growth, cell
Cell nuclei, DNA counting, growth evaluation, 78-79
Cell proliferation assay, FGFs, pituitary and brain, 169-170
CHEF 18 cells, 158-160, 163-165
CHO cells, selenium requirement in F12 medium, 256
Cholesterol, kidney cell line LLC-PK_1, 26
Chromatography
FGFs, carboxymethyl ion exchange, 168, 172-173, 183, 185
brain, 173-178
pituitary, 174-177, 179
FGFs, pituitary acidic
BioGel, 215
CM-Sephadex-C-50 gradient elution, 203, 204, 206
CM-Sephadex-C-50 stepwise fractionation, 201
FGFs, pituitary and brain
exclusion gel on Sephadex G-75, 173, 182, 183
Sephadex G-75 column, 201-203, 205-207
gas liquid, fatty acid analysis, 131
Pedersen fetuin and embryonin purification
affinity-dye binding, 134
CM cellulose, 133, 136
HPLC, 133, 137-139
metal (zinc) chelate affinity, 133, 134, 139, 140
Pedersen fetuin, separation of lipids and phospholipids, 130-131, 147
Clonogenic assay, small cell lung cancer, 58
Cloudman S-91 melanoma, 244
Collagen
coating solution, colon carcinoma tissue culture, 76-77

FGFs, pituitary and brain, 192
production, embryonin, α_2-macroglobulin, and Pedersen fetuin, 130, 143-145
Collagenase, colon carcinoma, 75-76
Colon carcinoma (human), primary tissue culture, 73-80
cell-aggregate cf. single-cell suspensions, 77
cf. cell lines and soft agar colony-forming assays, 73-74
cell preparation, 74-76
collagenase and hyaluronidase, 75-76
collagen-coating solution for dishes, 76-77
growth evaluation, cell nuclei DNA counting, 78-79
in vitro drug and radiation sensitivity, 73, 79-80
solutions, 76-79
insulin, transferrin, EGF, T_3, 77
Concentration
FGF and MGF, increasing, 194-195
and growth, kidney cell culture ($LLC\text{-}PK_1$), 27
Corneal endothelium
BCE-ECM, 188
FGF mitogenic, 184-185, 192, 193
Creatine kinase, small cell lung cancer, 58, 63
C_6-3D rat glioma cell line, 210, 218-219
C3H 10T1/2 cells. *See under* Fibroblasts, mouse embryo, NIH 3T3 and C3H 10T1/2
Cyclic AMP, 18, 27
Cycloheximide, 161-162
Cystamine, derivatization of insulin, 237-238
Cysteine inhibition, fibroblast culture in SFM, 254-255
4′-Deoxydoxorubicin, colon carcinoma sensitivity assays, 79
Dexamethasone, fibroblast culture in SFM, 250
Differentiation regulation, embryonal carcinoma cell culture, 107
Diphtheria toxin. *See under* Hormone-toxin conjugates, cell variant hormone responses
Disulfide linkages, hormone-toxin conjugates, 233, 236, 238
Dithiothreitol, fibroblast culture in SFM, 250, 260
DME:F12 medium
A431 cell culture, 96-103
embryonal carcinoma cell culture, 109
mouse embryo fibroblasts, 266
SV40 Balb/c-3T3 cell culture, 222-228, 230
see also Fibroblast culture in SFM; Ham's F12
DME H-16 medium, FGFs, pituitary and brain, 167-168
DMEM
bronchogenic epidermoid carcinoma cell culture, 84-93
colon carcinoma, 74-77
hormone-toxin conjugates, 240, 241
metanephros, organ culture, 34, 39
DNA
cell nuclei, counting, growth evaluation, 78-79
content, small cell lung cancer culture in HITES, 64, 67
synthesis assay, pituitary and brain FGFs, 170-171
Dome formation, MDCK established cell culture, 5-7, 21, 22
L-Dopa decarboxylase, small cell lung cancer culture in HITES, 58, 63
Drug sensitivity assays
colon carcinoma, 79

small cell lung cancer, 58

Eagle's MEM, fibroblast culture in SFM, 249
EDTA, pituitary FGF, 201, 205
EGF, 143, 144
A431 cell culture
inhibition, 99-103
reversal, 95-97, 102
bronchogenic epidermoid carcinoma cell culture, 90
colon carcinoma, 77
distinguishing regulators from nutrients, 158, 160, 161, 163, 165
embryonal carcinoma cell culture, 113, 121
fibroblast culture in SFM, 249, 250
hepatocytes, primary culture, 47
hormone-toxin conjugates, 234, 241-244
kidney cell culture, 14
mouse embryo fibroblasts, 267, 270
Electrophoresis
acid gel, pituitary FGF, 199, 203-204, 206
PAGE, Pedersen fetuin, 131-132, 134-137, 139
SDS-PAGE
embryonin, 141
FGFs, pituitary and brain, 168, 171, 182, 185, 189, 190
Elution, pituitary FGF
CM-Sephadex-C-50 gradient, 203, 204, 206
gel filtration, 202
Embryo. *See* Fibroblasts, mouse embryo, NIH 3T3 and C3H 10T1/2; Metanephros, mouse embryonic, organ culture
Embryonal carcinoma cells, F9, 111-114
Embryonal carcinoma cells (mouse), growth in defined SFM, 107-122
development and preparation of media (EM-3), 108-111
modification for differentiation, 117
differentiated cell line growth, 117-120
from myogenic cell line T984, 118
PSA-5E, 117
PYS-2, 117-118
differentiation regulation, 107
DME:F12, 109
EGF, 113, 121
endogenous GF release, 121
fatty-acid free BSA, 110, 111
fatty acids, 110, 111
fibronectin, 109-111, 115-116, 121
hexamethylene-bisacetamide, 111, 118
insulin, 110
laminin, 108-110, 115-116, 121
lipoproteins
HDL, 110, 112-120
LDL, 110, 112, 114
VLDL, 110, 112, 114
modification for differentiation, 117
monitoring cell lines, 119
multipotent, 107, 122
differentiation into parietal endoderm, 114-116
OC-1551 and -1003, 114-117
P-19, 6050AJ, PCC-3, PSA-1, 116-117
nullipotent, 107
OTT-6050, Pedersen fetuin and, 129, 130, 134-137, 142
restricted, 107, 120
restricted, growth and differentiation
F9, 111-114
nulli SSC-1, 113-114
PC-13, 112-114
retinoic acid, 110-111, 121
selection of cell line, 119

variation within cell lines, 119
Embryonin, 147-149
and cell growth, 142-143
collagen production and protease inhibition, 143-145
fatty acid carrier function, 147, 148
immunologic properties, 142
modulation of capillary endothelial cell migration, 145-147
physiochemical properties, 139-142
purification from fetuin, 133-134
affinity-dye binding chromatography, 134
carboxymethyl cellulose chromatography, 133, 136
HPLC, 133, 137-139
metal (zinc) chelate affinity chromatography, 133-134, 139, 140
SDS-PAGE, 141
see also Fetuin, Pedersen, α_2-macroglobulin contamination
Endocytosis, receptor-mediated, 244-245
Endothelial cell growth factor. *See* Myoblast growth factor
Epidermoid lung cancer cells (LE), 65-66; *see also* Lung cancer
Ethanolamine
A431 cell culture, 96, 100
hormone-toxin conjugates, 241

Fatty acid(s)
analysis, Pedersen fetuin, 131
carriers, embryonin, α_2-macroglobulin, Pedersen fetuin, 147, 148
embryonal carcinoma cell culture, 110, 111
-free BSA. *See under* BSA
SV40 Balb/c-3T3 cell culture, 222, 225-226, 228
FCS
A431 cell culture, 96, 98-99, 101, 103
FGFs, pituitary and brain, 167-168
hormone-toxin conjugates, 240, 241, 242
pituitary acidic FGF, 210, 211
SV40 Balb/c-3T3 cell culture, 222-223, 225, 229
Fetuin, Pedersen, α_2-macroglobulin contamination, 125-149
amino acid analysis, 132-133
analysis of fetuin preparation, 134-139
embryonin isolation, 137, 138
analytical PAGE, 131-132, 134-137, 139
assay for collagen stimulating activity, 130
assay for mitogenic activity, 129-130
biological activities, 126, 127
capillary endothelial cell migration modulation, 145-147
and cell growth, 142-143
chromatography, 127
collagen production and protease inhibition, 143-145
cf. Deutsch and Spiro fetuin, 127, 133, 134, 138-139
embryonin (HMW glycoprotein), 128
family of polymorphous glycoproteins, 126-127
fatty acid analysis, GLC, 131
fatty acid carrier function, 147, 148
FCS, 126
IEF, analytical and preparative, 132, 134, 135, 136
immunologic properties, 142
isolation of growth-promoting activity, 128
mouse OTT-6050 EC cells, 129, 130, 134-137, 142
NRK cells, 129, 130, 143

Ouchterlony double immunodiffusion, 133
physiochemical properties, 139-142
RME cells, 129, 130, 134, 142, 143, 144
separation of lipids and phospholipids, 130-131, 137
see also Embryonin; α_2-Macroglobulin
Fibroblast growth factors. *See* Pituitary acidic FGF
Fibroblast overgrowth, kidney cell culture, 9-10, 14, 20
Fibroblasts, Balb/c 3T3, 159, 163, 200; *see also* SV40 Balb/c-3T3 growth
Fibroblasts (human diploid) culture in SFM, 249-263
DMEM-F12, 250, 252
chicken egg white ovalbumin, 250, 257
trace elements, hydrocortisone, insulin, EGF, transferrin, 250
MCDB 110 and 202 media, 250, 253, 254, 256, 261, 262
dexamethasone, 250
DTT, 250, 260
liposome B, 250, 257, 261
PGE_1 and $F_{2\alpha}$, 250, 260
phosphoenolpyruvate, 250, 260
reduced glutathione, 250, 260
nutrient requirements, 250, 254-256
calcium and selenium, 255
cf. CHO cells, selenium requirement, F12 medium, 256
cysteine inhibition, 254-255
trace elements, 255-256
cf. other species, 261-262
RITC medium, 249-251, 254
BSA, 250
Eagle's MEM, 249
fibronectin, 250, 260
insulin, T_3, EGF, transferrin, 249
serum-replacing supplements, 256-261
ferric iron, 260
glucocorticoids, 256
lipid oxidation, 259-260
PDGF, 260-261
preformed lipids, 257-259, 262
trypsinization, 261
water-soluble synthetic lipid carriers, 259
"weeds," 249
Fibroblasts, mouse embryo, NIH 3T3 and C3H 10T1/2, 265-274
cf. BSC40 monkey kidney cells, 266, 270, 273
growth in SFM, 268-269
supplements listed, 269
SV40 propagation, 266, 268-269
growth in SFM, 269-273
C3H 10T1/2, 269-271
dose response curves, 270
growth curves, 271, 272
NIH 3T3, 272-274
SV40 infection, 270-274
medium and supplements, 266-268
antibiotics, 266-267
DME:F12, 266
EGF, 267, 270
fibronectin, 267, 270
HDL, 267-268, 272, 273
insulin, 269, 270
sodium selenite, 267
transferrin, 267, 269
SV40 transformation, 265
T antigen, 265, 271, 272
trypsinization and cell plating, 268
Fibronectin
A431 cell culture, 95, 97, 98, 100

embryonal carcinoma cell culture, 109-111, 115-116, 121
FGFs, pituitary and brain, 192
fibroblast culture in SFM, 250, 260
mouse embryo fibroblasts, 267, 270
SV40 Balb/c-3T3 cell culture, 222, 227-230
F9, embryonal carcinoma cell culture, 111-114
F12. *See* DME:F12 medium; Ham's F12

Glassware, hepatocytes, primary culture, 52-53
Glioma cell line C_6-3D, rat, 210, 218-219
Glucagon
bronchogenic epidermoid carcinoma cell culture, 84, 87, 89-91
hepatocytes, primary culture, 47
Glucocorticoids, serum replacement, 256; *see also* specific glucocorticoids
γ-Glutamyltranspeptidase, 35
Glutathione, reduced, fibroblast culture in SFM, 250, 260
Glycyl-histidyl-lysine, A431 cell culture, 100
Granulosa, FGF mitogenic, 184-185, 192, 193
Growth, cell (in culture)
A431 cells, 99-101
bronchogenic epidermoid carcinoma, primary culture, 87-89
and cell concentration, kidney cells ($LLC\text{-}PK_1$), 27
embryonin, 142-143
evaluation, cell nuclei DNA counting, 78-79
α_2-macroglobulin, 142-143
metanephros, organ culture, 37, 39
Pedersen fetuin, 142-143
reinitiation assays, hepatocytes, 53-54
Growth factors
embryonal carcinoma cell culture, endogenous, 121
epidermal. *See* EGF
fibroblast. *See* Pituitary acidic FGF
myoblast (endothelial). *See* Myoblast growth factor
nerve, 143
ovary, 209
platelet-derived, 143, 260-261
small cell lung cancer culture in HITES, autocrine production, 66-67
Growth factors, regulators cf. nutrients, 157-165
arrest test, 158-160
availability test, 161-163
cycloheximide, 161-162
Balb/c 3T3 cells, 159, 163
BHK cells, 161
cell cycle and timing of regulatory events, 158
CHEF 18 cells, 158-160, 163-165
choice among several, 163-164
EGF, 158, 160, 161, 163, 165
FGF, 158, 161
insulin, 158, 160, 161, 163, 165
somatomedin C, 163, 164
SV40-3T3 cells, 161
thrombin, 163-165
transferrin, 158, 161, 165
transformation test, 160-161
mutation, 160

Ham's F12
bronchogenic epidermoid carcinoma cell culture, 84-93
colon carcinoma, 74-77
hormone-toxin conjugates, 240, 241
metanephros, organ culture, 34, 39
HCG, hormone-toxin conjugates, 234

HeLa cells, 143
Henle's loop, kidney cell culture, 14, 15
Hepatocytes, primary culture (rat), 43-54
  arginine-free medium, stock solutions listed, 45
  biological and inorganic reagents, 51-52
  buffer solutions listed, 47, 48
  cell counts, 47-48
  cell isolation and plating, 45-47
    adult, 49
    fetal, 48
    glucagon and EGF, 47
  differentiated properties, listed, 44
  growth reinitiation assays in SFM, 53-54
  hydrocortisone, 44, 45
  inosine, 44-45
  insulin, 44, 45, 47
  media and digestion buffer construction, 43-45
    steps, outline, 46
  morphology, 50
  L-ornithine, 43, 44, 46
  tissue culture equipment, plastic, and glassware, 52-53
Hepatoma, rat H35 cells, hormone-toxin conjugates, 243
Hexamethylene-bisacetamide, embryonal carcinoma cell culture, 111, 118
Hormone-toxin conjugates, cell variant hormone responses, 233-246
  A431 cells, 244
  cell variant selection, 240, 242, 244
  Cloudman S-91 melanoma, 244
  disulfide linkages, 233, 236, 238
  DME and/or F12, 240, 241
    BSA, 241
    EGF, 241, 242, 244
    ethanolamine, 241
    insulin, 241-242, 244
    supplements, 241
    transferrin, 241
  dose-dependent profiles, cell killing, 239
  EGF, 234, 243
  FCS, 240-242
  HCG, 234
  human placental lactogen, 234
  insulin-DTa conjugate, synthesis, 233, 237-238, 242-245
    derivatization of insulin with cystamine, 237-238
    diagram, 236
    diphtheria toxin, A fragment (DTa), 233, 234, 235
    disulfide bridge formation and purification, 238
  insulin-RICa conjugate, synthesis, 233, 238-240, 243-245
    crosslinking and purification, 239-240
    derivatization of insulin with SPDP, 238-239
    ricin, A subunit (RICa), 233, 234, 235, 237
    ricin, B subunit, 237, 245
  rat hepatoma H35 cells, 243
  receptor-mediated endocytosis, 244-245
  resistant variants, 243
  schema, routes of entry, 234
  3T3-L1 preadipocytes, 234-235, 243, 244
    differentiation in SFM, 242
    growth in SFM, 240-241
H35 cells (rat hepatoma), 243
HUE cells, FGF and, 185, 191, 194-195
Human placental lactogen, 234
Hyaluronidase, colon carcinoma, 75-76
Hydrocortisone, 218
  fibroblast culture in SFM, 250
  hepatocytes, primary culture, 44, 45

kidney cell cultures, 11, 13, 14
cell line LLC-PK_1, 26
MDCK established cell culture, 5, 6
metanephros, organ culture, 35, 39

IEF
FGFs, pituitary and brain, 168-169, 173
brain, 177, 181-186
pituitary, 185, 190
Pedersen fetuin, 132, 134-136
pituitary acidic FGF, 214-218
Inosine, hepatocyte primary culture, 44-45
Insulin
A431 cell culture, 95, 97, 100
bronchogenic epidermoid carcinoma cell culture, 84, 87, 89-91, 93
colon carcinoma, 77
distinguishing regulators from nutrients, 158, 160, 161, 163, 165
EGF, fibroblast culture in SFM, 250, 261
embryonal carcinoma cell culture, 110
fibroblast culture in SFM, 249, 250
hepatocytes, primary culture, 44, 45, 47
metanephros, organ culture, 35, 39, 40
mouse embryo fibroblasts, 269, 270
SV40 Balb/c-3T3 cell culture, 222-226, 229
-toxin conjugates. *See under* Hormone-toxin conjugates
-transferrin synergism, 222, 223
Interferon, colon carcinoma, sensitivity assays, 79
Iron, ferric, 260
Iron oxide, rabbit proximal tubule cell culture, 15, 16
Isobutylmethylxanthine, kidney cell line LLC-PK_1, 27

Keratinization. *See under* Bronchogenic epidermoid carcinoma, growth and differentiation
Keratinocytes, human skin, 93-94
Kidney cell culture, established, in K-1 defined medium, 3-9
LLC-PK_1, 3, 8-9, 20, 22
properties listed, 5
MDBK, 8
MDCK, 3-7, 11-12, 18, 20-22, 34
dome formation, 5-7, 21, 22
hydrocortisone, 5, 6
PGE_1, 5, 6, 26, 28
plastic substratum, 5
properties listed, 4
salt and water transport, 21-22
transferrin, 5, 6
T_3, 5, 6
medium components, 6
see also Metanephros, mouse embryonic, SF organ culture
Kidney cell culture, primary, defined SFM, 9-21
adult mouse, K-1 medium, 13
antibiotics, 10
baby mouse, K-1 medium, 10-13
EGF, 14
fibroblast overgrowth, 9-10, 14, 20
growth and maintenance in medium K-1, various species, 13-14
Henle's loop, 14, 15
hormonally defined, 9
hydrocortisone, 11, 13, 14
cf. MDCK cells, 11-12
methodologies for culturing different subpopulations, 14-15
nephron segments (tubules and glomeruli), 10-11, 14

PGE_1, 11, 13, 14
rabbit kidney
epithelial cells from microdissected nephron segments, 18, 20
cf. other species, 18
proximal tubule cells, medium RK-1, 15-19
T_3, 14
trypsin, 10-11, 15
Kidney cell line BSC40, monkey, 266, 268-269, 270, 273
Kidney cell line LLC-PK_1 (proximal tubular), defined SFM, 25-30
cAMP, 27
cholesterol, 26
defined medium used to isolate primary cultures, 28-29
growth dependencies and concentration, 27
hydrocortisone, 26
isobutylmethylxanthine, 27
in K-1 defined medium, 3, 8-9, 20, 22
cf. MDCK, 25-28
properties listed, 5
T_3, 26
vasopressin, 26-28

Laminin
embryonal carcinoma cell culture, 108-110, 115-116, 121
FGFs, pituitary and brain, 192
Linoleic acid, SV40 Balb/c-3T3 cell culture, 224-226, 228-230; *see also* Fatty acid(s)
Lipid(s)
carriers, water-soluble, serum replacement, 259
oxidation, serum replacement, 259-260
and phospholipid separation, Pedersen fetuin, 130-131, 147
preformed, serum replacement, 257-259, 262
Lipoproteins
high density (HDL), mouse embryo carcinoma, 110, 112-120
fibroblasts, 267-268, 272-273
low density (LDL), 110, 112, 114
very low density (VLDL), 110, 112, 114
Liposome B, fibroblast culture in SFM, 250, 257, 261
Liver. *See* Hepatocytes, primary culture
LLC-PK_1 cells. *See* Kidney cell line LLC-PK_1 (proximal tubular), defined SFM
Lung cancer
adenocarcinoma, 65-66, 69-70
A431. *See* A431 human epidermoid carcinoma cell culture
A549, 59, 65
bronchogenic. *See* Bronchogenic epidermoid carcinoma, growth and differentiation
epidermoid, 65-66
small cell, incidence, therapy, survival, 58
Lung cancer, small cell (human), growth in HITES SFM, 57-70
APUD, 58-59
bombesin, 63, 66
characterization of cell growth in HITES, 63-65
DNA content, 64, 67
clonogenic assay, in vitro drug and radiation sensitivity, 58
creatine kinase BB, 58, 63
cytogenetic analysis, 69
L-dopa decarboxylase, 58, 63
growth factors, autocrine production, 66-67
growth of cell lines, 59-61
HITES medium, 60-61

cf. other lung cancers, 65-66, 68
adenocarcinoma (LA), 65 66, 69-70
A549 cell line, 59, 65
epidermoid (LE), 65-66
NCI-H23 and -H125, 65
selectivity of HITES medium, 65
cf. serum-supplemented, 57, 61-62

α_2-Macroglobulin, 137, 147-149
biological activity, 127
and cell growth, 142-143
collagen production and protease inhibition, 143-145
fatty-acid carrier function, 147, 148
immunologic properties, 142
modulation of capillary endothelial cell migration, 145-147
pathophysiology, 149
physiochemical properties, 139-142
subunit structure, 141
see also Embryonin; Fetuin, Pedersen, α_2-macroglobulin contamination
MCDB 110 and 202 media. *See under* Fibroblast culture in SFM
MDBK cells, 8
MDCK cells
cf. LLC-PK_1 (proximal tubular), 25-28
cf. primary kidney cell culture, 11-12
see also Kidney cell culture, established, in K-1 defined medium
Melanoma, Cloudman S-91, 244
Metanephros, mouse embryonic, organ culture, 33-40
at explantation, 36, 37
growth and viability, 37, 39
hydrocortisone, 35, 39
insulin, 35, 39, 40
medium preparation and storage, 34-35, 39, 40
morphogenesis, 37
PGE_1, 34, 35, 39, 40
after six days of culture, 38, 39
T_3, 34, 35, 39, 40
tissue harvesting and processing, 35-37
tissue isolation and culture, 33-34
transferrin, 34, 35, 39, 40
see also Kidney cell culture, established, in K-1 defined medium
α-Methylglucoside, 11, 18, 29
Mitogenic activity assay, Pedersen fetuin, 129-130
Myoblast (endothelial cell) growth factor, 167
extraction from brain and pituitary, 187
see also under Pituitary and brain FGFs
Myogenic cell line T984, embryonal carcinoma cell culture, 118

NCI-H23 and -H125 cells, 65
NG and NG-2 antigens, 117
NGF, 143
NRK cells, Pedersen fetuin, 129, 130, 143

OC-1551 and -1003 embryonal carcinoma cell culture, 114-115, 117
Organ culture. *See* Metanephros, mouse embryonic, organ culture
L-Ornithine, hepatocytes, primary culture, 43, 44, 46
OTT-6050 EC cells, Pedersen fetuin, 129, 130, 134-137, 142
Ouchterlony double immunodiffusion, Pedersen fetuin, 133
Ovalbumin, fibroblast culture in SFM, 250, 257

Ovary growth factor, 209

Parathyroid hormone, A431 cell culture, 100
PC-13, embryonal carcinoma, 112-114
PCC-3, embryonal carcinoma, 116-117
PDGF, 143, 260-261
Pedersen fetuin. *See* Fetuin, Pedersen, α_2-macroglobulin contamination
PGE_1, 143, 146
  fibroblast culture in SFM, 250, 260
  kidney cell cultures, 11, 13, 14
  MDCK established cell culture, 5, 6, 26, 28
  metanephros, organ culture
$PGF_{2\alpha}$, fibroblast culture in SFM, 250, 260
pH
  A431 cell culture, 98
  FGFs, pituitary and brain, 171-172, 182
Phlorizin, 22
Phosphoenolpyruvate, fibroblast culture in SFM, 250, 260
Phospholipid and lipid separation, Pedersen fetuin, 130-131, 147; *see also* Lipid(s)
Pituitary acidic FGF (bovine), 209-219
  isolation, 210-211
    assays, 210-211
    C_6-3D cell line, 210, 218-219
    FCS, 210-211
    3T3 fibroblasts, 210, 219
  multiple forms, 218
  precipitation, 215-218
  preparation, 211-215
    cf. basic FGF, 212
    BioGel chromatography, 215
    CM-Sephadex column, 212
    DEAE-Sephadex column, 212-213
    FGFa, 213, 215
    FGFb, 213
    fractions listed, 213
    IEF, 214-218
  stimulation of colony growth in agarose suspension culture, 218-219
    assay, 211
    hydrocortisone effect, 218
Pituitary and brain FGFs (bovine), 167-195
  ABAE cells, 185, 191, 194-195
  cell proliferation assay, 169-170
  CM ion-exchange chromatography, 168, 172-173, 183, 185
    brain, 173-178
    pituitary, 174-177, 179
  collagen, 192
  concentration, effect of increased, 191, 192
  exclusion gel chromatography, 173, 182, 183
  extraction, 171-172
  F_4 and F_5 fractions, 175
  F_3 fraction, 172-173
  FGF-1, 173, 175, 180-181, 183-185
  FGF-2, 173, 175, 183-185, 188, 189
  fibronectin, 192
  gel filtration, 168
  HUE cells, 185, 191, 194-195
  laminin, 192
  media (DME H-16) and supplements (FCS), 167-168
  and MGF, use in cell culture, 167, 187, 188-195
    increasing concentrations, 194-195
    senescence delay, 192-193
    vascular endothelial cells, 189, 192
  mitogenic for various bovine tissues, 184-185, 192, 193
    listed, 193

pH, 171-172, 182
preparative IEF, 168-169, 173
brain, 177, 181-186
pituitary, 185, 190
SDS-PAGE, 168, 171, 182, 185, 189, 190
stimulation of DNA synthesis assay (^{3}H thymidine), 170-171
temperature, 172
tissues, 167
Pituitary carcinoma, 103
Pituitary FGF (bovine), preparation, 199-207
Balb/c 3T3 cell ^{3}H bioassay, 200
cf. brain, 207
BSA, 200, 204, 205
inactivation by heat, acid, SDS, 206
isolation procedure, 200-204
acid gel electrophoresis, 199, 203-204, 206
ammonium sulfate precipitation and extraction, 201
CM-Sephadex-C-50 gradient elution, 203, 204, 206
CM-Sephadex-C-50 stepwise fractionation, 201
EDTA, 201, 205
gel filtration elution profile, 202
homogenization and extraction, 200-201
PMSF, 201, 205
recovery from columns and dialysis, 205
protein determination, 200
source of pituitaries, 200
stable to lyophilization, 205
Plastic substratum, MDCK established cell culture, 5
Platelet-derived growth factor, 143, 260-261
Plating
A431 cell culture, 97-99
hepatocytes, primary culture, 45-47
mouse embryo fibroblasts, 268
SV40 Balb/c-3T3 cell culture, 226-230
PMSF, pituitary FGF, 201, 205
P-19 embryonal carcinoma cell culture, 116-117
Preadipocytes, 3T3-L1, 234-235, 240-244
Protease inhibition, embryonin, α_2-macroglobulin, and Pedersen fetuin, 143-145
PSA-1, embryonal carcinoma, 116-117
PSA-5E, embryonal carcinoma, 117
PYS-2, embryonal carcinoma, 117, 118

Rabbit kidney cell culture
epithelial cells from nephron segments, 18, 20
cf. other species, 18
proximal tubule cells, 15-19
Radiation sensitivity assays
colon carcinoma, 79
small cell lung cancer culture in HITES, 58
Receptor-mediated endocytosis, 244-245
Retinoic acid
bronchogenic epidermoid carcinoma cell culture, 84, 87-93
embryonal carcinoma cell culture, 110-111, 121
Retinoids, bronchogenic epidermoid carcinoma cell culture, 91, 94
antikeratinization, 92
Ricin. *See under* Hormone-toxin conjugates, cell variant hormone response
RITC medium. *See under* Fibroblast culture in SFM
RME cells, Pedersen fetuin, 129, 130, 134, 142-144

Salt transport, MDCK established cell culture, 21-22
SDS, pituitary FGF, inactivation, 206; *see also under* Electrophoresis
Selenium, 84, 87, 89-91, 93, 255
  CHO cells, 256
  sodium selenite, 267
Senescence delay, FGF and MGF, 192-193
Serum
  components, 125
  spreading factor, A431 cell culture, 100
  supplementation cf. SFM
    bronchogenic epidermoid carcinoma cell culture, 84, 85, 92
    small cell lung cancer culture in HITES, 57, 61-62
  see also FCS
Skin, keratinocytes, human, 93-94
Sodium selenite. *See* Selenium
Somatomedin C, 163, 164
Somatostatin, A431 cell culture, 100
SPDP, derivization of insulin, 238-239
Squamous cell carcinoma, rat bladder, 93; *see also* Bronchogenic epidermoid carcinoma, growth and differentiation
SSC-1, embryonal carcinoma cell culture, 113-114
Storage
  metanephros, organ culture, 34-35, 39, 40
  SV40 Balb/c-3T3 cell culture, 222
SV40 Balb/c-3T3 cell growth in SFM, 161, 221-230
  antibiotics, 222, 223
  colony formation, 230
  DME:F12 medium, 222-228, 230
  dose response to supplementation, 227
  fatty-acid free BSA, 222, 224-226, 228-230
  fatty acids, 222, 225, 226, 228
  FCS, 222, 223, 225, 229
  fibronectin, 222, 227-230
  infection, 270-274
  insulin, 222, 224-226, 229
    -transferrin synergism, 222-223
  linoleic acid, 224-226, 228-230
  passage and plating of cells, 226-230
    trypsin, 227-230
  serum proteins and detoxification, 222
  storage, 222
  transferrin 224-226, 229
  cf. untransformed cells, 224, 225
  see also under Fibroblasts, mouse embryo, NIH 3T3 and C3H 10T1/2

T antigen, SV40, 265, 271, 272
Temperature
  A431 cell culture, 98
  FGFs
    pituitary and brain, 172
    pituitary, inactivation by heat, 206
Teratocarcinoma. *See* Embryonal carcinoma cells, growth in defined SFM
T_3
  bronchogenic epidermoid carcinoma cell culture, 84, 87-93
  colon carcinoma, 77
  fibroblast culture in SFM, 249
  kidney cell culture, 14
    cell line LLC-PK_1, 26
  MDCK culture, 5, 6
  metanephros, organ culture, 34, 35, 39, 40
T_4, bronchogenic epidermoid carcinoma cell culture, 91
3T3 fibroblasts, 69
  pituitary acidic FGF, 210, 219

see also Fibroblasts, mouse embryo, NIH 3T3 and C3H 10T1/2; SV40 Balb/c-3T3 cell growth in SFM
3T3-L1 preadipocytes, 234-235, 240-244
Thrombin, 163-165
^{3}H-Thymidine assay, FGFs, pituitary and brain, 170-171
Tissue harvesting, metanephros, organ culture, 35-37
T984 myogenic cell line, 118
Trace elements, 250, 255-256; *see also* specific elements
Transferrin
A431 cell culture, 95, 97, 100
colon carcinoma, 77
distinguishing regulators from nutrients, 158, 161, 165
fibroblast culture in SFM, 249, 250
hormone-toxin conjugates, 241
MDCK established cell culture, 5, 6
metanephros, organ culture, 34, 35, 39, 40
mouse embryo fibroblasts, 267, 269
SV40 Balb/c-3T3 cell culture, 224-226, 229
-insulin synergism, 222, 223
Transformation test, regulators cf. nutrients, 160-161
Triiodothyronine. *See* T_3
Trypsin/trypsinization
A431 cell culture, 97-99
kidney cell culture, 10-11, 15
mouse embryo fibroblasts, 268
serum replacement, 261
SV40 Balb/c-3T3 cell culture, 227-230
T222 tumor transplant in nude mice, 86-87
Tyrosine aminotransferase, 243
Umbilical cells (HUE), 185, 191, 194-195

Vascular endothelial cells, FGF and MGF, 189, 192
mitogenicity, 184-185, 192, 193
Vascular smooth muscle, FGF mitogenic, 184-185, 192, 193
Vasopressin, kidney cell line LLC-PK_1, 26-28
Virus. *See* SV40 Balb/c-3T3 cell growth

Water transport, MDCK established cell culture, 21-22

Zinc chelate affinity chromatography, fetuin and embryonin, 133, 134, 139, 140